BEVERAGES AND THEIR ADULTERATION

WILEY

BEVERAGES

AND

THEIR ADULTERATION

ORIGIN, COMPOSITION, MANUFACTURE,
NATURAL, ARTIFICIAL, FERMENTED,
DISTILLED, ALKALOIDAL AND
FRUIT JUICES

BY

HARVEY W. WILEY, M.D.,

AUTHOR OF "FOODS AND THEIR ADULTERATION."

WITH 42 ILLUSTRATIONS

PHILADELPHIA
P. BLAKISTON'S SON & CO.
1012 WALNUT STREET

PREFACE

The continued favor which has been shown to my book entitled "Foods; and Their Adulterations" inspires the hope that this volume which is a companion to the food volume will receive equal support.

In view of the fact that the sentiment for prohibition has gained so rapidly in the country as to result in the submission of a constitutional amendment, for national prohibition, to the legislatures of the several states for ratification, many have suggested that a description of alcoholic beverages may be out of place. Just the contrary is the case. The American citizen who desires all the information possible in making up his mind on this question will certainly be helped by a knowledge of the origin, manufacture, chemical composition and geographical distribution of the various forms of alcoholic beverages, both fermented and distilled.

The rapidly increasing use of fruit juices demands a prominent place in this volume, for their description and composition. The many types of so-called soft drinks, which will undoubtedly have a greater vogue as the area in which alcoholic beverages are manufactured and sold decreases, warrant a rather full description of them here. The so-called medicines which consist chiefly of alcohol, and which are held by the Bureau of Internal Revenue as non-medicinal but alcoholic, are fully described.

Proper space is also given to a discussion of our most popular beverage, coffee, and its related products, tea, cocoa and chocolate.

Water is the typical beverage and in the hope of its coming into more general use it occupies the place of honor in this volume. With potable water, naturally, mineral waters both artificial and natural are included. Milk, another national beverage, has already been described in the volume on "Foods; and Their Adulterations."

With each subject treated are described the common adulterations and misbrandings which may be practised therewith.

Every endeavor has been made to secure accuracy of statements and data. The latter are taken chiefly from official sources. It should be understood that this book was not written for the scientific investigator but for the average, sober-minded, reasonably well-educated American citizen, who is daily taking a greater and deeper interest in what he eats and drinks.

HARVEY W. WILEY.

WASHINGTON, D. C.

TABLE OF CONTENTS

LIST OF ILLUSTRATIONS

INTRODUCTORY

BEVERAGES

Definitions.—Beverages, as distinguished from foods, are those substances which are consumed in a liquid state and which may or may not have food value. There is no legal distinction between a food and a beverage. Beverages in the food law are defined as part of food. In general it may be said that all products which require mastication are foods in the common acceptation of that word. All liquid products which are consumed as or with foods, or for quenching thirst, requiring no mastication, may be classed under the common appellation of beverages. Water is a beverage and so classed by the food and drug act and by regulations and court decisions thereunder.

Beverages of a Two-fold Character.—It follows from the above that beverages are conveniently divided into two great classes, the bases of which depend upon the point of view of the utility of the article consumed. Such products as milk, sweet cider, fruit juices, extracts containing food properties, and similar condimentals, serve as foods; while at the same time they may be condimental, in other words, appealing to the taste. On the contrary, things which are consumed in a liquid state which have little or no food value, belong to the purely condimental class of beverages. As types of these bodies there may be mentioned tea, coffee, ginger ale, mineral waters, potable waters, and others of similar character.

A sub-division may also be made of those beverages which contain toxic properties, either natural or added. Among these in the first rank are all alcoholic beverages, especially distilled. In addition to these, coffee and tea as well as wine and beer may be classed in this sub-division because of the toxic alcohol or caffein which they contain. Among soft drinks, those to which caffein or cocaine or other deleterious substances have been added would be included in the sub-classification.

Beverages for Quenching Thirst.—The primal use of the beverage is to secure a sufficient liquid content in the fluids of the body. Whenever for any reason there is need of more water in the body, a sensation of thirst is developed. The common expression of thirst is "dry." This term, however, has been almost completely appropriated by the fictitious idea of what thirst signifies. In other words, the desire for an alcoholic beverage is now very commonly referred to as expressing the idea contained in the word "dry." The phrase "I am so dry" which occurs in the drinking song does not refer to any lack of moisture in the body. The use of alcohol in excess develops thirst instead of satisfying it. In like manner the use of excessive quantities of common salt or other mineral substance of that nature requires a larger quantity of water in the system to properly dilute the saline contents of the liquids of the body.

There is only one beverage for real thirst, and that is water. The character of water to be used in the satisfaction of thirst will be the subject of special consideration.

Difficulties of Description.—It is very difficult to undertake a description of the beverages in common use without digressing into the realm of hygiene and physiology, and it is still more difficult to do so without considering the ethical aspects of the question. I wish it to be understood, first of all, that the object of this book is not to discuss chiefly questions of health or hygiene, nor of ethics, although this may be done occasionally by way of incidental remarks. The object of this volume is to describe the common beverages which we consume. It is important to know their origin, their methods of manufacture, their composition, and the proper conditions attending their transportation and use. It is important that the informed citizen should know something of the nature of these beverages other than that presented by their commercial aspect. Especially should he know the difficulties which attend the securing of a water supply suitable for potable purposes. The various methods which have been practised of securing this water supply, as well as the causes of contamination, will be developed to a considerable extent because of their vital importance to human welfare.

In so far as possible the personal experience of the author has been called upon to furnish the principal data of this volume.

It is not difficult, however, to imagine that no one person could, even if he made a life-long specialty of the study, become personally acquainted with all phases of the great industries connected with the beverages of mankind.   I have therefore endeavored, in so far as I could, to draw my information other than that obtained from personal experience and investigation from reliable sources.

PART I

WATERS

Pure Water.—Pure water, that is water which is not mixed with any other substance, is practically unknown. To obtain pure water artificially requires the highest skill of the chemist's art combined with the most delicate technique. Water has been described as the universal solvent. Not only does it dissolve many other liquids, but acts upon most solid bodies producing a greater or less degree of solution, and particularly does it dissolve the gases with which it may come into contact. For this reason water also dissolves the materials of its containers, especially if they be glass, the degree of solution being very marked in some kinds of glass and less marked in others. By reason of its tendency to dissolve matters of this kind water can only be secured in a pure state with the greatest precautions and by preparing it in vessels to which it is inert. Pure water, in the strictest sense of that term, is therefore a chemical curiosity and does not concern us in any other way. The art of preparing pure water is strictly chemical and does not form any part of the present discussion. The distilled water of commerce, when properly prepared is free of living organisms and contains only minute quantities of foreign bodies.

Water as a Beverage.—In this manual water will be considered soleily as a beverage. The term "beverage" has a wide application and is often construed to apply only to artificial preparations. Th s, however, is not a correct interpretation of the term. Under the Food and Drugs Act of June 30, 1906, all beverages are regarded as foods. Water is the most important of all beverages. It is also an indispensable food. It forms a large part of all living organisms—in fact the principal part of many of them, such as the very succulent vegetables. The human body contains approximately 70 percent of water. The importance of water as a food, or in other words as a beverage, is therefore evident. It is practically the one beverage which is universal in its consumption. The abundant presence of water is necessary to life, both animal and vegetable. The great deserts of the surface of the earth are made so principally by the lack of water.

Classification of Water as a Beverage.—Water is used as a beverage in various forms. The principal source of drinking water is found in the springs and wells. Spring water or well water consists of the water which falls upon the surface of the earth and either enters the surface of the earth directly or subsequent to its forming streams of greater or less size. Passing through the earth the water absorbs the soluble principles with which it may come into contact. When the quantity of these dissolved matters is not large there is formed ordinary spring or well water. In water of this kind the amount of dissolved matter is not sufficient to give any characteristic taste. Spring or well water of this kind forms by far the greatest part of the water used as a beverage in all except crowded centers such as the towns and cities. When the water comes to the surface through natural causes it produces the spring. When water is secured by digging or boring into the earth there is produced what is known as the well. Unless the material through which water percolates is exposed to some infection spring and well water is universally regarded as the most wholesome form of water as a beverage.

Rain Water.—Rainfall is often collected and stored in cisterns and used for beverage purposes. As will be explained further on, rain water is subject to many sources of pollution, particularly by entangling matters floating through the air or collecting débris off the roofs or other surfaces upon which it first falls. This is particularly apt to be the case with the first portions of the rain. In heavy and long-continued rains the portions which fall toward the end of the precipitation are almost pure. They contain very little solid matter in solution but usually are saturated with the nitrogen and oxygen of the atmosphere and contain more or less carbon dioxid; as rain water must also be regarded melted snow either collected upon the roofs of houses or otherwise. The water which comes from the melting of the snow has all the characteristics of the water precipitated directly as rain. When these waters have come in contact with the soil they begin at once to dissolve mineral and organic matters and lose their distinctive character as rain water. The melted snows of high mountain peaks are believed to be the purest forms of rain water that exist.

Stream and River Water.—When the water flowing from the surface of the earth or from springs collects into streams or rivers it is often used as a source of the water as a beverage. Such water is often taken directly for consumption or pumped into reservoirs and distributed to crowded centers of population. Waters of this kind form the principal supply of the large cities.

Distilled Water.—Distilled water, as recognized in the Pharmacopœia, ninth decennial revision 1916, is described as "A colorless limpid liquid, without odor or taste and neutral to the official indicators . . . . It should give no reaction for ammonia, nitrates, nitrites, sulphates, chlorids, carbon dioxid. When 100 mils are evaporated on a water bath to dryness not more than 0.001 gram of residue should remain."

Further, the Pharmacopœia requires that distilled water should contain no organic matter. Distilled water is often used for beverage purposes, but only in limited quantities. It is rarely distributed for public consumption, but is usually prepared on the premises where it is consumed, except in the case of bottled waters or those which are regarded as imitations or artificial waters. Distilled water has toxic properties for plants probably due to lack of plant food, and except for the reason that it is likely to be sterile and free from all soluble impurities is not highly regarded as a beverage. It is usually quite free of dissolved gases, has a flat and insipid taste, and does not equal in potable properties water from springs and wells.

Mineral Waters.—Mineral waters are those which have extracted from the soil and rocks through which they have passed sufficient quantities of mineral matters to give them a distinct character. These dissolved matters are of the most varied character. They consist largely of the salts of the ordinary bases, namely, soda, potash, lime, magnesia, and iron and more rarely lithia and radio-activity. The rarer elements are sometimes found in these waters, but only in minute quantities. Occasionally poisonous principles, such as arsenic, alum and copper, are found dissolved in waters of this character. In general, it may be said that the use of mineral waters as a regular beverage is not to be commended. These waters are useful for special purposes, which will be described further on.

Chalybeate Waters.—Chalybeate waters are those which contain a sufficient quantity of iron to give them a distinctive ferruginous character. They are highly prized in many forms of disease, the iron being supposed to be of a tonic nature. In chalybeate water the iron which is in solution usually becomes precipitated on standing, either by oxidation or by the escape of carbon dioxid from the solution.

Sulphur Waters.—Sulphur waters are those mineral waters which contain sufficient quantities of compounds of sulphur, chiefly hydrogen sulphid, to give them a distinct character. These waters are usually recognized very easily by the unpleasant—to most persons at least—and characteristic odor of hydrogen sulphid pervading them. Elemental sulphur is found separated from waters of this kind forming excretions of greater or less extent around the orifices of the springs. Mineral waters may at times contain sulphur dioxid in solution, but in general the term sulphur water indicates a water saturated to a greater or less extent with hydrogen sulphid.

Alum Waters.—These are mineral waters in which the sulphates of aluminum, either as such or in combination with other bases, exist. These waters are extremely styptic in character and have the characteristic taste and effect of weak solutions of alum. This whole class of mineral waters can only be regarded in the light of remedial agents rather than that of beverages.

Alkaline Waters.—Alkaline waters are mineral waters in which there is dissolved sufficient quantities of carbonate or bicarbonate of soda or similar salts to give the water a distinct alkaline taste and reaction. Alkaline waters are found particularly in regions where the rainfall is deficient. The natural rain not being sufficient to dissolve all the soluble matters in the soil, these alkaline bodies, products of decomposition of rocks, accumulate. During periods of heavy rainfalls which frequently occur in such regions large quantities of these alkaline salts go into solution and are carried into the earth. From this source are fed springs and streams which furnish the alkaline waters at intervals between the rains. These alkaline waters are esteemed in some localities for their medicinal properties, but are not suitable for consumption as a beverage. Laxative waters usually contain large quantities of sodium sulphate (Glauber's salts).

Gypsum Water.—This is a form of mineral water in which there is dissolved a large quantity of sulphate of lime, which when it exists in a certain form of crystallization, is known as gypsum. These waters have a peculiar, bitter unpleasant taste, and while used largely in some localities as a beverage they cannot be regarded as wholesome when intended for constant use.

Hard Water.—The term hard is applied to those waters which form a heavy curd with soap. The hardness is due to the solution of excessive quantities of the salts of magnesia and other similar mineral substances. Since these salts are found held in solution by the carbon dioxid which the water carries, the boiling of hard water will precipitate part of the mineral matters thus held. The original hardness of such water is called the total hardness, and the hardness which remains after the precipitation of the bodies above mentioned the permanent hardness. The hardness of water is more important from its technical use perhaps than from its use as a beverage. Hard waters, however, on account of the large quantities of mineral contained therein, are not suitable for permanent consumption. Technically they are objectionable, especially for use in boilers because of the coating of the boilers produced by the precipitation of the mineral matters, often in very firm layers, upon the surfaces of the boiler. This precipitation is often seen in vessels used in the kitchen where hard water is employed. Various precipitants have been used in order to diminish the hardness of water and make it suitable for beverages and for technical purposes.

Treatment of Hard Water.—Inasmuch as the carbon dioxid which a water naturally holds adds very greatly to the solvent power of the water for mineral substances, it is evident that all those substances which are held in solution in water containing carbon dioxid may be precipitated by simply boiling the water. This form of hardness is caused chiefly by the sulphates of lime and magnesium, bodies which are more or less soluble in water not containing carbon dioxid. The permanent hardness is cured usually by treating the water with carbonate of soda, which converts the calcium sulphate into calcium carbonate and precipitates it.   The carbonate of soda also acts upon the Epsom salts, that is, magnesium sulphate, converting it into carbonate and thus precipitating it.   The quantity of carbonate of soda employed should be adjusted to the amount of lime and magnesia sulphates which are present.

If waters are excessively alkaline the treatment with bicarbonate of soda may not leave the water in a fit condition either for drinking or for washing purposes.   If there is a large quantity of magnesium sulphate in the water the amount of sodium sulphate formed is so large as to interfere with the emulsification of the soap, thus interfering with the washing process.   In such conditions the use of freshly slaked lime is to be preferred to that of carbonate of soda.   The slaked lime converts the magnesium sulphate into magnesium hydrate, which is precipitated, leaving an equivalent amount of calcium sulphate in solution.   If the amount of calcium sulphate is considerable, however, a part of it will be precipitated.   Thus, it is generally advisable to treat hard water both with carbonate of soda and slaked lime.   When lime and carbonate of soda are both used in softening water, it is advisable that the lime should be used first.

According to some authors the permanent hardness is caused chiefly by magnesium chlorid and calcium sulphate.

Method of Removing Hardness.—One of the common methods of removing hardness is by means of soap.   When a soap, which is made with potash and soda as a base, is added to the water containing calcium and magnesium salts the soap is decomposed, and the fatty acid is set free in an insoluble state.   It is upon this reaction that the old method of determining hardness, known as Clark's method, was based.   This process has been varied in many ways by chemists in different countries, but the original principle has not been departed from.

The hardness is usually estimated in the number of milligrams of calcium carbonate in a standard volume of water—usually 100,000 parts.   In England and America it is usually determined by the number of grains of calcium carbonate in an imperial gallon in England and a United States gallon in America, which means, in England, the number of grains of calcium carbonate in 70,000 parts, and, in America, in 58,381 parts.

The methods of determining hardness of water will be found in all works on water analysis.

Importance of Determining Hardness.—In so far as drinking water is concerned it is not important to know the hardness apart from the quantity and kind of mineral matters contained. In other words, for potable purposes, the hardness of water *per se* is of no consequence. The effect of the mineral matters in solution upon the health and digestion are the principal things, in fact, the only things to be considered in so far as drinking purposes are concerned. For technical purposes, however, the matter is quite different. Here it is of the utmost importance to know the character of the dissolved mineral matter. If, for instance, water is to be used in a steam boiler, the rapidity with which it coats the tubes and sides of the boiler is a matter of grave consequence. Often the tubes of the boiler become so coated with the deposited mineral matters as to make it almost impossible to generate steam therein. The boilers are rapidly burned out, great damage is caused, and danger of explosion engendered.

Wherever water is to be used for technical purposes, it should be carefully examined, especially for its temporary hardness.

Salts which are very soluble, such as the chlorids of soda, lime and magnesia, are of little consequence, but those which are easily precipitated, like the carbonates, sulphates, etc., are of great consequence. In such cases complete chemical analysis would be required to determine whether or not a water is suitable for purposes of this kind.

Water used for Irrigation Purposes.—It is well known that certain mineral compounds, when they accumulate in the soil, are extremely harmful to vegetation. Common salt is one of those substances; carbonate of and sulphate of soda are also of this kind. On the other hand carbonates of lime and magnesia and the sulphate of lime are not, as a rule, harmful, but often beneficial. When a water supply is to be used for irrigation purposes, it is of the utmost consequence that a chemist be called in to determine whether or not the water has enough of objectionable mineral constituents to render it unsuitable for the proposed purposes. In this connection it must be borne in mind that the salts may accumulate, and do accumulate, in soils which are subjected to irrigation. Thus, if the water at first does not have a sufficien quantity of these objectionable salts to render it injurious, the accumulation of these salts in the soil in a few yea s may render it unfruitful. In such cases artificial drainage coup with excess of water applied between crops will preserve the . from injury.

Soap may also be used for softening hard wat. To determine how much of the soap is necessary a solution of s ap is made of a definite strength, usually in the proportion of 20 grains to a tablespoonful of rain or distilled water. This solution may be made in a larger quantity and a tablespoonful of it added to a pint of the water whose hardness is to be tested. After thorough shaking the water is set aside, and the foamy lather which has resulted should be permanent for at least five minutes. When the correct quantity of soap solution is determined by experiments of this kind, then the required quantity of soap can be added to large quantities of water.

For ordinary hard waters, where facilities for making the foregoing tests are not at hand, or the skill to make them is lacking, very good results may be obtained usually by assuming a certain average degree of hardness and proceeding as follows: Five hundred gallons of water are treated with a solution of five pounds of carbonate of soda in two or three gallons of water. Numerous so-called boiler compounds are offered for sale, but they consist essentially of some one of the ingredients just described.

In many cases water is sufficiently softened for bathing purposes by the addition of a proper quantity of borax.

Mineral Matters Left in Softening Water.—Waters may often be of such a character that the process of softening necessarily leaves in the water a number of precipitated materials. Water of this kind should always be allowed to settle; in other words, be carried to sedimentary basins before it is used. In this way gravity will speedily clear the water, since the precipitated matters are always considerably heavier than the water itself. This process, however, does not care for the soluble materials which are left in the water, chiefly compounds of sodium. Thus the actual state of the water in some cases after softening may be worse for potable purposes than it was before.

Objections to Softened Waters.—Where waters are to be used in steam boilers this is not so important, but where it is to be used in the house for all kinds of domestic purposes, including drinking, it is an important matter. No process of chemical treating can be regarded as restoring water to an ideal condition of potability. No amount of sedimentation, for instance, will remove soluble sodium salts and completely remove sulphate of lime from solution. These bodies often give a distinct alkalinity to the water itself, which may amount to from thirty to forty parts per million. In such cases it is better not to attempt to secure a complete softening of the water, but only a partial one, since the original materials which make the water hard may be less injurious than the residual materials which come from a complete softening.

Method of Correcting Alkalinity.—It may sometimes be desirable to eliminate, to some extent, the alkalinity in the process of softening water. According to Johnson[1] this alkalinity should be limited to the smallest possible amount and rigid precaution should be taken to overcome it. This is done by various methods. The most common is that of saturating the water with carbon dioxid. The carbonic acid unites with the excess of alkali and forms a bicarbonate, which is quite effective in restoring the water to neutrality or even in making it slightly acid. Any process by means of which large quantities of carbon dioxid may be blown through the water is effective. This process is necessarily expensive and unless the source of carbonic acid is abundant and cheap cannot be used in a commercial way.

If these devices for an excess of caustic lime or caustic soda are not applicable, sometimes a good result may be reached by mixing the treated with an equal amount of the untreated water, which usually carries a considerable quantity of carbon dioxid in solution.

Artesian Water.—This is a term applied to water which is secured by deep borings into the earth. An ordinary well is usually under 100—and at most 150 feet—in depth. If it is required to go 200 to 500 feet in depth or more the water is usually termed artesian. The term artesian is said to be derived from the town of Artois in France where a deep well of this kind was first known. Often in artesian wells the water rises above the earth's surface of its own force; in other words, it forms a flowing well. Any very deep well, whether flowing of its own accord or not, is now regarded

[1] Water Supply Paper 315, U. S. Geological Survey, page 81.

as an artesian well. The waters of artesian wells usually have different properties from those which are derived nearer the surface, because of their contact with different characters of soluble matters. Many artesian wells yield sulphur, chalybeate, or alum waters; in others the water is so impregnated with common salt as to be unsuitable for consumption; while in many others the water is of excellent quality.

Driven Wells.—There are certain localities in which the under layers of the soil consist largely of sand, through which water percolates with more or less facility. The water which may be secured in such localities by digging into the sand may be regarded as a supply from Nature's sand filtration plant. The most economical method of securing a water supply from such sources is by means of a driven well. A "driven well" is an ordinary iron pipe driven into the surface by means of being attached to a steel point, and as each section of pipe is added to that which has gone below the whole tube is forced deeper into the soil until it finally penetrates the water-bearing sand layers. The first section of pipe which is driven into the soil is furnished with finely perforated sides through which the water can enter and all the particles of sand be excluded. The water usually rises to near the surface of the earth and is removed by ordinary pumping. Hundreds of wells of this kind can be put down under favorable circumstances and be kept supplied to their full capacity by the natural seepage of the water through the sand layers. The quantity of water which can be secured in this way is of course limited to the area and character of the water-bearing sand. This method of supply is usually resorted to only for very small cities. A very good quality of water is usually secured by this method, due to the complete filtration to which it is subjected in the ground.

Pure Water.—It has been stated that pure water can only be regarded as a chemical curiosity. It is important in the consideration of water as a beverage, to understand what the term "pure water" as ordinarily used implies. It is evident that it does not relate to chemical purity, nor can it be applied in any sense to limit in any particular way the various kinds of substances which water used as a beverage may contain. The term must be applied therefore in its common signification, namely, to mean water which is free from any infectious or harmful germ or other living substance, free from contamination with sewage or excreta of any character, and containing no greater quantities of mineral substances than give it character as a potable water. A pure water therefore may contain large numbers of other substances without losing its right to that name. If the use of a water does not tend to produce infection of any kind nor to load the system up with undesirable mineral substances, nor to derange in any way the metabolic processes, it may be regarded as pure.

The Usual Constituents of Drinking Water.—Pure drinking water includes water drawn from the ordinary uncontaminated sources of the spring, the well or the running stream. Such water contains always the essential gases of the atmosphere in solution with greater or less quantities of carbon dioxid, traces of other gases, traces of common salt, of lime, of magnesia, of soda, of iron, and many other mineral substances. It is free from any excess of organic matter, of ammonia or nitric acid, nitrous acid, and chlorin derived from any other source than common salt. It will hardly be contested by physiologists and hygienists that water of this description properly handled is better for constant use as a beverage than any of the other forms of water described. In general, therefore, the term "pure water" may be applied to water which does not carry any diseases, is free from contamination with sewage or excreta, and carries only a moderate amount of harmless and even helpful natural mineral constituents.

Peaty Water.—Water which has collected over or passed through deposits of decaying vegetable matter such as the peat deposits which exist very abundantly in all parts of the world becomes saturated with organic matter and usually has a brownish tint. Such waters are called peaty. In many localities they are regarded as of special value, but it cannot be claimed on any scientific grounds that they are in any way superior or equal to the pure waters described above. Peaty waters are highly valued for some technical purposes and are supposed to add a certain flavor to certain products made therewith, as for example, whisky, especially Scotch whisky. There is no real ground, however, for the belief that the peaty flavor as such is of any very great value in such cases. Peaty water is usually remarkably free of mineral constituents, and for this reason it probably extracts a larger quantity of the mineral constituents of barley or other grains with which it is brought into contact. To this extent it is perhaps better suited than the ordinary spring or well water for use in the preliminary brewing and fermentation precedent to distillation.

Location of Springs and Wells.—In selecting water as a beverage the location of the spring or well is important. The purest waters are those which, other things being equal, are least exposed to the contamination incident to human life. The spring which bursts from the mountain side far from any human habitation is as a rule less likely to be infected with sewage or excreta than if it is found in thickly populated districts. Wells existing in the open country where there is only one farm house very near their location afford a better supply of water than wells existing in the city. Although cities at the present time are usually supplied with perfect systems of sewage, which would seem to prevent any contamination, yet the water from such wells is looked upon by the highest medical authorities with a certain degree of suspicion. To such an extent is this suspicion carried that during the year 1907 all the wells in the City of Washington were ordered closed by the Health Officer. This was done in spite of the protests of thousands of inhabitants who were using these wells and who preferred the water thereof to the City supply derived from the Potomac. The Health Officer, however, felt justified in closing these wells on general principles and also by reason of the examination thereof which disclosed a degree of purity far less than that which should be demanded from waters from these sources. In the establishment of a country home where the natural supply of water is to be depended upon special care should be exercised in so locating the house, the barns, the stables and the out buildings as to render it practically impossible to contaminate the water supply of the family thereby. The same precautions should also be observed in the digging of a well. It should be so situated as not to be subject to contamination from the sources indicated.

The Divining Rod.—In the early history of the country where settlements were made so far from springs as to make it desirable to locate a well, there was a very wide-spread popular superstition relating to the location of the proper site for the well.

The "water witch," so called, or "wizard," and his divining rod, were important adjuncts to the location of the farm house. By means of the divining rod the so-called water wizard was supposed to locate a subterranean stream of water so that in digging the well there would be no possibility of not striking a vein of flowing water. It is not surprising that the water "witch" achieved great credit as being accurate in his prognostications, since it is well known that it is almost impossible to dig a well anywhere in ordinary geological formations without striking water at a reasonable depth. The fresh branches of various trees were supposed to possess the magic property of indicating water. Witch hazel was the tree whose branches were usually preferred, though the apple was very much in vogue. The divining rod was a twig with two equal forks which were cut at convenient lengths. The forks were held in the hands of the "wizard" and the branches extended in a horizontal direction. It was the popular superstition that in passing over a stream of subterranean water the branch or twig would be pulled toward the earth, thus indicating to the "wizard" the proper locality for the well. An interesting description of the use of the divining rod is furnished by Professor F. H. King (Rural New Yorker, September 14, 1907). This is an account of an experiment conducted by Professor King together with Professor T. C. Chamberlain of Chicago. The "wizard" with whom the experiment was made was regarded as an intelligent, conscientious and honest man. He was interested in nature and particularly in geology, and he thoroughly believed that he possessed the power of locating veins of subterranean water with the aid of the divining rod. Before starting upon his search it was his custom to look carefully over the field and he would fix his eyes upon a certain spot in the field. As he neared this objective point the stick and his eyes changed their angle together, so that as he crossed what in his judgment was a vein of water both his eyes and the twig were looking downward. After locating a vein of water by crossing and recrossing at various points he believed himself able to locate where the flow of water would be more or less vigorous. To the scientists who accompanied him it appeared that he was disposed to locate the veins of water in localities where his trained eye had learned that they were most apt to exist, and his predictions of the veins of water and especially of where they would give the strongest flow usually coincided with the conditions of topography which would tend to concentrate the underground flow. Although the twig was repeatedly observed to bend it was attributed to the unconscious muscular effect of the operator, since both observers were of the opinion that this motion was entirely unconscious on the part of the "wizard" himself. The " wizard" in question was very confident of his ability to predict the places where water was to be found. He had much less faith in his ability to predict the distance below the surface. Mr. J. C. Hodges of Tennessee takes issue with Chamberlain and King in regard to the divining rod being altogether deceptive, and makes a strong argument in its favor. (Rural New Yorker, October 12, 1907). Mr. Hodges is quite of the opinion that the explanation given that the man by his experience would naturally locate a stream in a given position is erroneous. He believes that the various streams, especially those which are found in the Appalachian regions, can be found by the man with the divining rod with almost absolute certainty. This is common every-day experience. Not only are there places where wells may be bored, but the depth and size of the streams are told with certainty. Mr. Hodges himself claims that the divining rod in his hands is practically reliable. He relates an instance where he was blindfolded and led in this condition until he located a small stream. In succession a large number of streams was located, each being marked with a peg when the divining rod indicated the presence of the water. In this way he was led three times to the same place and a peg was driven each time. After the bandage was removed it was found that the three pegs which were put down in this way were less than three feet apart. Mr. Hodges claims that the force with which the divining rod is drawn toward the earth is a measure to the magnitude of the stream. The witch hazel, in his opinion, is the only divining rod which can be relied upon. The term "water witch" therefore should be applied rather to the divining rod itself than to the man who carries it. It would be if applied to him, a water wizard. He records other instances to justify his belief in the efficacy of the water witch. He confesses that he is not able to explain the phenomenon but he is sure that it is of an electric character, some men being charged with electricity as others are not. Mr.

Hodges further asserts that the belief in the divining rod has not been confined to the ignorant alone and many instances of its power have been recorded by scholarly men. In spite of the eloquent argument of Mr. Hodges it must be admitted that modern science can have little or no faith at all in the divining rod. It is one of the superstitions which has had a long vogue but which is rapidly dying out.

Common Names Used in Describing the Characteristics of Natural Waters.—Unfortunately the use of adjectives in the English language is attended with so much latitude that the same term has often different meanings to different individuals. The character of adjectives describing natural or polluted waters has never been fixed by any national or international congress. Investigators and authors usually follow their own inclinations in descriptive terms, so that much confusion exists therefrom.

The term "pure" is generally applied in common language to all clear waters that show no sign of color nor turbidity. This common distinction of course is very apt to be misleading. Waters may be pure in the above sense and yet carry disease-producing germs which threaten health and even life.

The term "muddy" is applied to waters in which visible silt is suspended in quantities that make the water opaque. This condition is illustrated in the small streams after heavy rains in localities where the soil washes readily.

The term "bog" is applied to waters which are in contact with organic matter to such a degree and for such a length of time as to acquire a brown tint.

The term "putrid" is applied to all waters, usually stagnant, which acquire for any cause a bad smell.

From the sanitarian's point of view water is "polluted" when it contains any extraneous substance, derived especially from the surface, which tends to give it any property threatening to health. These pollutions may come from natural causes, that is, the decay of organic matter naturally in the soil or on its surface, or they may arise from excrementitious matter derived from dwellings, barns, poultry houses or pigpens.

The water is said to be "infected" particularly if it carries any pathogenic germ of a character suitable to reproduce the disease in the person consuming the water.

Finally, from the sanitary point of view a water is "pure" if it contains only the ordinary mineral substances which are extracted from the soil by the water and which are not of a harmful character, and when it is suitable for drinking and other domestic uses. In this sense the word "pure" has quite a distinct signification from the same word used from the chemical point of view.

Composition of Water.—Chemically pure water is composed of only two substances, hydrogen and oxygen. These materials in the free state exist only in a gaseous condition. Oxygen forms practically one-fifth of the volume of the atmosphere. Hydrogen does not exist very largely in a free state, though its occurrence in nature as hydrogen is sometimes observed. In nine pounds of pure water in round numbers eight pounds are oxygen and one pound hydrogen. If the composition of water is referred back to the gaseous state of its two components, it is found that two volumes of hydrogen combine with one volume of oxygen to form water. The union takes place on the application of a flame, with tremendous violence. The strongest containers of mixtures of hydrogen and oxygen in proportions to form water are ruptured with the greatest violence if the combination takes place. The composition of water may be shown very beautifully and very easily by an experiment which produces its decomposition. If water containing an electrolyte such as diluted sulphuric acid is subjected to the action of the electric current hydrogen will be given off at the negative pole and oxygen at the positive pole of the electrical apparatus. If the negative pole be placed in one arm of a bent tube and the positive pole in the other, the gases may be collected separately. This is illustrated in the simple apparatus shown in the figure. The positive pole is marked + and is found to be giving off oxygen; the negative pole is marked − and is found to be giving off hydrogen. The gases rise in the tube and are collected. It will be seen that the volume of the hydrogen collected is just double that of the oxygen. This form of simple apparatus is called by chemists an eudiometer. If pure water with an electrolyte which cannot be decomposed is used the resulting gases will be pure. Pure water is so poor a con-

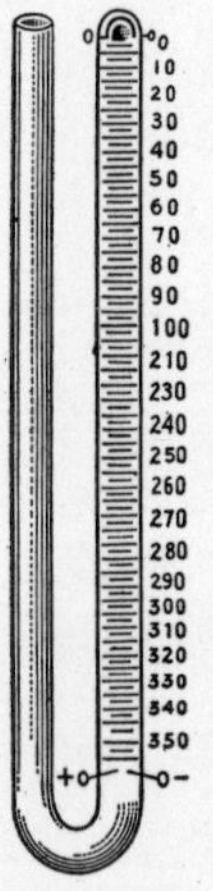

FIG. 1.—Eudiometer.

ductor of electricity that no decomposition will take place in the apparatus described because of the inability of the current to pass through the connecting arm filled with pure water between the positive and negative poles. The addition of a few drops of sulphuric acid enables the current to pass, while at the same time the acid is not in danger of being acted upon appreciably by the electric current itself. The result is that nothing is decomposed except pure water, so that pure hydrogen will be found in one arm of the apparatus and pure oxygen in the other. If the operation be conducted over an inverted tube in which both poles are in the same tube, the oxygen and hydrogen will escape together gradually forcing the liquid out at the bottom of the tube.

Rare Elements Present in Water.—Attention has been called already to the fact that water is a great solvent. It acts with vigor upon many mineral substances in the soil and dissolves gases and other materials in the atmosphere. Water thus charged with other re-agents becomes more active on certain mineral substances. Thus the solvent power of water may be increased for some substances by dissolving other substances therein. This is notably illustrated in the case of limestone which is only slightly soluble in pure water but very soluble in water carrying carbon dioxid in solution. Such water passing through layers of limestone becomes heavily impregnated with dissolved carbonate of lime. On the water emerging from the spring or the well the carbon dioxid escapes and the excess of lime is precipitated. When such water is boiled in the tea kettle all of the carbon dioxid escapes and a part of the carbonate of lime is precipitated. In families where such water is used the tea kettle may often become incrusted with a layer of carbonate of lime. In addition to the ordinary substances found in water, those which are of more rare occurrence are frequently found. This is illustrated in the fact that some waters contain small quantities of lithium, in fact lithium seems to be a very widely distributed element and is found in many waters. As it will be seen further on, many waters claimed to be lithia waters are not so in fact, but have their lithium added after being taken from the source. There are genuine lithium waters. These waters are usually highly valued for medicinal purposes, though the value of the lithium therein as a remedial agent has been very much exaggerated. Some still rarer elements, or at least those less known, are often found in water. Among these may be mentioned argon and other rare constituents of the air. Certain waters also exhibit phases of radio-activity and have thus obtained a vogue as healing agents far beyond their merits.

Interpretation of Water Analyses.—It is not expected that the general reader of this work will have expert knowledge of the method of interpreting analytical data. There are some fundamental principles of interpretation which can be applied even by the layman, and a knowledge of which will help in the interpretation of the data which are given.

Purpose of Analysis.—The purpose of an analysis of water varies with the use to which the water is to be put. From a scientific point of view the purpose of an analysis is to give accurate information concerning the composition of the sample. From a potable point of view it is to determine whether or not the water is suitable for consumption. From a technical point of view the desired knowledge lies in the direction of the utility of the water for the technical purposes to which it is to be applied. Analyses are made of water, also, with a view of ascertaining what action, if any, the water may have on the containing vessels carrying it or in which it is to be held. For beverage purposes, and that is the principal purpose of this book, the analysis of water would be particularly interesting from a sanitary point of view. The presence of certain substances in water is connected more or less intimately with the potability of the sample, and the following principles of interpretation may be applied with that point in view.

Chlorin.—The presence of chlorin in a sample of water indicates the existence therein of certain soluble chlorids. Among these common salt is the most frequent. The presence of chlorin in drinking water, due to a small content of common salt, is normal and unobjectionable. Experience has shown that the water derived from the springs and wells in places remote from the ocean and from centers of salt deposits contains only a very small quantity of common salt. If a larger quantity of chlorin than would be due to this source is found in the water, the first inquiry to be made is whether or not the source of it is near the ocean or a salt mine. If this is the case, the excess of chlorin is not open to suspicion. If, on the other hand, the source of the water is not contaminated by natural salt deposits, then the presence of the excess of chlorin is indicative of contamination with sewage or waste. The urine secreted from human and other animals is extremely rich in common salt, and hence the justification of the suspicion, where a high content of chlorin suddenly appears, of contamination with sewage. In the interpretation, therefore, of analyses, where a large amount of chlorin is indicated, the point of most importance is whether or not this is a normal condition or one which is produced by the abnormal conditions above described.

Ammonia.—The presence of ammonia in water in small quantities is normal. This nitrogenous substance exists either in a free state or what is known as albuminoid ammonia, that is, soluble protein matter. Whenever a water is found to contain an exceptionally high content of ammonia in either one of these forms, a suspicion of contamination is justifiable. The contamination is of an organic character, that is of sewage or animal wastes or decaying vegetable or animal matter.

The presence of free ammonia in small quantities of course is not objectionable to health. That is not the point which is important. The significant feature of the presence of excessive amounts of these bodies is in the probability of contamination just mentioned.

Nitrites and Nitrates.—Nitrites and nitrates, the salts of nitrous acid and of nitric acid, are normally found in water, especially the nitrates. The presence of these bodies in minute quantities is therefore not objectionable. When, however, there is any considerable increase in the quantity of nitrites especially, pollution of an organic nature may be suspected. Either in the decay of this organic matter nitrous acid may be produced, or the organic matter present may work upon the nitrates normally present and reduce them to a lower state of oxidation, namely, nitrites. The nitrites are extremely transient in character. They are likely to oxidize rapidly to nitrates when the water is brought in contact with free oxygen.

If a water contains more than 0.003 part of nitrites per million, it is sufficient indication of contamination to warrant an investigation.   The normal quantity of nitrites and nitrates occurring in the water should be taken as a basis of comparison in each locality. Any sudden increase in the amounts of these bodies is almost a certain indication of contamination.

Total Solid Matter in Waters.—When a potable water is evaporated to dryness, it leaves a residue.   All of the materials forming this residue are grouped together under the term "total solids."   It is an important question to know just how much of this material may be present in a water without impairing its value for potable purposes.   The quantities of total solids in drinking waters vary greatly.   Some waters are almost as pure as distilled water, containing only mere traces of total solids. Others contain considerable quantities of mineral salts in solution. Sanitarians differ in regard to the total quantity of these bodies which a water may contain without being condemned for potable purposes.   As little as 60 parts per million in some localities would be regarded as excessive; while in many other localities, where waters are highly impregnated with mineral salts, as, for instance, gypsum waters, as much as 1,200 parts of solids per million may be tolerated.

Kind of Matter.—The kind of mineral matter is perhaps of more importance in this relation than its total quantity.   If the mineral substances present have medicinal qualities, as, for instance, sulphate of soda or sulphate of magnesia, less quantities of total solids are tolerated.   If, on the other hand, they are of very little physiological effect, as, for instance, carbonate of lime and the sulphate of lime, a larger quantity may be tolerated. When a water has a total mineral content, of all kinds, exceeding 1,200 parts per million, it may be regarded as ceasing to belong to the potable class and to pass over into the realm of mineral waters.

Sanitary Analyses.—When a water is examined solely to determine whether it is contaminated attenton is paid, aside from a bacteriological analysis, especially to the determination of free and albuminoid ammonia and nitrites and nitrates.   There are certain points that should be borne in mind when considering a sanitary analysis of a water, viz., when all the figures are high, or certain ones are extremely high, the chemist can usually say with a reasonable degree of certainty that the water is contaminated; but in some cases all of the figures may be low, thus indicating to the chemist that the water is perfectly healthful and yet bacteria may be present which would have a very injurious effect upon the health of the consumer. In such cases a sanitary analysis serves only as an indication and not as a proof, and should be accompanied by a bacteriological examination to give results of a positive nature.

The following is an analysis made in the Bureau of Chemistry of a sample of water, which one can say with a reasonable degree of certainty is contaminated.

Parts per million

Free ammonia	2.04 (high)
Albuminoid ammonia	0.17 (high)
Nitrogen as nitrates	Trace
Nitrogen as nitrites	Faint trace
Oxygen consuming capacity	4.75 (high)
Chlorin	12.00
Total solids	55.00

Following is an analysis of a water which one can say with a reasonable degree of certainty is pure.

Parts per million

Free ammonia	Trace
Albuminoid ammonia	0.02
Nitrogen as nitrates	Trace
Nitrogen as nitrites	Trace
Oxygen consuming capacity	0.10
Chlorin	4.00
Total solids	128.00

Following is an analysis of another water which one can say with a reasonable degree of certainty is pure.

Parts per million

Free ammonia	0.03
Albuminoid ammonia	0.06
Nitrogen as nitrate	0.106
Nitrogen as nitrite	Trace
Oxygen consuming capacity	0.175
Chlorin	4.800
Total solids	355.00

A bacteriological examination of this water, also indicated it was pure. It showed only 38 organisms per mil[1] (c.c.) and no gas producers in either 0.1 or 1.0 mil.

[1] Cubic centimeter.

Following is an analysis made in the Bureau of Chemistry of a water, which is extremely doubtful. One cannot judge of its purity without a careful bacteriological examination and also a further chemical investigation.

Parts per million

Free ammonia.................................. 0.80
Albuminoid ammonia.......................... 0.108
Nitrogen as nitrates........................... 5.00
Nitrogen as nitrites........................... 0.0005
Oxygen consuming capacity..................... 0.80
Chlorin....................................... 14.50
Total solids................................... 399.00

Value of Bacteriological Examination.—Until within a few years it was customary to judge of the purity of a water by a chemical examination. The quantity of organic matter which a water contained, the amount of free ammonia therein, the amount of so-called albuminoid ammonia which could be developed, the presence of chlorin not coming from the natural element of common salt, the amount of oxygen, which organic matter therein contained, consumed, were all regarded—and justly so—as marks of purity or impurity. There is no desire in this case to throw any doubt upon the value of these sanitary analyses. Such examinations are of the utmost importance. Since, however, the evil effects produced by water are rarely due to any inorganic substance, except in the case of mineral waters, nor to mere organic matter it is evident that a chemical examination is not sufficient to judge of the purity of the supply. In the present state of our knowledge we may say safely that a bacteriological examination is more important from a sanitary point of view than a chemical. The only doubt which attends a bacteriological examination is that as a rule the pathogenic germs which are in water are so few in number that they may escape detection. In all cases a very small quantity of the water is used for examination. This quantity is usually diluted necessarily in order that in seeding a sterilized medium therewith the number of bacteria which grow therein are not too crowded to be counted; hence, a very small portion even of the small portion originally used actually finds its way to the media used for growth. Hence in a water, say, containing 10,000 bacteria to the cubic centimeter and only 10 typhoid germs it is quite possible that in seeding the growth media not one of the 10 germs will find a lodgment. In order to avoid this in all cases of suspicion large numbers of bacterial counts and cultures must be made. In this way it is quite likely that the deadly germ for which the search is made will not long escape detection.

Determination of the Pollution of Water.—Reference has already been made to the sanitary analysis of water by chemical means and to the sanitary analysis by means of bacteriological investigations. The question therefore arises as to how the pollution of a water may be accurately determined, and also can any conclusion be drawn as to the character of the pollution. In the cases of greatest pollution, that is where the water carries visible fragments of detritus of different kinds, a microscopic examination will usually reveal the character of the contamination. The various kinds of débris may be identified by structural or other features and thus easily traced to their source. One of the most common contaminations of running water is particles of silt. These are of greater or less size, according to the force of the current and the time which has elapsed since the fragments of soil have been detached. The examination of the water will reveal the character of the silt, especially where the particles are still large enough to have definite physical characteristics. The silt, however, may exist in a state of such fine sub-division as to remain in suspension for many days or even indefinitely. Such silt has no physical properties by means of which it can be identified. Fragments of food or excreta or particles of paper or other matters will indicate at once gross pollution with sewage or refuse from the dwellings. Fragments of wood pulp will indicate contamination from a paper factory. Discoloration due to tannin and particles of tan bark or fragments of hair or leather will indicate contamination from a tannery, and so on through the list of possible contaminations. When the character of the gross contamination is not sufficient to determine the source of pollution recourse must be had to sanitary, chemical and bacteriological analyses. Often the ordinary chemical analysis will reveal the source of pollution. This is the case wherever soluble matters are poured into the water supply from any source whatever and which are of a nature to enable the chemist to determine their character. Thus, colorless solutions, such as salts of sulphite or sulphate of soda or other soluble salts can be easily identified by analytical means.

If the matters producing the contamination are colored a particular tint or at least some shade thereof borne by the original matters will reveal the character of the pollution. Peaty and swamp waters are often revealed also by the color, which is usually caused by humus substances in solution or suspension. The character of the bacterial flora will also be an indication of the possible sources of pollution. Not only the nature of the substance found, but also the amount in which it occurs, has to be taken into consideration. This is not true of pathogenic bacteria, which if present even in very small numbers indicate a water unfit for consumption. Bacteriologists regard particularly as an indication of dangerous pollution the presence of bacillus coli and streptococci and many others which can be found described in standard works on the bacteriology of water.

Kinds of Bacteria in Water.—The number of bacteria of all kinds in water varies enormously. After a long rain the last part of the precipitation may be almost free from impurities, including bacteria. The same is also true of water from snow collected on high mountain peaks, yet even in these cases absolute freedom from bacteria is not secured. The water that comes from great depths after filtration through rocks and beds of sands often has its bacterial content reduced to almost zero. The same is true of the waters of certain springs. When, however, water comes from any source near the surface of the soil its bacterial content is always likely to be high. The number of bacteria in the same water may vary from day to day. Warm and sultry days cause a great increase in the rapidity of bacterial growth, so that in changing from cold to warm weather suddenly the number of bacteria in any given water may rapidly change.

Rapidity of Multiplication.—The rapidity with which bacteria increase in number is sometimes phenomenal. Even in water where the bacterial food is not present in great abundance the increase is very rapid. In milk, which is a most favorable medium for rapidity of growth, the rate of increase is sometimes quite incredibly large. At ordinary room temperature the number of bacteria in any given amount of water increases at different rates. Franklin found, as quoted by George Newman (Bacteriology and the Public Health) that a water which contained in round numbers 1,000 organisms at the time of the collection of the sample had 6,000 at the end of 6 hours, 7,000 at the end of 24 hours, and 48,000 at the end of 48 hours. In another case where water was collected from a deep spring the number of organisms per cubic centimeter was found to be 7. In 24 hours the number had increased to 21. After three days the enormous number of 495,000 were found. This was at the usual temperature of the month of April. The kind of bacteria has also to be considered in this connection. Some varieties multiply much more rapidly than others.

Causes which produce Rapid Growth.—There are two principal causes which produce very rapid growth of bacteria. The first of these is the presence in the medium of quantities of material upon which the organisms can feed; the second important factor is the temperature. All organisms, however, do not behave the same when at the same temperature. It has been found that organisms which grow in cold storage plants do not develop rapidly if at all at blood heat. On the other hand, organisms which develop normally at blood heat do not develop rapidly at a freezing temperature. It is believed that all these organisms, rapidly adapt themselves to their environments, so that after a few generations the organism which is normally grown at blood heat may develop at a low temperature, and *vice versa*.

Periodic Increase or Decrease.—All observers have noticed that the rate of increase or decrease in the bacterial count in water, as well as in other media, is irregular. There may be a rapid increase up to a very large number, then a sudden change showing no increase, or even a great falling off in the number of bacteria. This change may be due to various causes; for instance the amount of food which is at the disposal of the bacteria may become exhausted. In the second place, the bacteria by their decay and by the excretions of poisonous matters may render the medium germicidal and thus destroy their own life. Of course changes in temperature are likely also to increase or decrease the rate of multiplication.

Presence of Streptococcus in Water.—Authorities differ respecting the significance of that variety of organism known as streptococcus in food products and water. That it does occur in water is obvious from the findings of numerous observers. On the other hand, most waters of ordinary purity—in fact nearly all which can be so considered—are practically free from these organisms. They are often entirely absent in water and even in soils. Their presence in water is regarded by most authorities as an indication of pollution of a very dangerous kind. Fortunately the strepto-cocci themselves do not have any very great vitality. They are easily killed and usually do not persist for any very long period. Many authorities regard some forms of streptococci as wholly innocuous. When a bacteriological examination reveals the presence of this organism in water it is always advisable to make a search to see if there is not some source of pollution that can be discovered and eliminated.

Pathogenic Bacteria.—As has already been stated in the general discussion, there are many forms of diseases which are believed to be largely propagated through water. Among these the two which have been mentioned as most important are typhoid fever and cholera. The character of the organism which produces cholera is not so well understood as that which produces typhoid. The skilled bacteriologist will have no difficulty in recognizing the presence of typhoid germs and by isolating and cultivating them can reproduce them at will. Of course any water containing the typhoid germ is totally unfit for consumption. While, as has been intimated, such water may be used with impunity by some persons in vigorous health, it may at any moment seize upon one who consumes it, producing dangerous and even fatal results. The typhoid bacteria, when they do exist in water, are usually in very small numbers. It is therefore difficult to discover them. It is considered sufficient to discover other organisms which constantly accompany the typhoid germ, such as the colon bacillus and when these are found to assume that the typhoid germ may also be present and treat the water accordingly. There are many methods by which the typhoid germs enter the water. They may enter before consumption even when they are not present in the original source. When typhoid is epidemic all parts of the house and the vessels which may be in the house or near it may become infected. The water which may be brought into the house entirely free of typhoid germs, by being placed in vessels which are contaminated or polluted or otherwise exposed to infection, may become the vehicle to carry the germ in a danger-ous way. The cholera bacillus is even more dangerous than that

of typhoid. Fortunately it is rarely present in waters and therefore need not be feared except in cases of an epidemic. In all
such cases no water should be regarded as above suspicion.

Effect of Boiling.—If the consumption of water-bearing pathogenic germs is a necessity, that is, if no other source of water can
be had, such water should always be boiled a sufficient length
of time to kill every germ which it contains before being used.
The boiling of water is a precaution which should be practised
whenever there is any danger of infection, whether the pathogenic
germ has really been discovered in water or not. It is related
of the Japanese medical corps that during the war with Russia
every advance of the Japanese army was preceded by a corps
of physiologists and bacteriologists. Every source of water which
the Japanese soldiers might be called upon to drink in their advance was carefully examined and placards posted so that the
soldiers might know whether it was safe to drink this water raw
or only after boiling. As a result of these precautions it is well
known that typhoid fever was infrequent among the Japanese
soldiers during the whole two years of the war. For domestic
use the water used for drinking purposes should always be boiled
in case of doubt or during the prevalence of typhoid fever in the
neighborhood. The boiled water is conveniently kept in a sterilized glass-stoppered bottle placed in the refrigerator.

What is Bad Water?—This is a question which comes naturally
after a study of pure water and may be answered in many ways,
and from many points of view. The most recent contribution
on this subject, by Gehrmann[1] gives a review of the general
methods of judging of water as to its character, and especially
as to what constitutes water unfit for human consumption, or
bad water.

Bacteriological Data.—From the bacteriological point of view,
the count of the number of colonies produced in the usual method
of investigation is an index of the number of bacteria which may
be found. The number, however, is not any more important
than the kinds, and if the bacterial examination reveals the
presence of a number of species allied with those of the pathogenic
character, the water would be pronounced bad, even if no germs of
any given specific disease were discovered. If the investigations

[1] First Report of the Lake Michigan Water Commission, 1909, page 68.

of the analysis show the presence of the colon bacillus or colon group of organisms, it is considered as satisfactory evidence of sewage contamination. It is possible that there may be instances where the colon bacillus exists which has not been due directly to a growth from the excreta of an animal, but even if these bacteria be transmitted through several stages of growth outside of the animal and afterward enter the water, it is evident that the contamination is one of degree only, and is to be avoided.

Bacteria due to Spores.—The English bacteriologists have also regarded the group of intestinal bacteria, which is propagated by spores, as of equal significance with the colon bacillus. Whether or not every sample of water should be condemned where organisms of this kind are found, as is usually the case, is thought of differently by experts. On the side of safety, it seems to me the only course to pursue is to determine at once by careful investigation the character of the water in which any of these organisms is found. Their mere presence excites a grave suspicion which should be resolved in all cases in favor of the consumer.

Effect of Number.—In regard to the actual number of bacteria in water, experts also differ as to the point at which the water should be condemned. Some authorities fix the limit at 300 organisms per cubic centimeter, while others think this is too high, and fix a much smaller number, namely, from 50 to 100. Others of perhaps equal authority extend the limits and say that the water may contain as much as 1,000 bacteria per cubic centimeter. It would be advisable if the lawmakers, after consulting with competent and unprejudiced experts in this matter, should fix upon a maximum limit and regulate it by law. Otherwise, there must of necessity be continual discussions respecting the number of bacteria the water may have before it should be condemned. There is abundant evidence to show that when water is properly filtered through sand, the number of bacteria which it contains will fall even below the lowest limit mentioned above. Water that contains not to exceed 25 organisms per cubic centimeter may be regarded as typical of what water should be, though it may contain much larger numbers without being open to condemnation.

Color Tests.—Many tests have been proposed, of a chemical and organoleptic character, as well as the color tests, for judging of the purity of water and as a means of classifying the water as bad. All of these tests have more or less value, without doubt, but none of them, alone, can be considered as authoritative. If water is colored, it is evidently due to contamination, which ought to be avoided. Many persons consider bog water, which has a distinctly brown tint, as not being thereby disqualified for drinking or other purposes. Water which is colored by the refuse of industrial operations can hardly be regarded as suitable for human consumption, even if the coloring agents themselves be not distinctly of a poisonous or deleterious character. In other words, the inorganic coloring matters, to which the color of water may be due, while less open to suspicion and more readily identified, are not for that reason to be classed other than the organic coloring matters, which the water may extract from the soil through which or over which it passes.

Chemical Tests.—Among the chemical tests, the determination of the quantity of nitrates and nitrites has been uniformly held to be the most significant, respecting the judgment of the water in regard to contamination. The presence of common salt also, in localities where removed from the ocean and from salt springs, in any than the ordinary small quantity, is regarded as a source of suspicion; but of not so important a character as the nitrites and nitrates.

Free Ammonia.—The presence of free ammonia in water is an index that it has been in contact with decaying organic matter containing nitrogenous elements, and thus it is laid open to suspicion. Ammonia itself, in the small quantity and in the very dilute condition in which it is found in water, would not be considered injurious, but only as an index of contamination.

Basis of Judgment.—The real value of the judgment of the water cannot be secured until a careful examination of the sources of supply has been made. This examination ought to include the whole watershed from which the supply comes. Otherwise, there will be a defect in the data which cannot be easily supplied. In addition to this, also the geological structure of the rocks and the character of the soil are of importance, since often, especially in limestone regions, due to the subterranean passages, water supplies may be contaminated from a great distance, without the contour of the soil leading anyone to suspect that such a contamination would take place.

Quantity of Water in the Body.—The tissues of the animal body as well as those of the vegetable are made up chiefly of water. The quantities of water found in the principal tissues of the human body are as follows:[1]

Percent

In the blood } corpuscles.................55.00 } whole blood
              serum....................90.00 } 76.00
In the muscles..........................76.00
In the fat..............................8.00
In the bones...........................14–44
In the brain and nerves.................70–77
In the hair.............................5.00
In the nails............................5.00

The average percentage of water in the body may be given as approximately 73 percent. In a man of average weight—150 pounds—there is an average of 110 pounds of water. The importance of water as a food is at once evident.

Functions of Water.—The principal functions of water in the physiological activity of the human body are based chiefly upon its use as a vehicle. It is a carrier or solvent of the nutrients, as well as of the products of catabolism, that is the breaking down of the tissues, and of the excreta. The average content of water for instance in the urine is 94 percent, and in the feces 76 percent. No physiological action could take place in the absence of water. Desiccation always means death, both to the animal and to the vegetable. There is an old Latin adage: *Corpora non agunt nisi soluta,* which signifies that bodies do not perform any of their peculiar functions unless they are in solution. This is particularly true of the nutrients which are active in the process of building up, that is what is known by physiologists as anabolism, and in the processes of decomposition which produce the phenomena of catabolism. Water in some form is the universal necessity of the human organism. Not only is water taken as such into the system in the ordinary process of eating, but in all beverages, none of which are entirely devoid of water. Tea, coffee, and milk are composed very largely of water; even distilled alcoholic beverages are usually more than half water, while the fermented beverages are from 85 percent to 95 percent water. Experience

[1] Data furnished by Surgeon General's Office, Public Health Service.

has shown that of the five pounds of food eaten by a man of average weight in a day, fully three-fourths or four-fifths are water, the total quantity of dry food in a day being only about one pound and a half, while the total quantity of moist food is 5 or 6 pounds. Water therefore forms approximately four-fifths of the diet of the human race. Hence it is the most important beverage, both in quantity and in function, of all those to which man is addicted.

When Water Should be Used.—There are many differences of opinion among physiological writers respecting the proper time for the use of water. There is one rule, however, which when properly followed leads to the best result, namely, that water should be drunk when one is thirsty. The quantities of water, for instance, which are required in winter when evaporation from the surface of the body is restricted, is very much less than that which would be used by a man in summer, especially if engaged in hard labor. A good rule is to drink frequently, rather than to drink much at a time.

There is much difference of opinion respecting the use of water during meals. Formerly physiologists were almost unanimous in their belief that the quantity of water used with meals should be restricted, if used at all. The later investigations by Professor P. B. Hawk and his assistants favor the use of a reasonable quantity of water with meals. The secretion of saliva containing the enzyme ptyalin, which converts starch into sugar, does not appear to be restricted by the use of water while eating, but rather promoted. There should, of course, be a sufficient quantity of liquid in the masticated food to favor the action of the digestive enzymes in the stomach. When the food is too dry the action of these enzymes must be to that extent restricted. Theoretically too great concentration of the food in the stomach would interfere more seriously with digestive action than too great dilution. The present status of the problem, therefore, favors a somewhat generous use of liquids, such as milk and water, at meals. This liquid should be taken when no food remains in the mouth, in other words "between bites."

Temperature at which Water Should be Drunk.—The proper temperature for drinking water is a matter of some interest. In many cases physicians advise the use of hot water or warm water, especially in the morning before breakfast. This is doubtless at times advantageous. There is no hygienic objection to drinking warm or even hot water, if not hot enough to produce any pain or injury. There is a natural tendency in man to prefer cool water, especially when thirsty and in hot weather. Before the invention of the methods of preserving ice this natural taste of man was gratified by the natural refreshing coolness of the water at its source. Many springs and wells, even in the hottest summer, give water of rather low temperature. With the advent of the modern "ice age," which furnishes even to those in most moderate circumstances plenty of ice during the hot months, the habit of drinking water ice-cold has come into general use. This habit should be avoided or restricted and the people should learn to drink water at its natural temperature as furnished by the spring or well. I should strongly advise against the drinking of water, especially in hot weather, at a temperature lower than 50°F. Temperatures of 55°, 60° and 65° are agreeable and even refreshing, and these can be obtained either at the natural source or by judicious cooling. The addition of ice directly to water should never be practised. It makes the water too cold and invites infection. Ice is often cut from rivers and ponds and is found to carry very often more germs and more dangerous germs than the water from which it is taken. The reason of this is that the ice may collect débris from those who pass over it before and during cutting and thus contain more germs than the natural water from which it was formed. The chief objection to drinking ice-cold water is in the sudden chill which it gives to the coats of the stomach, which cannot be otherwise than injurious. The matter is well put in these verses:

"Full many a man both young and old
  Has gone to his sarcophagus,
By pouring water icy cold
  Adown his hot esophagus."

Artificially Cooled Water.—It is possible easily to obtain water of a reasonable degree of refrigeration, say, 50°F., without the addition of ice directly. This can be done by the very simple device, which is in common use, of placing a moderate amount of ice external to the exit tube through which the water supply is [drawn. The Fig. 2 shows a simple construction which permits the cooling of water in a moderate manner without bringing the ice into contact therewith. As is shown in the figure, the water is contained in a glass bottle, which is inverted into the funnel-shaped receptacle, filling it so as to seal the mouth of the bottle. As the water passes from the mouth of the bottle through the conduit, it is cooled by the ice exterior thereto. The quantity of ice may be so adjusted as to prevent the water from reaching a temperature lower than 50°. As the water is drawn off air enters and takes the place of the water in the bottle. This principle is well known to chemists, and the apparatus is known as Mariotte's bottle. The common habit in this country of placing cracked ice in the glass is greatly to be deplored. Any one will enjoy drinking water far more by having it at a medium temperature, and the risk of injury is far less.

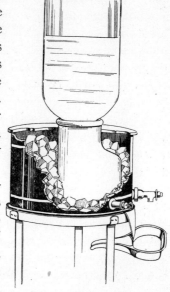

FIG. 2.—Water-cooler.

Purification of Drinking Water.— Since even with great care the sources of drinking water may from time to time become contaminated, it is of supreme importance to know what treatment is necessary in order that the waters may be made fit for consumption. To this subject chemists, bacteriologists and engineers have directed much attention. An interesting summary of the history of water purification has been published by the Geological Survey,[1] to which I owe the data following.

Historical Sketch.—The ruins of antiquity show that large storage reservoirs were common in ancient times, and it is well known that the Chinese for thousands of years have used alum as a coagulant in muddy water in order to accelerate clarification. Perhaps the earliest literary reference to filtration appears in the "Ousruta Sanghita," a collection of medical lore written in Sanskrit probably 4,000 years ago. In a letter to the British Journal of Preventive Medicine, Mr. Francis E. Place, of Jaipur, Rajpu-

[1] Geological Survey, Water-Supply Paper 315.

3

tana, India, calls attention to this reference, in which the following statement appears: "It is good to keep water in copper vessels, to expose it to sunlight, and to filter it through charcoal."

Modern history does not record any attempt at filtration until 1829, when the 1-acre slow sand filter was built by James Simpson for the East Chelsea Water Co., at London, England. The germ theory of disease was then unknown, and the filter was built to perform the offices of a mechanical strainer for the purpose of removing the turbidity from the water. This filter is still in service, however, and is doing work of a nature far exceeding the purpose for which it was designed and built.

Typhoid fever as a specific disease was discovered in 1829, but it was not until 1849 that the germ theory of disease was seriously advanced. An act of the British Parliament of 1852 made compulsory the filtration of the entire water supply of the metropolitan district. This action was the result of the severe cholera epidemic of 1849 and was the first of a series of attempts to purify water for hygienic reasons.

The first noteworthy movement in this country for the purification of a public water supply was made in 1866, when the city of St. Louis sent James P. Kirkwood to Europe with instructions to investigate the art of the purification of water as there practised. On his return Mr. Kirkwood made an elaborate report, which will always remain one of the classics on the subject. His recommendations for St. Louis were not adopted, however, apparently for sound reasons, as none of the purification works in Europe which came under Mr. Kirkwood's observation had a water to treat that was similar to the water at St. Louis. The waters of western Europe are almost uniformly clear, whereas that of the Mississippi is extremely turbid.

In 1872, about five years before Mr. Kirkwood's death, a plant was built at Poughkeepsie, N. Y., in accordance with his plans. This was the first practical attempt at purification of a municipal water supply in America. Plants of a type similar to that built at Poughkeepsie were built somewhat later at Lowell, Mass., Columbus and Toledo, Ohio, and elsewhere, but most of them failed of the purpose for which they were intended.

Quite extensive experiments on slow sand filtration were also made at Boston, Mass., Louisville, Ky., and elsewhere.

The classic investigations of the Massachusetts State Board of Health at Lawrence, Mass., were begun in 1887 and are still in progress. Up to a few years ago the work at the Lawrence Experiment Station, so far as water purification is concerned, was limited to studies on slow sand filtration. The construction of the Lawrence city filter, first placed in operation in 1893, was one of the results of these investigations.

In 1893 the first carefully conducted experiments with the newer process of mechanical water purification were made by Edmund B. Weston on the water supply of Providence, R. I., and in 1895 the much more elaborate studies in the same line were begun at Louisville, Ky., and continued through 1897. These Louisville experiments, conducted under the direction of George W. Fuller, formed the beginning of practical demonstrative investigations into the various methods of water purification. Similar studies followed successively at Pittsburgh, Pa.; Cincinnati, Ohio; Washington, D. C.; New Orleans, La.; Philadelphia, Pa.; and elsewhere.

All of this experimental work gave a great impetus to water purification in this country, and not only has the number of cities installing water-purification works increased rapidly during the last 10 years but the design and construction of such works has now reached a high plane of excellence. The more advanced ideas in this regard were first manifested in the works at Albany, N. Y., designed by Allen Hazen, and in the works of the East Jersey Water Co., designed by George W. Fuller and built at Little Falls, N. J. The former plant was first used in 1899 and the latter in 1902. At Albany the filters are of the slow sand type, and at Little Falls of the mechanical or rapid sand type.

In 1900, according to Hazen, 1,860,000 people, or 6.3 percent of the urban population of the United States were being supplied with filtered water. In 1904 the number of people so supplied had increased to 3,160,000, or 9.7 percent of the urban population of the country. Since that time many large cities have installed filter plants until now (1918) about 15,000,000 people in the United States are being served with filtered water.

Principal Methods of Modern Purification.—The methods employed for purifying water for potable purposes are classified as follows: First, boiling; second, treatment with chemicals;

third, filtration; fourth, treatment with chemicals and subsequent filtration. By means of one or a combination of the above methods, it is possible to convert the water supply which has become contaminated into a safe and palatable beverage.

In describing these methods of purification, it is not the intention to approve of them as a substitute for the securing of pure, uncontaminated water. That should be the object of every householder and municipality. Methods of purification should not be permitted to interfere with, or delay in any way, attempts to secure pure water at the source. In order that the sources be uncontaminated, the general practice of pouring waste and sewage into running streams must be regarded as entirely reprehensible. Many years will elapse before the running waters of the country can be protected from such contamination. Meanwhile we must be content with doing the best possible to purify the contaminated waters.

The day will doubtless come when all pollution of the streams of the country will be forbidden by national, state or municipal laws. Already the state of Ohio has made a start in this direction, and the state of Indiana, as well as some other states, has instituted a careful sanitary survey of running streams for the purpose of determining the extent and character of the pollution. Among the great cities Baltimore has already established a sewage disposal plant, which will be described further on. International questions also arise in connection with this subject. The Great Lakes are becoming contaminated, thus presenting a problem which can only be dealt with by international means. Attention will be called in coming pages to some of the systems of sewage disposal, notably that of Chicago, which places upon individuals, towns, corporations and states, below the city of Chicago, a serious problem in connection with the diluted Chicago sewage. These problems will be discussed only in so far as is necessary to convey to the public the knowledge of how pollutions occur and the necessity of remedying or preventing them.

Sanitary Control of Water.—The pollution of the water supplies threatens injury to the people of the United States in two principal ways. In the first place the pollution of the streams is destroying the fish supply, thus removing a very important source of food. Further than this, the pollution of the streams entering the sea, especially near large cities, is so great that the oyster and clam supplies are polluted, thus threatening the health and life of the citizens of the country.

The second threat to the prosperity of the country is in the pollution of the water supply for human consumption. The United States Government, through the Department of Agriculture, and the public health service and the state governments in many states is endeavoring to minimize the pollution of the running streams. As the population grows denser and manufacturing industries are more widely scattered, the threat of pollution becomes greater. In looking to the future control of the running streams it is evident that the states and the government must continue to coöperate, by reason of the fact that the state can control the stream only within its borders, while the United States can control the streams only as they flow from one state to the other. Thus, for complete control both sets of authorities must act.

Flow Through the Chicago Canal.[1]—The authorized flow through the canal was fixed by the War Department at 4,167 cubic feet per second, but at the present time and for a number of years past, the flow has averaged approximately 8,000 cubic feet per second. This diversion has of course, lowered the lake level to a certain extent, probably not over a few inches, but owing to the natural fluctuations of the level of Lake Michigan the amount of such lowering cannot be determined without elaborate calculation. A very complete discussion by the U. S. Lake Survey of the probable effect on lake levels of the diversion of their waters is given on page 5401, Annual Report, Chief of Engineers, 1900 and pages 4120–4131, Annual Report 1904.

The level of Lake Michigan is constantly fluctuating. Local and temporary oscillations of irregular amount and duration occur frequently. These oscillations which are due to the influence of wind and changes in barometric pressure, sometimes amount to as much as five feet, but average considerably less than one foot per day. In addition to these temporary oscillations, there is a consistent seasonal change in level during each year, the highest stage occurring during the summer months. Since 1860, the greatest annual change in elevation between the highest and

[1] Letter from War Department District Engineer, Chicago, April 26, 1918.

lowest monthly mean of any year is given by the U. S. Lake Survey as 2.23 feet and the least annual fluctuation as 0.57 foot. The mean annual level varies from year to year but without any regularity. Since 1860 the maximum variation between the highest and lowest monthly mean during the whole period was 4.59 feet.

St. Louis vs. Chicago.—One of the most important illustrations of the necessity of control has been found in the celebrated contest between St. Louis and Chicago respecting the diversion of the drainage waters of Chicago from the lake through the drainage canal to the Mississippi River. The sending of such a huge body of sewage into the Illinois River and thence into the Mississippi River is evidently a menace to the inhabitants who live along the banks of these two streams. While it is true that the pollution of a stream is cared for in course of time from natural causes, it is also true that in flowing water very great injury may be done before these natural causes have had an opportunity to completely purify the water. The term "completely purify" may be too strong, for in natural purification the end results of the oxidation processes still remain in the water. While the nitric acid and the nitrates formed therefrom and the ammonia and the common nitrates which remain in the water are in themselves no very great threat to health, the esthetic point of view is offended, by offering for human consumption the purified sewage of large cities in which all the end results of purification remain.

The principle of the natural purification of the water rests upon the oxidation of the organic matter, which is the principal carrier of dangerous infection, by means of organisms which produce a solution of the solid bodies and a change in the chemical nature practically of all the bodies contained in the sewage. These organisms act more rapidly in warm than in cold weather. Other things being equal, the sewage would be more rapidly purified naturally during the summer than during the winter. During the summer, also, the volume of the carrying waters is less, and hence the necessity for purification is greater. These organisms not only attack the carbohydrate bodies and destroy them, but they also convert the nitrogenous bodies into nitric acid and ammonia.

It is not safe to depend upon these natural causes for purification, as they act irregularly and at varying speeds of activity and in varying completeness of final results. The sewage of cities and of towns, should be impounded and purified before being allowed to enter the running streams.

Baltimore Methods.—The sewage of Baltimore, formerly, was poured into the Chesapeake Bay, polluting the waters almost beyond the state of toleration in the vicinity of the city, destroying the fishing interests and polluting the oyster beds throughout the whole length of the Chesapeake Bay. The sewage of Baltimore is now carried to large areas of filter beds, distributed over them in a fine spray in the most convenient form to be acted upon by the organisms which grow in the purifying receptacles. Only the clear, purified, effluent water is allowed to enter the Bay.

The city of New York has already taken steps toward conducting the immense sewage of that great metropolis under the East River and onto the sandy plains of Long Island, where it may undergo a similar purification. The State of Ohio has enacted legislation looking to the speedy prohibition of pouring the sewage of Cleveland and Cincinnati into Lake Erie and the Ohio River respectively. The States of Indiana and Illinois particularly have undertaken local surveys of their own waters and have prosecuted them with fine results. The citizens of those two states will soon be able to learn from official sources what streams carry water suitable for potable purposes and what do not. This problem is of supreme importance to the people from the aspect of sanitation and health.

Filtration.—The most obvious method of removing gross materials from water is by filtration. All solid particles—silts, sand, clay, and other mechanical impurities—may be removed in this way. In water taken from running streams the most abundant mechanical impurity, especially after heavy rains, is silt or clay. In many sources of water supply these impurities are never absent. As an example, it may be said that the waters of the Missouri River are never clear, and the Ohio River is probably muddy three-fourths of the year. Many large cities on the banks of these streams take their water supply from the Rivers. This is true also—or was true—of the Potomac, the waters of which are muddy for a large portion of the year. Until the installation of the filtration plant at Washington it was a rare thing to have clear water unless the river waters were first collected in reservoirs and allowed to settle for a long time. Two principal methods of filtration are practised: the first may be described as the chemical method. In this process the suspended matter is coagulated by the addition of alum, ferrous sulphate or lime, or all three combined. These bodies usually produce the coagulation of the suspended silt, by the process technically known as flocculation. By some the suspension of the silt in water is supposed in one sense to be an electrical phenomenon, and the addition of an electrolyte like some of the soluble salts mentioned above is sufficient to produce coagulation and precipitation. Whatever be the cause, however, the fact remains that to remove suspended silt from water by any kind of filtration coagulation can be effected as described. It is extremely difficult to remove suspended silf from water by any kind of filtration except that of the sand bed, to be described hereafter, without previous coagulation. After the coagulation the particles of coagulated material can be separated by passing through filters usually made of sand or sand and charcoal, which retain the precipitated matters. These filters, as in fact all others, require frequent washing in order to remove the occluded silt. The filtration may be accomplished by gravity, that is by throwing the water on the filtering bodies and allowing it to percolate, or by mechanical force, that is by forcing the water through the filters by pumps. For small supplies, such as for factory use or for small cities, the filtration by means of pumps and small filters can be practised. For large supplies, however, more economical methods are required.

Disadvantages of Chemical Filtration.—The objection to filtration such as just described is found in the introduction of chemicals into the water supply, which in themselves are objectionable. Neither alum nor ferrous sulphate is desirable in water for domestic consumption. Both are injurious and in any considerable quantity are often poisonous. When used in conjunction with lime and judiciously manipulated none of these substances remain in solution; they are all removed with the material which is separated by filtration. Where applied, without care, as is often the case, in larger quantities than are necessary, some of these materials remain in the water supply; hence this form of filtration should only be used under the constant control of a skilful chemist who by his careful oversight and frequent examinations may assure the consumer that the water is free from the objectionable added materials. For these reasons the chemical process of filtration will probably never command universal confidence.

Simple Mechanical Filtration.—By far the more desirable method of separating gross particles from water is the sand method of filtration in use in many cities. In this country extensive sand filtration plants have been installed for the use of the City of Philadelphia, Washington, and many other cities. It is not the purpose in this manual to go into any description of the methods of building sand filters. The principle of the sand filtration rests upon the fact that where muddy water or water polluted with gross materials is brought upon a bed of sand so arranged as to allow free percolation from the lower surface the water may be secured in the effluent perfectly bright when the filter is properly working. Such sand filters do not give a clear filtrate when first put in use because the sand particles are too coarse and allow some of the silt to pass. There is gradually formed upon the upper surface a layer of silt or mud which fills up the grosser pores of the sand and finally becomes thick enough to retain all of the suspended particles, even without previous coagulation. When the filter is thus covered with its mud covering the water in the effluent is perfectly clear. The sand bed remains in operation until the mud layer becomes so thick that the water ceases to percolate. The mud with the adhering sand is then removed mechanically, washed if necessary to recover the sand, which is replaced upon the bed, when it is again ready for use. It is evident that a sand filter of this kind must cover a large area where the water supply is for a populous community. Many acres of filter surface are necessary for a city of 100,000 inhabitants. The original construction of such a sand filter is extremely costly, but when once finished its operation is comparatively economical.

Slow Sand Filtration.—One of the best examples of slow sand filtration of a water supply is that of Philadelphia, which has four plants, namely, at Lower Roxboro, with a daily capacity of 12,000,000 gallons; Upper Roxboro, with a daily capacity of 16,000,000 gallons; the Belmont plant, with a daily capacity of 67,000,000 gallons; and the latest and newest one of Torresdale, with a daily capacity of 240,000,000 gallons. The total cost of the Torresdale slow sand filtration plant in Philadelphia was $9,208,000.

In general it may be said that the cost per acre for building a slow sand filter is about $60,000. The gross daily capacity of a 1-acre filter is approximately 3,000,000 gallons, although its average capacity is only about 2,500,000 gallons. If the daily consumption be 125 gallons per capita, an acre of filter would supply 20,000 people. The first cost would therefore be about $3 per person, and the interest on the investment amounts to 18 cents per annum for each person. It is estimated by the experts of the Geological Survey that if the water be boiled as a substitute for sand filtration, the cost in fuel for 6 weeks will be about the same as the annual cost for supplying water from the slow sand filtration, from which the bacteria are so thoroughly removed that it is almost as safe as boiled water. On the basis of the consumption of 125 gallons per person per day, the actual cost of slow sand filtration would be about 36 cents per person per annum.

Rapid Sand Filtration.—Diminishing the area of the filter and rendering it more rapid in its operation has attracted a good deal of attention in the last few years. I have visited a plant of this kind at Columbus, Ohio, which has a capacity of 30,000,000 gallons daily. The rapid filtration is preceded always by a chemical treatment, which flocculates and precipitates the undesirable matters in the water and also attacks and destroys a large part of the living bacterial organisms and renders the subsequent filtration more easy and more rapid. Naturally the sand grains forming the filter bed of the rapid filtration are much coarser than in the slow filtration, and this renders the previous precipitation necessary in order that the effluent water may be bright, clear, and free of undesirable matters.

One great advantage of the rapid sand filtration is in the arrangement of the machinery in such a way as to force water back through the filter when it is desired to be cleaned, which loosens the coagulant from the sand and carries it off with the wash water to any desirable exit. The cost of cleaning the filter is reduced to a minimum both of expense and of time. The largest rapid sand filtration plant in the country is that at New Orleans, which has a daily capacity of 40,000,000 gallons.

Chemicals Used in Coagulation.—The most common chemicals which are employed in coagulating waters previous to rapid sand filtration are compounds of aluminum, lime, and iron. The alums, namely, ordinary potash or ammonia alum, or sulphate of alumina, are perhaps the most efficient coagulants known. If the alum is omitted and sulphate of iron (copperas) is used only, certain modifications are necessary, and especially the introduction of the lime compound, which is a necessary ingredient when the coagulation is produced by ferrous sulphate (copperas) alone.

Attention is called by the report of the Geological Survey to the fact that to obtain satisfactory results from the use of lime when used with iron sulphate as a coagulant it is necessary to make use of sufficient lime to neutralize and precipitate the iron. The use of too little lime results in poor coagulation, caused by the incomplete precipitation of the iron, some of which is usually left in solution and appears in the effluent of the filters. The use of too much lime results in the formation of lime incrustants, which deposit in the air and strainer systems and cause much trouble through clogging.

Further, water which is treated with lime and iron will show an increase in permanent hardness, as compared with the effect of the use of compounds of aluminum. Aside from sanitary reasons the aluminum salts are considered more satisfactory as coagulants; they remove color from water more rapidly and completely and make it possible to obtain by filtration a more brilliant water than do iron salts.

Removal of Bacteria by Sand Filtration.—One of the most important points in connection with filtration as above described is the fact that bacteria which themselves possess definite size are too large to pass through the fine pores of the mud cover of the sand filter. The result therefore is that when the sand filter is perfectly used and the effluent water perfectly clear the bacteria which the water first contained are left almost entirely upon the mud film. Examinations of the effluent water from the sand filters in Washington have shown as low as 8 or 9 organisms only per cubic centimeter, and rarely do they go above 20. The experience in other cities is doubtless comparable to that in Washington. Thus the sand filtration serves a double purpose; it not only removes the suspended matter and makes the water clear, but it also retains the principal part of the bacteria which the water originally contained. It is evident that a water which is never muddy would not be greatly benefited by passing through a sand filter unless the sand was fine and compact because it would be impossible for it to form the mud cover necessary to its perfect function. If the sand is fine enough the bacteria may be largely removed. There is always an element of danger in the use of sand filters. Imperfections or breaks may appear and through these the filtered water may be reinfected; also in cleaning the filters, and immediately after cleaning pollution may take place. In spite of all of these faults, the sand filters have generally resulted in lowering the death rate from typhoid. The sand filters have also been used with previous sedimentation or precipitation. In these cases rapid and effective work has been done. In Cincinnati, notably, the death rate from typhoid has been greatly lessened. It does not exceed 4 to 6 for each 100,000 of population. In Washington, on the contrary, the deaths from typhoid have not been materially decreased by the installation of the same filtration plant. Possibly the presence of the disease in Washington may be attributed to some other cause. Since writing the above the death rate at this time (June, 1918) has materially diminished and the number of cases of typhoid fever in the city has fallen below 30.

The Septic Tank.—Another method of purifying highly polluted water—which is not employed however for furnishing water for domestic consumption—is the use of the septic tank. The septic tank is a device employed for purifying sewage so as to prevent the contamination of the streams below a city by the sewage of the city. The theory of the septic tank is to bring the sewage into contact with highly active bacterial organisms. These organisms have the property of attacking the organic matter in the sewage and oxidizing it and thus purifying the water. There is of course left in the effluent water the products of oxidation, for instance the effluent contains considerable quantities of nitric acid, or nitrates, due to the action of nitrifying organisms. The cellulose and starchy matters in the polluted water are also attacked by organisms which have the power of destroying them by fermentation or oxidation and converting them into harmless forms. The septic tank is composed usually of alternating layers of charcoal and coarse sand or gravel. The materials must be coarse enough to permit the purifying organism to find a nidus in the interstices, otherwise the septic tank would serve merely as a mechanical filter. This would operate to defeat its purpose, namely, to destroy the organic matter of all kinds which the sewage contains. This does not contemplate the destruction of particles of any noticeable size. For this reason the sewage which is to be purified must first be strained so that only the finer particles of its suspended matter remain therein. The sewage of Baltimore is now purified in a manner substantially as just described.

Rapidity of Oxidation.—The rapidity of oxidation which takes place in the septic tank is quite remarkable. In very favorable circumstances in the course of an hour or two the most objectionable sewage may be transformed into an effluent which is apparently not disagreeable either to the sense of smell or taste. For sentimental reasons it is not desirable to drink this effluent water, though a person drinking it might not be reminded of the fact that an hour before it had been nothing but most objectionable forms of sewage. The principle of the septic tank is found applied on a large scale by distributing the sewage over naturally sandy plains and thus depending upon the natural arrangement of the soil particles for the oxidizing purposes. A large part of the sewage of Paris is treated in this way, and the sandy plains thus irrigated not only produce an oxidation of the sewage, but utilize the fertilizing materials therein for the growth of crops. The principal objection to this method of purification of sewage is that in winter it acts slowly, so that it is difficult to purify the sewage in an open field at temperatures at or below freezing point. Naturally the efficiency of the septic tank depends upon the development therein of those organisms capable of rapidly destroying organic matter. These organisms exist in the sewage, and the arrangement of the layers of charcoal and sand in the septic tank afford them an opportunity of exerting their maximum activity. The surfaces of the charcoal especially become covered with these nitrifying organisms and they even penetrate into the interior of the porous substances, thus affording a maximum surface and a maximum intensity of action.

Activated Sludge.—The belief that the disposal of sewage by turning it into running streams is only a temporary expedient which will in the near future be abandoned, is growing rapidly. Our cities are constantly increasing in population. The quantity of sewage is growing larger, and the degree of contamination of running streams is becoming so great as to be practically beyond control. In addition to the septic tank, which is a rather expensive method of disposing of sewage, it has lately been discovered that the deposits of sewage known as sludge may be utilized largely for promoting purification. By innoculating the sludge with the organisms which destroy protein and cellulose, under certain conditions which are controllable, the purification of the sewage resting upon the sludge is greatly promoted. The principles upon which this destruction depends are the activating of the sludge by the introduction of the proper bacteria and the securing of proper aeration and movement to bring the sewage into contact with this activating material. This method has been exhaustively studied by the Water Commission of Illinois.

One of the chief objections to the purification of sewage in Northern cities is due to the fact that the low temperatures of winter interfere seriously with bacterial activity. If the pools in which the sewage is stored can be protected from the freezing weather either by cover or by sinking them deeper into the earth, the temperature suitable to the destruction of organic matter is more easily maintained. In addition to this the bacterial activity itself is always attended with the evolution of heat, which may be so conserved as to continue to promote the activity of the sludge even in cold weather.

Professor Bartow of the University of Illinois has experimentally determined the high value of an activated sludge in hastening the purification of sewage. I have had the opportunity of seeing this process near Urbana. His reports are found in the publications of the State Water Commission. His work is now temporarily suspended while he is engaged in field duty with our Army in France. Pearse has studied the results of utilizing activated sludge on the sewage of the Chicago packing houses and concludes that:[1]

The activated sludge process offers the best promise of a solution of the problem of treating packing-house waste of any suggested up to the present time. For the study of conditions in Chicago the construction of a unit plant has been recommended to handle 1.5 million gallons per 24 hours, to obtain working results on a large enough scale to determine the design of the final plant."

[1] American Journal of Public Health, Vol. VIII, No. I, Jan., 1918, page 55.

Effects of Different Methods of Filtration.—Not only is it desirable to have water purified to remove colors and suspended matters which offend the eye, though perhaps not strictly inimical to health, but also, and this is most important, it is desirable to have a water supply which does not threaten health. In nearly all cases of purification of water for municipalities the death rate of water-carried diseases, typical of which is typhoid fever, has decreased. In some instances this decrease has been most marked, while in others it has been slow and not so striking. The report of the Committee on Water Supplies of the American Public Health Association, for 1913, shows the following effect upon typhoid fever death rates for the following cities using mechanical and sand filters:

Mechanical filters:	Before filtration Per hundred	After filtration thousand of population
Binghamton, N. Y.	47	15
Cincinnati, Ohio	50	12
Columbus, Ohio	78	11
Hoboken, N. J.	19	14
Patterson, N. J.	32	10
Watertown, N. Y.	100	38
York, Pennsylvania	76	21
Sand filters:		
Albany, N. Y.	74	22
Lawrence, Mass.	114	25
Washington, D. C.	57	33

From the above tabulation it is apparent that mechanical filters are more effective than sand filters in diminishing the death rate from typhoid. Whether this is a mere coincidence, or whether it is based upon the essential and radical differences of the two methods, I am unable to say. Very often the good effect of filtration in the diminution of the typhoid death rate is not exhibited at once, but only after a number of years. This, I think, is probably due to the fact that the water pipes and distributing pipes of the cities often contain deposits which tend to retain the typhoid bacillus so that it may be constantly regenerated. Thus, it requires several years to remove the infection from the distributing system. It would be advisable, I think, in cases of this kind, to force through the distributing system water heavily charged with hypochlorite of lime or some other germicidal body, and allow it to flow freely from the spigots in every house. The residual water could then be driven out by the freshly filtered water for two or three days, until the germicide is eliminated.

An instance of the slow effect of diminishing typhoid death rate is seen in the case of the city of Washington, where the diminution of the death rate has been extremely slow. At this date (Jan. 1918) the morbidity of typhoid has been so greatly reduced that only 30 cases are under treatment. Another instance is that of New Orleans, where filtration was first established in 1909. The previous death rate per hundred thousand in New Orleans had ranged from 55 in 1907 to 29 in 1908. The first year after filtration the death rate remained at 29; in the second and third years it rose to 32 and 31 respectively; while in 1912 the effect of the filtration began to be distinctly felt, when the death rate fell to 14.

In general it may be said that the improvement of drainage, the establishment of efficient sewage systems, the installation of sanitary privies, and the purification of the water, have added very greatly to the average life of the inhabitants of our cities. But the full measure of benefit has not yet been accomplished. Many cities are expending vast sums of money in seeking unpolluted sources of water supply. Seattle, San Francisco and Los Angeles, on the Pacific Coast, have spent, or are to spend, hundreds of millions of dollars to secure a pure water supply from the mountains. New York City has just expended a vast sum to get its water supply from an uncontaminated source. In my opinion, no kind of a debt which promises more benefit to future generations can be contracted than that which looks to the sanitary betterment of the water supply.

Causes of Stagnant Water.—When water is confined so as to be practically motionless except as it may be moved by the wind or by the application of other forces to the surface, there is apt to be an accumulation of débris, which manifests itself in various ways. When the depreciation of the water is so marked as to become visible it is called stagnant. The condition which arises is revealed to the senses in various ways; the water may become foul smelling and thus the stagnant condition be revealed through the sense of smell. The growths of various flora, such as algæ in various forms, is also a mark of stagnancy. In such cases the water becomes coated with a scum, usually green, which indicates the abnormal growth of organisms of the kind mentioned. The débris of organic life, either of the bacterial flora or the flora represented by the type of the algæ by decay also adds additional matters to render the water less desirable. This condition of stagnancy often occurs in reservoirs of water which are intended for potable purposes, especially during the hot months. The mere growth of simple forms of algæ cannot in any sense be considered as a contamination in the sense of being injurious to health. These forms of life as a rule are perfectly harmless. They become objectionable solely by their abundance and by their decay. The appearance of the water is not attractive and no one would be willing to consume a water thus infected, even if convinced of its harmless character. During the winter months the low temperatures prevent the growth of the different kinds of organisms mentioned, and thus conditions of stagnancy are not so often revealed. Especially upon farms where the water for the animals is contained in ponds the stagnant conditions are apt to obtain during the summer months. In city supplies where the water is held in reservoirs for the purpose of settling or otherwise, the condition may also obtain.

Treatment of Stagnant Water.—When possible stagnant water should not be used for human consumption, but there are occasions when it is the only source of supply at hand. In such cases it is important that the objectionable conditions ber emoved. Various methods of removal are practised. Mechanical agitation which would naturally prevent the growth of these organisms is not applicable as a rule. Therefore recourse is usually had to chemical treatment. Various kinds of chemicals as will be shown further on added to stagnant water poison the cells of the organisms and thus apparently purify the water. The purification consists in the death and decomposition of the organisms growing upon the surface and in the body of the water. It is seen that no real purification takes place in one sense, since the remains of these organisms are simply deposited and may continue to decay as soon as the poisonous principle is sufficiently removed to permit the renewal of bacterial activity.

DESTRUCTION OF GERMS

Chemical Sterilization.—The destruction of pathogenic germs in water by chemical processes has long been practised and with

4

various re-agents. Among those which have been used particularly for this purpose are sulphate of copper and so-called hypochlorite of lime. On account of the great expense of the sulphate of copper and its extremely poisonous effect upon animal life, especially fish, the greater use of the lime compound has of late years come into vogue. It is now used in many parts of the United States and Canada to such an extent that when one speaks of the chemical treatment of water for the destruction of pathogenic germs by chemical means, the chlorin-lime compound is the one indicated unless otherwise specified. It is a cheap and effective method for destroying certain of the threatening organisms which are usually present in polluted water.

The destruction of germs in water must not be confounded with the process of eliminating silt. For this purpose certain electrolytes, like alum or ferrous sulphate and lime, are more generally employed and are much more effective than the hypochlorite of lime. Also when vegetable organisms are to be destroyed, such as algæ, there is nothing so effective as copper sulphate. Hence, hypochlorite of lime is not the ideal substance to use as a precipitating agent when the waters are subsequently to be filtered through sand or other filters. The treatment of water with hypochlorite is also not a decolorizing process.

In order that the germicidal action of the hypochlorite may be most effective, it is highly important that suspended and organic matters be previously precipitated. There is one point of excellence in the use of hypochlorite found in the fact that any excess of it is extremely disagreeable both to the taste and to the nose, and hence reveals itself directly in the water. When an excess of the re-agent has been employed, the water becomes to that extent unpotable. For this reason the use of hypochlorite of lime requires a careful chemical supervision, in order that the quantities which are necessary for purification may not be exceeded. In all these cases the only excuse for the use of chemicals of any kind is the choice between two evils. There is always a danger of using too little or too much of the chemical. In the first case the water, which is supposed to be free of danger, is not so; and in the second place, the excess of the chemical may prove injurious. The ideal solution of the problem in all cases would be to seek a source of unpolluted water, but as this is not always obtainable, we must expect that the use of purifying re-agents of the kind mentioned is a necessity or at least a precaution which should not be omitted.

Liquid Chlorin and Ozone.—Another purifier of water is ozone. Great claims have been made for ozone as a germicidal agent for water in the last few years, and many forms of apparatus for generating and using it have been applied. The schemes for promoting the use of ozone are hard to dissociate from commercial exploitation. Ozone undoubtedly has the power to destroy the life of pathogenic organisms if it can be brought into contact with them. The inefficiency of the ozone generators and the difficulty of thoroughly bringing ozone into contact with the germs to be destroyed are practical objections to the general use of this re-agent. It also is evidently very much more expensive than the simple treatment with hypochlorite, which remains today the most efficient and most economical method of chemically treating water for germicidal purposes. Violet rays, radiant emanations and other forces of a mysterious or little understood character, have also been recommended for the sterilization of water. These recommendations have only an academic value, inasmuch as their efficiency has not been demonstrated in a practical way.

Liquid Chlorin.—Since the introduction of the so-called hypochlorite treatment of waters, in 1908, many complaints have been made by reason of using an excess of the re-agent. Many difficulties have been encountered in properly mixing this re-agent with the water supply to be purified, leading to the complaints above mentioned. To avoid this trouble and to have a more controllable supply, efforts have been made to use liquid chlorin instead of the hypochlorite. Chlorin gas is easily liquefied and kept in strong cylinders in a liquid state in the same manner as is practised with carbon dioxid. Liquid chlorin has been used with good effect in combatting typhoid fever in the waters of Lockport, Tonawanda and Buffalo, New York, Chicago and Milwaukee. The use of it is easily controlled and it mixes with great facility and evenly with the whole water supply. Liquid chlorin does not remove turbidity nor color from water, and therefore it is not a complete substitute for the ordinary forms of precipitation and filtration.

Efficiency of Hypochlorite.—The committee on water supplies of the Sanitary Engineering Section of the American Public Health

Association has recently submitted its report.[1] In answer to requests for information regarding the extent and efficacy of the hypochlorite treatment 110 replies were received. About 75 percent of the waters treated were from rivers, 20 percent from lakes, and the rest from wells. The total daily output of treated water was nearly 2,000 million gallons. The average cost of treatment is 25 cents per million gallons. In very bad waters, such as the Bubbly Creek in Chicago, 60 pounds of hypochlorite per million gallons of water are used. The minimum quantity reported was four pounds per million gallons. Only about 30 percent of the plants reported tested the treated water for free chlorin. Great variations are found in the strength of commercial hypochlorites. The maximum percent of available chlorin was 44 and the minimum 15, and the average, 33 percent. The percent of bacteria removed by the chlorination process is found to be about 98.

Cost of Hypochlorite.—The cost of hypochlorite varies with locality and quantity used. From the Atlantic Coast to Indiana the average cost is $1.40 per hundred pounds. Thence to the Mississippi, $1.78. Thence to Denver, $2.60. The small plants pay greater sums than the large—varying from $1.46 to $2.84 per hundred pounds. These costs at the present time (1918) are much higher.

Application of Sulphate of Copper.—Mr. W. A. Taylor, Chief of the Bureau of Plant Industry, has kindly furnished the photograph of the simple method of applying sulphate of copper. He says: "This reservoir, shown in the figure, was seriously affected with an abundant growth of a small green alga, which imparts a very pronounced piquant odor to the water. The reservoir was so badly infested that the disagreeable odor could be detected during the summer at a distance of half a mile. An application of copper sulphate at the rate of two-tenths of a part of sulphate to a million parts of water by means of the bags shown in the figure was sufficient to completely eradicate this trouble and during the succeeding years one or two applications during the summer have been sufficient to keep the reservoir in good condition."

Restricted Use of Sulphate of Copper.—In spite of the fact that these experiments demonstrated in a most convincing way

[1] American Journal of Public Health, September, 1915, page 918.

the practicability of purifying water reservoirs in this manner, no great practical use has been made of the process. This has been due, doubtless, to the fact that many experimenters were not satisfied respecting the desirability of purifying a water supply by the addition of a substance known to be of such a poisonous character as sulphate of copper. The present high cost of the reagent and the danger of injury to fish have also had a limiting influence.

Sterilization with Permanganate.—The polluted water according to the French method is treated with a mixture of potassium

Fig. 3.—Distribution of copper sulphate. (*Courtesy of W. A. Taylor, Bureau of Plant Industry, U. S. Department of Agriculture.*)

permanganate, sodium carbonate and slaked lime in equal proportions. After five minutes eight parts of anhydrous ferrous sulphate are added for each seven parts of the mixture used in the preliminary treatment. The permanganate oxidizes the organic matter and destroys microörganisms, the sodium carbonate precipitates any calcium sulphate which may occur naturally in the water and the calcium hydrate precipitates any bicarbonate of lime which may be present. On the addition of the ferrous sulphate the excess of permanganate is removed in the form of a dense precipitate. The water drawn off from the precipitate is very pure and limpid and contains only a very small amount of the sulphates of potassium and sodium. This method, when practised upon the waters of the Seine, which contained a great number of bacteria, including the colon bacillus, produced a perfectly sterile water. There is very little danger attending the use of an excess of either of the re-agents. An excess of the first precipitating mixture would impart to the finished product a noticeable pink color, while an excess of iron, due to the addition of the ferrous sulphate, would be precipitated partly by the sodium carbonate and partly from the calcium carbonate resulting from the first reaction.[1]

TYPICAL CITY WATER SUPPLIES

Protecting Lake Michigan.—The chief problem in the Chicago water supply is to protect the waters of Lake Michigan by changing the flow to the Mississippi and at the same time not to divert enough water to seriously disturb the level of the Lakes and the flow of the St. Lawrence. Whether this can be accomplished by using the maximum quantity of 10,000 cubic feet per second, only the future can determine. At the same time international relations must not be strained and the friendship and collaboration of our neighbors on the North are necessary to accomplish the best results. In the end it will doubtless be found that the citizens of all the towns and cities bordering the Great Lakes and their connecting rivers will have to purify their sewage before it is returned to the lacustrine circulation.

Purification of the Chicago Water.—Even with the deflection of the Chicago River, causing it to flow into the Mississippi instead of Lake Michigan, the Chicago water is not entirely safe. Especially is it found that after heavy storms the deposits which have settled for many years on the bottom of the lake are stirred and the water becomes turbid and impure. The Board of Trustees of the sanitary district of Chicago recently summarized the condition of the water as follows:

"The general quality of the Chicago city water supply during

[1] Lancet, 1908, Volume 174, page 1024

the last few months of 1913 was good. Only storms of unusual severity agitated the bottom of the lake sufficiently to cause the water to appear turbid. The slight increase of the Chicago death rate of 1913 over that of 1912 was not due to the water supply."

In a statement made to me by the Department of Health of the city of Chicago, on January 8th, 1916, the following information is given as to the methods of purifying the Chicago drinking water, and also the quantity thereof consumed:

The water supply of Chicago is obtained from Lake Michigan by means of six crib intakes located from two to four miles from shore, along the water front. Two of these crib intakes, *e.g.*, the extreme north and south cribs, which have in the past shown a slight amount of contamination, are being treated by means of chlorid of lime in a dosage of two-tenths part available chlorin per million. The water from the harbor intake, two mile crib, is now being treated by means of liquid chlorin in a dosage of two-tenths part per million.

The total pumpage during the year 1915, as given by the Department of Public Works, is 221,654,000,000 gallons. During the year 1916, to date, the average pumpage has been 630,000,000 gallons daily.

Water Supply of New Orleans.—One of the most interesting problems connected with the water service of large cities is that presented by New Orleans. The City of New Orleans is built upon a deep bed of silt which has been deposited in past ages from the silt-charged waters of the Mississippi. The curious fact has been discovered that at a considerable depth beneath the surface, reached only by artesian wells, a water is supplied which is practically a bog water. It is extremely soft, as bog waters usually are, and very deeply colored with the organic matter which is held in colloidal suspension in water supplies. Large quantities of this brown water are found in the hotels and other public buildings and in use for technical purposes in different parts of the City. Another important supply for New Orleans has been cistern water. A dwelling house in the residential portion of New Orleans is not considered complete without a large wooden cistern. Owing to the character of the soil and the nearness of the water line to the surface it has been found impracticable to build cisterns in the ground. They are therefore constructed as wooden tanks above the surface. All of these supplies, however, were found to be insufficient as the city increased in size. It therefore became necessary to study the possibilities of supplying the water from the Mississippi River.

Character of the Mississippi River Water at New Orleans.—The ability of water both to dissolve matters with which it comes in contact and to carry them off mechanically, has already been referred to. In this sense water is the universal scavenger: it flushes the streets; it carries off the refuse of factories, of dwelling houses, and of stables; it transports mechanically large quantities of mineral and organic matter, and dissolves in the earth through which it passes all the soluble constituents of the soil. The water, therefore, when it reaches the spring, the stream, and the river, is in one sense Nature's sewage, and carries either in solution or in suspension all of the refuse of Nature's activity to which it can gain access. In this country the lower Mississippi River represents the greatest sewer of the world. It drains the entire area between the crest of the Alleghanies and the crest of the Rocky Mountains. This area is in round numbers 1,260,000 square miles. The city population on the banks of the Mississippi and the streams tributary thereto is about 10,000,000. The rural and town population is probably double that number. It may be said that fully 40,000,000 people, or nearly half of the population of the United States, reside in the Mississippi Valley. The river is the *cloacum magnum* of the United States. The character of the water which flows past the City is a matter of great interest. A study of the Mississippi water was carefully made by a committee appointed for the purpose of proposing a system of water supply and sewerage for the City. The results of a long series of studies extending over three years show that the Mississippi River is a very muddy stream containing about 600 parts per million suspended matter, consisting chiefly of silt and clay. If this suspended matter were separated from the water it would amount to nearly three tons of solid matter in each million gallons of water. During very dry seasons the quantity of suspended matter is very much diminished, but during periods of flood there has been found to be as much as 2,400 parts per million.

Pollution.—Although draining such an immense area on which so many millions of human beings as well as domesticated animals live, the Mississippi River water just above New Orleans has been found to be remarkably free of pollution. This is due to the fact that the volume of the water is very great in proportion to the number of inhabitants on the water shed. The purity of the Mississippi water is largely due to dilution as well as to the chemical and mechanical processes of purification which are constantly going on within the flow of the stream. In spite of the vast amount of sewage which is poured into the river, only the merest traces of it can be detected at New Orleans.

Fineness of the Suspended Matter.—It is evident from the conditions which exist that the suspended matter in the river water at New Orleans must be of extreme fineness, in other words, approaching the colloidal state. The erosion which produces the turbidity of the water, to be sure is constantly going on, but the coarse sediments which are fed into the Mississippi River through its principal affluents, namely, the Missouri and the Ohio, have about 1,000 miles of distance in which to settle before reaching New Orleans. The relative amount of suspended matter which will be deposited in a given time from Mississippi water at New Orleans is not very great. Experiments have shown the following data in regard to the amount of suspended matter deposited in 24 hours in different localities tributary to New Orleans: for instance in the water of the Missouri River at Kansas City it is found that 82 percent of the suspended matter, and in the water of the Ohio River at Cincinnati 62 percent will be deposited in 24 hours; while in the water of the Mississippi River at New Orleans only 45 percent of the suspended matter will be deposited in the same time.

Bacteria in the Mississippi River Water.—Bacteria examinations of the river water extending over about three months in the summer, when the bacteria are the most numerous, show the maximum number of bacteria per cubic centimeter of 16,000 and a minimum of 2,300, and an average of 8,000 per cubic centimeter. After filtration in the effluent water there was found a maximum of 300, a minimum of 21 and an average of 110 per cubic centimeter. The maximum bacterial efficiency of the filter was found to be 99.4 percent, its minimum 96 percent, and its average 98.2 percent.

Present Methods of Purification.—The New Orleans water purifying establishment is now completed and is delivering to the city of New Orleans 17,000,000 gallons of pure water per day.

The plant has a capacity of 60,000,000 gallons, and this could be easily extended to supply 168,000,000 gallons per day. The method of operation is as follows:[1]

Water is taken from the Mississippi River, three-quarters of a mile away, through a 48-inch suction line, which is laid level and with its top two feet below extreme low water in the river, and lifted by the low-lift pumps, through the head or controlling, house, into one of the two grit reservoirs, in passing through which it deposits about 10 percent of the heavier suspended matter which it contains. From the grit reservoir the water passes through the head house again, and, in proportion to the amount of water passing and its condition, lime and sulphate of iron are added to it, after which it passes through one of the two sets of chemical mixing passages, which afford a runway back and forth and up and down of about a mile, during which any deposit is prevented by the motion of the water. This keeps the chemical solution thoroughly mixed with the water until the full chemical reaction required has taken place. Then once more the water passes through the head house, going this time to one of the two sets of settling reservoirs, through which its passage is very slow, for the express purpose of causing the deposit of the now coagulated masses of clay and precipitating lime and magnesia which were originally contained in solution in the water, as well as of the iron and lime which were added to bring about this precipitation. The result is that when the water has reached the outlet of the settling reservoirs and is ready once more to pass through the head house to the filters nearly all of the suspended matter and a large amount of the chemicals which were in solution in the river water, together with those which were added in the treatment, have been left behind in the settling reservoirs, and the remainder, about fifty parts per million out of an average of 750 parts of suspended matter contained in the raw river water, are in such changed condition that they are easily removed by the filters, which yield an effluent entirely free from suspended matter and containing less than half as much lime and magnesia and only about 1 percent as many bacteria as the river water originally contained. The resultant water is bright and sparkling in appearance and safe and desirable for every use.

The filters are of the so-called rapid type. They are, however, merely gravity sand filters, designed to handle large quantities of water and to be very easily and cheaply cleaned. In filtering, the water enters above the sand layer, passes through it and is collected by a system of drains into one effluent pipe for each of the ten filter units, which effluent pipe is automatically throttled to prevent a too rapid flow through the filters. As the suspended matter in the water accumulates in the sand layer, the latter is gradually choked up, until finally the filter will not pass as much water as is required of it with the throttle on its effluent pipe wide open. This stage is reached

[1] The Journal of Industrial and Engineering Chemistry, Volume 7, No. 5, May, 1915, page 444.

in from 100 to 300 hours of service under present operating conditions in New Orleans, and then the filter has to be cleaned. The process of cleaning consists in closing the inlet and outlet of the filter from its connection with the operating portion of the system and in forcing filtered water into the effluent pipe, through the drains, and up through the sand layer, at a high velocity, which stirs up the sand layer, loosens the mud and causes it to overflow through troughs placed at a higher elevation than that to which this velocity will raise the sand of which the filtering material is composed. It takes less than 10 minutes to go through the entire operation of washing a filter, and only about ½ to 1 percent of the filtered water is required for filter washing.

The entire cost of treating and filtering the water and pumping it into the distribution system is not over two cents per 1,000 gallons, and the cost of water delivered to domestic consumers through meters by the city is now usually less than one-fourth of what these same consumers had to pay for raw river water pumped direct from the river before the city constructed and operated its own waterworks plant.

The methods of purification used in Chicago and New Orleans are typical of the two kinds of water supply for cities—viz., natural reservoirs, such as lakes and running waters such as rivers.

Water as a Carrier of Disease.—While the medicinal value of water has perhaps been exaggerated, the possibility of its carrying disease, on the other hand, has been minimized. There are many diseases which are transmitted by germs of some kind or other or by direct infection. It is now established beyond doubt that the diseases which were known so long as malarial were misbranded. Of all the various forms of malarial diseases, none of them is caused by bad air. In earlier times when so many people were suffering from ague (chills and fever), it was noticed that in building houses in such regions the higher lands were selected—not only on account of beauty of view but for salubrity. Those who dwelt in the lowlands and near the waters were more subject to the "malarial" disease than those living at higher altitudes. This was popularly supposed to be due to the fact that the air in the lowlands was polluted. Since the epoch-making discovery of the activity of the mosquito in carrying diseases of this kind, it is now easily seen that those who lived in the lowlands and near the water supplies were very much more likely to be bitten by mosquitoes than those living on the highlands. Water therefore is only a side issue in the transmission of diseases by means of the mosquito in that it supplies the breeding places for the larvæ of these insects. What is true of ague and chills and fever is also true of yellow fever, which is the result of a direct infection transmitted by a mosquito of another kind. The mosquito which produces the malarial fever is the variety known as anopheles, while that which produces the yellow fever is the stygomia. Another disease of a very dreadful character which is transmitted largely by water, in the opinion of most experts, is typhoid fever. Here, there is a direct infection through the water itself. The germs of typhoid fever live indefinitely in water, and even in ice. When these germs are taken into the alimentary canal they as a rule produce no effect. The perfectly healthy organism is either able to kill these germs or to prevent them from securing a hold. When, however, these germs come in contact with the organism in a certain state of fatigue or inability to resist, then they enter the intestine, propagate, and produce the well-known eruption on the small intestine which is characteristic of typhoid fever. Another disease which is supposed to be largely carried by water is cholera. Just in what way the germ is transmitted does not seem to be clear, but that water is the principal vehicle is believed by many. The many forms of diarrhœa, both infantile and adult, are doubtless produced often by water. In fact very often the diarrhœa which attacks travelers is ascribed to a change of water, though perhaps this is not always the case. But at least the possibility of transmission of diseases in this way through water must be considered. There are perhaps many other forms of diseases which depend more or less for their distribution upon the water supply, but those mentioned above are the most deadly and the most common. If therefore we accord to certain waters therapeutic or remedial effects, we must accord to certain others pathologic or disease effects. The object which must be kept in view in such cases would be to increase the consumption of waters that have therapeutic effects, and to decrease or eliminate completely the consumption of waters which are dangerous to health. Both of these events are capable of accomplishment by the application of industry, knowledge and skill. It seems only reasonable to believe that the advance of hygienic, therapeutic and engineering knowledge and skill will in the course of a few years eliminate most of the dangers incident to the drinking of water.

Epidemics Due to the Water Supply.—There are many incontestable evidences that epidemics of the most fatal sort have often been traced to the water supply. An alarming epidemic of typhoid fever occurred at Cornell University a few years ago. A very large number of the students of this University were ill with typhoid fever and unfortunately quite a large number lost their lives. The epidemic was of such a serious nature that it came near closing the doors of the institution.

The Troy Epidemic.—A more recent outbreak of typhoid fever, due to a polluted water supply, occurred in Troy, Pennsyl-

FIG. 4.—Studying contamination of water at Troy, Pa. (*Pennsylvania State Dept. of Health.*)

vania, in October, 1912. In a town of 700 persons there were 232 cases, practically 33 percent of the population. Fortunately the outbreak was not very virulent. The mortality for this reason was low, amounting to only 8.19 percent. This infection was due to a connection between the West Branch of Sugar Creek and one of the reservoirs supplying the town with water. According to the official report by the Health Commissioner of the state, a direct communication had been made between the Creek and collecting system, both through a perforated barrel and also along the upper side of the diverting wall beneath the bed of the Creek. A coloring matter, fluorescein, was placed in the perforated barrel 550 feet above the collecting reservoir, and the color occurred in the reservoir within 20 minutes. When placed in the Creek just above the diverting wall, it appeared in the reservoir 55 feet distant in 33 seconds. The illustration shows the placing of the coloring matter in the West Branch of Sugar Creek, by means of which the connection with the reservoir was demonstrated.

The Cost of the Troy Epidemic.—The Health Commissioner of Pennsylvania, the late Dr. Dixon, made an interesting study of the cost to the town of Troy of the epidemic of typhoid fever which attacked so large a proportion of the citizens of the town in the autumn of 1912. He finds that the total cost of the epidemic to the town was $28,000. Compared to this it is estimated that the cost of installing a complete system to supply pure water for the town would not have exceeded $22,000. The above represents the direct cost in money. The incidental cost, however, is much greater. The cost of the services of the health officers of the state, the estimated loss due to the decrease in trade, and the estimated minimum value of lives lost based upon the usual figures for determining such value, shows a total additional cost of about $105,000.

When it is considered that this loss practically fell upon 700 people, the magnitude of it is more readily comprehended. From all this it follows that one of the most economical expenditures which a town could make would be devoted to placing pure water, entirely free from any threat of disease, at the disposal of every citizen.

The Fly and the Mosquito.—It is not just to place the whole blame for typhoid on the water supply. In the city of Washington the typhoid rate remained quite high several years after the completion of the sand filters. Bad milk, the fly and the mosquito are enthusiastic allies of bad water in increasing the typhoid death rate. A higher death rate from typhoid during the summer and autumn months is probably due largely to the above causes. Fuller has shown[1] that the typhoid rates in Richmond and in Washington have followed practically the same curves, and that

[1] Journal of the Franklin Institute, July, 1915, page 55.

therefore the conditions in the two cities are similar. Better sanitary inspection of milk, a more eager campaign against the fly, the filling up of swamps and low places and thus combating the mosquito, have done a great deal in both Washington and Richmond, according to Fuller, in curbing the death rate from typhoid.

Reduction of Typhoid Fever in Rural Communities.—Typhoid fever is one disease which, taken the country over, is more prevalent according to population in rural than in urban localities. The data collected by the United States census show a remarkable preponderance of this disease in rural districts. In a circular issued by the Public Health Service January, 1916, attention is called to the fact that during the past year 16,369 rural homes in 8 different states were inspected by the officials of the Service. A previous survey, made in 1914, had shown that in rural communities less than 1 percent of the homes had sanitary toilets and more than 50 percent of the people were using water from contaminated sources. The typhoid fever rate, which is an excellent index of sanitary conditions, was therefore remarkably high in these regions.

The officials of the Public Health Service feel encouraged in the belief that as a result of the campaign of education which they have undertaken among rural communities the morbidity rate of typhoid fever, as well as the mortality rate, has already been considerably reduced. A similar service has been rendered to the State of Indiana by the Board of Health of that state, under the direction of Dr. John N. Hurty. Over 5,000 farmers' homes have been inspected in that state, and a campaign has been vigorously prosecuted in that state to remedy the startling insanitary conditions which have been found. It is evident that by cleaning up the water supplies of country homes and installing modern sanitary toilet systems the typhoid death rate of rural districts will be reduced even to a lower figure than that which prevails in the cities.

State Control of Waters.—A number of the states have lately been much interested in water control, to which reference has already been made, especially the states of Indiana and Illinois. The state alone cannot control its water supply, only in so far as it is produced within the state. Any efficient state control must therefore be in collaboration with adjoining states and with the Government. Pennsylvania is another one of the states that is taking a great interest in the purity of its water supply. Dr. Dixon, the former Commissioner of Health, has emphasized the impoitance of a pure water supply in the following words:

We can live longer without food than we can without water. It makes up 66 percent of our bodies. In its purity as Nature causes it to fall from the atmosphere it is the most healthful of all liquids and the only one which will satisfy the thirst. Without it our blood could not circulate through the vessels of our anatomy. The infant in embryo is fed by blood of which water is such an important part. It must be fed by blood which is clean and pure in all its parts in order to build up a healthy youngster.

He further calls attention to the fact that the Schuylkill River has become a gigantic sewer and that the Delaware is fast reaching the same condition. As a result of this pollution of the water supply typhoid fever has become in many cases epidemic. In addition to the outbreak at Troy which has already been mentioned, a late outbreak at Nanticoke caused 445 cases of illness with 10 percent of deaths. This epidemic, as was the case in Troy, was traced directly to the water supply. Anthrax is also a disease which is carried by water. Many other diseases may also be water-carried, as, for instance, cholera and smallpox.

State Control in Illinois.—The State of Illinois has created a water commission which is established at the University at Urbana, and is doing perhaps more than any other state in the rigid and scientific investigation and control of the state waters. The President of the University is the head of the organization, and the Director is Dr. Edward Bartow. A report is made annually of the activities of this Survey. The last report shows a praiseworthy activity along all the lines of investigation looking to improving the water supplies of the state. It is the purpose of this Survey to cxamine all the principal waters, and in this particular report there is an extended study of the water of the Sangamon River and its tributaries. Another interesting part of the report is that relating to the disposal of sewage along the northern branch of the Chicago River and its tributaries.

Canning Factory Wastes.—Important to the technical industries is the chapter on the disposal of canning factory wastes.

As a result of this investigation of cannery waste it has been possible to illustrate that by the use of broad irrigation beds, that is, distributing the waste over superficial areas, the waste of the canning factory can be effectually disposed of. It is of course evident that for such disposition the soil must be suitable, that is, sandy and pervious to water, and of a contour suitable to the creation of large beds in which the level of the sewage will not greatly vary from place to place, in connection to settling basins by which the greater part of the solid matter may be removed. This method of disposal of the waste may be safely considered as insuring the purity of the running streams, which up to this time have been practically the sole recipients of sewage waste.

Summary.—By way of summary, it may be pointed out that the dweller on the farm now has at his command all of the household conveniences of the city dweller, and morever he can be fortified in the same degree, or even to a greater degree, against a contaminated water supply or offensive and insanitary final disposal of sewage, and this without excessive cost. It is impossible in speaking of sanitation on the farm, especially as it relates to water supply and disposal of sewage, to lay down hard and fast rules which may be followed blindly. On the contrary, it will be necessary in arranging for any of the conveniences outlined in the foregoing to take full cognizance of the influence of local factors. The problems presented are not, however, difficult or at all complex, and with a moderate amount of study the up-to-date and intelligent farmer can with but moderate assistance from the agricultural engineer design and install water supplies, sewerage systems, plumbing systems, heating and ventilating systems, and all that is necessary of a material nature to render the comfort of life on the farm complete and its appointments even luxurious.

PART II

MINERAL WATERS

Definition.—Mineral waters, as has already been explained, are natural waters which by reason of the acquirement of large quantities of some element, by solution or otherwise, are removed from the sphere of common potable waters. This denaturing, so to speak, of a potable water may be accomplished either by the acquirement of an excess of the ordinary soluble salts which spring and well waters contain in minute quantities, or by the solution of an odorous gas, such as hydrogen sulphid, or by the acquirement of some poisonous or medicinal body, such as arsenic or iron.

There is no definite limit of separation which can be established between a potable and a mineral water. In point of fact, the division of waters into these two classes is not attended with a great deal of difficulty. Medicinal waters are often potable, and mineral waters of a low degree of mineralization are constantly used for potable purposes. In fact, the use of mineral waters is not by any means confined to the persons who make pilgrimages to the places where they are produced, but the principal consumption is in the way of bottled waters, which are sent sometimes to far distant countries.

Number of Mineral Waters.—The number of mineral waters is legion. There are many hundreds of them known in the United States, and in foreign countries they are almost, if not quite, as numerous. Only a few waters have acquired world-wide reputation, but many waters that have not acquired this reputation are of equal excellence. The classification of mineral waters has been arranged by Haywood[1]. In some respects it is the best classification which has been published, especially by reason of its compactness. It is as follows:

[1] United States Department of Agriculture, Bureau of Chemistry, Bulletin No. 91.

Group	Class			
Thermal Non-thermal	I. Alkaline	Carbonated or bicarbonated Borated Silicated	Sodic Lithic Potassic Calcic	Non-gaseous Carbon dioxated
	II. Alkaline-saline	Sulphated Muriated Nitrated	Magnesic Ferruginous	Sulphureted Azotized
			Aluminic Arsenic	Carbureted Oxygenated
	III. Saline	Sulphated Muriated Nitrated	Bromic Iodic Silicious	
	IV. Acid	Sulphated Muriated	Boric	

Haywood gives further explanation of some of the groups classified, as, for instance, thermal waters, which are those which issue from the earth at a markedly higher temperature than the surrounding atmosphere. He defines alkaline waters as those which have an alkaline reaction and contain carbonic or bicarbonic acid ions in predominating quantities. Saline waters are those which have an alkaline or neutral reaction and contain sulphuric, muriatic, or nitric acid ions in predominating quantities. Alkaline-saline waters lie between the alkaline and saline. They embrace those which have an alkaline reaction and contain (1) sulphuric, muriatic, or nitric acid ions along with carbonic or bicarbonic acid ions, both classes being present as predominating constituents or those which have an alkaline reaction and (2) those that contain sulphuric, muriatic, or nitric acid ions along with boric or silicic acid ions. Acid waters are those which have an acid reaction, and contain either sulphuric or muriatic acid ions in predominating quantities.

Apollinaris Water.—Among the waters imported into this country none has such vogue as the Apollinaris. A study of this water is more or less typical of other imported waters which are valued and used in the country. One of the first points which arises in the consideration of a mineral water is this: Is it a natural or artificial water? A natural water is a water bottled directly as it comes from the spring, without any treatment of any kind except the mechanical steps which are necessary to transfer it to the bottle and to cork it.   If the spring is naturally a carbonated spring the bottled water will be a naturally carbonated water, though of course it will not have the same content of carbon dioxid unless it is pumped from a closed tube in the spring and bottled directly without giving the carbon dioxid an opportunity to escape.   In so far as known, however, there is no such bottling apparatus at work.   Hence it follows that the naturally bottled sparkling water will probably contain considerably less carbon dioxid than the spring water itself.   The question, also, of whether this water is natural or artificial, has been raised in this country by reason of the classification of imported waters for dutiable purposes.   The revenue laws make a distinction between the duties laid upon a natural water and those which are laid upon an artificial water.   This distinction arises whenever the natural water is treated in any way by means of which any of its natural contents save carbon dioxid, as mentioned above, are eliminated, or any new contents are added for any purpose whatever.   An artificial water is also held to be one in which carbon dioxid is added under pressure, whether it be the carbon dioxid obtained from the spring itself or derived from other sources.   In regard to the classification of such waters for dutiable purposes, the opinion of Attorney General MacVeagh, issued on July 26, 1881, may be cited:

The Apollinaris water, on the contrary, is not bottled as it flows from the spring, but it is, in the first place, heavily surcharged with carbonic acid gas, and ten parts of salt are added to 10,000 parts of water. It is alleged on behalf of the importers, that not only is this the carbonic acid gas such as escapes from the spring itself or from fissures in the rock immediately around the spring, but that this water when being bottled is charged with no greater portion of such gas than it possesses at a depth of 50 feet below the surface, and that the process is merely to restore to it the gas which had thereafter escaped. This contention is, however, vehemently denied, and it is insisted that not only is there no evidence that the quantity of gas is the same, but that the evidence clearly establishes that the water as bottled contains a very considerable excess over the water in the spring.

As to the salt, the importers allege that it is simply added to preserve the water in its natural state, and to prevent contact with the cork from altering it. This, however, is also as earnestly denied, and it is insisted that the salt is added, like the alleged excess of carbonic acid gas, for the sole purpose of altering the natural character of the water as it flows from the spring, and of enhancing its value as a sparkling and palatable beverage.

In view of this conflict of testimony, and of the fact that Special Agent Adams, of your Department, Mr. Sherer, the chemist selected by him, as well as Appraiser Howard, and Collector Merritt, of the Port of New York, have, after thorough consideration, concurred in finding that the water in question is subjected to such alteration after it leaves the spring as to render it an artificial mineral water. I am of the opinion that it ought to be so regarded, and held to be liable to duty.

I cannot say that the question is free from doubt; but I have less hesitation in reaching the conclusion I have stated, because it will be subject to review, if the importers desire it, and the question will then reach the only tribunal which, in my judgment, is entirely competent to decide it.

The Secretary of the Treasury, however, did not follow the opinion of the Attorney General, but on the twenty-eighth of January, 1882, issued the following as the opinion of the Treasury:

"I conclude, therefore, that the water imported under the name of Appollinaris water is an article which is produced by nature, and is not the handiwork of man; that it is a natural and not an artificial water."

Under the rulings and practices of the Food & Drugs Act, Apollinaris is not classed as a natural water. Carbon dioxid has been added and iron abstracted.

Official Requirements.—The official regulations covering the subject of the proper labeling of medicinal and table waters is found in Food Inspection Decision 94. In the opinion of the Department of Agriculture all manufactured waters should be labeled as either artificial or imitation, the choice of words being left to the manufacturer. If no distinction is made in the labeling of a water it will be inferred that it is of natural origin. All waters which, though natural in the beginning, have anything added to them or abstracted from them after they come from source, should be labeled as artificial or should be so labeled as to indicate that certain constituents have been added to or abstracted from them. If lithia be added to a natural water, the label should be either "artificial lithia water," "water artificially lithiated," or "water treated with lithia." If carbon dioxid be added, whether the carbon dioxid be of the manufactured variety or collected from the spring itself, the water should either be labeled as "artificially carbonated water," "water artificially carbonated," "water treated with carbon dioxid," or "contains added carbon dioxid." No water should be labeled as a natural water unless it be in the same condition as at source, without additions or abstractions of any substance or substances. No water should be labeled as "medicinal water" unless it contains one or more constituents in sufficient amounts to have a therapeutic effect from these constituents when a reasonable quantity of the water is consumed. No water should be named after a single constituent unless it contains such constituent in sufficient amounts to have a therapeutic effect when a reasonable amount of the water is consumed. No water should be characterized by a geographical name which gives a false or misleading idea in regard to the composition of said water. For instance, it would not be correct to designate a water as "lithia water" merely because the water came from Lithia, Fla., or Lithia, Mass. No natural American spring water should be named after a foreign spring, unless the name of the foreign spring has become generic and indicative of the character of the water, except to indicate a type or style, and then only when so qualified that it could not be offered for sale under the name of the foreign spring.

In a later opinion respecting mineral waters, issued in May, 1914, it is held by the Bureau of Chemistry that if salts be added to a natural water the quantity of salts added need not be stated. Such words as "fortified," "concentrated," "added salts," etc., do not convey the proper information to the purchaser and are considered misleading and objectionable. The label should say: "Contains added sodium chlorid," "Contains added sodium bicarbonate," "Fortified with magnesium sulphate," or any truthful legend of a similar import which conveys the proper information to the consumer.

It is also held by the Bureau of Chemistry that water which has been artificially carbonated cannot ever be properly labeled "a natural mineral water," and if iron or calcium salts or other mineral constituents have been removed or have been allowed to separate and thus disappear from the water this fact should clearly appear on the label.

Mineral Waters of Saratoga.—The most famous mineral waters of the United States are those which occur in and near the village of Saratoga. Situated near the foot hills of the Adirondacks, the mineral waters of Saratoga are peculiarly favored as respects immunity from contamination. They flow through a series of

FIG. 5.—One of the several spouting springs in Geyser Park, Saratoga. (*Courtesy of the Commissioners of the State Reservation at Saratoga Springs.*)

soluble rocks, which give to them their saline character. They also evidently are brought into intimate contact with sources of carbon dioxid, as most of them are highly carbonated. Among the more famous springs of Saratoga may be mentioned the Vichy, Hathorn, Arondack, Geyser, Peerless, Magnetic, Star, Seltzer, High Rock, Lincoln, Chief, Victoria and Carlsbad. Inasmuch as these waters are owned by the State a somewhat detailed description without advertising any private interest may be permitted. The waters are quite typical of the principal mineral waters of the World.

Origin of the Name "Saratoga."—In the case of springs as famous as those of Saratoga, and which bid fair in the future to have their fame established on a scientific basis, all data relating to the springs are of interest. In the Report of the Commissioners[1] an interesting account is given of the origin of the name:

The aboriginal Saratoga, the Saraghtogha or Saraghtoghie of the red man, and to us—at this distance in time—the romantic Saratoga, dates from a very remote period. Actual history and enticing legend provide a most interesting narrative if one chooses to construct it from available authorities. The appearance of the oldest Saratoga spring is shrouded in the obscurity of geologic ages; but it is fairly certain that a fire was kindled by some aborigines who camped by that spring, during the Second Crusade, if over the charcoal remains disclosed at a depth of 16 feet the deposition of tufa was as rapidly made in those days by High Rock spring as in recent time by Geyser spring. It is indeed fascinating to think that while Conrad III, of Germany, and Louis VII, of France, were engaged in their unsuccessful attack on Damascus, a band of red men were making propitiatory offerings to the Manitou of High Rock spring about the year 1150. Brief mention should be made of the visit of the first white man, as far as we can conjecture, to the healing waters of Saratoga. From the *Jesuit Relations* we learn that a band of Mohawk Indians in 1643 brought with them to the Saratoga springs Father Isaac Jogues, who had been surprised and taken prisoner by a marauding band. Repairing to the "Medicine Waters of the Great Spirit" with their captive, they rested and secured refreshment and restoration to health and strength at the fountains whither they had been for ages accustomed to repair for relief from exhaustion and disease.

The composition of two typical waters of Saratoga has been determined by the Bureau of Chemistry[2] and is given below:

[1] State Reservation at Saratoga for 1914, pages 22 and following.
[2] Bulletin No. 91, 1905, pages 85 and 86.

Vichy Water, Saratoga, N. Y.

(Sodic muriated alkaline-saline, carbon dioxated)

Gases (number of cubic centimeters per 1,000 grams at 0°C. and 760 mm. pressure).—Carbon dioxid (free), 255; carbon dioxid (set free from bicarbonates on evaporating to dryness), 125.6.

Constituents	Parts per million	Percent of total inorganic material in solution
SiO₂ (silica)................................	14.7	0.60
SO₄ (sulphuric acid radicle)..................	16.7	0.68
HCO₃ (bicarbonic acid radicle)...............	684.3	28.04
NO₃ (nitric acid radicle).....................	Trace	
NO₂ (nitrous acid radicle)....................	Trace	
PO₄ (phosphoric acid radicle)................	None	
BO₂ (metaboric acid radicle).................	Trace	
AsO₄ (arsenic acid radicle)..................	None	
Cl (chlorin).................................	893.3	36.60
Br (bromin).................................	Trace	
I (iodin)...................................	Trace	
Fe ⎫ Al ⎭ (iron and aluminum)....................	4.2	0.17
Ca (calcium)................................	41.3	1.69
Mg (magnesium).............................	9.1	0.37
Ba (barium)	0.06	
Sr (strontium)..............................	None	
K (potassium)...............................	3.8	0.16
Na (sodium)................................	772.1	31.63
Li (lithium)................................	c.65	0.03
NH₄ (ammonium)............................	0.714	0.03
Total..................................	2,440.924	100.00
Free ammonia..............................	0.675	
Albuminoid ammonia.......................	Trace	
Nitrogen as nitrates........................	Trace	
Nitrogen as nitrites........................	Trace	

Hypothetical Form of Combination

Formula and name	Parts per million	Percent of total inorganic material in solution
NH₄Cl (ammonium chlorid)	2.11	0.09
LiCl (lithium chlorid)	3.94	0.16
KCl (potassium chlorid)	7.2	0.30
NaCl (sodium chlorid)	1,459.5	59.79
Na₂SO₄ (sodium sulphate)	24.7	1.01
NaBO₂ (sodium metaborate)	Trace	
NaNO₃ (sodium nitrate)	Trace	
NaNO₂ (sodium nitrite)	Trace	
Na(HCO₃) (sodium bicarbonate)	693.1	28.40
Na(HCO₃)₂ (barium bicarbonate)	0.11	
Mg(HCO₃)₂ (magnesium bicarbonate)	54.8	2.25
Ca(HCO₃)₂ (calcium bicarbonate)	167.3	6.85
Fe(HCO₃)₂ (ferrous bicarbonate)	13.4	0.55
SiO₂ (silica)	14.7	0.60
Total	2,440.86	100.00

Hathorn Water, Saratoga, N. Y.

(Sodic muriated alkaline-saline (ferruginous, bromic, and iodic))

Gases (number of cubic centimeters per 1,000 grams at 0°C. and 760 mm. pressure).—Carbon dioxid (free), 75.0; carbon dioxid (set free from bicarbonates on evaporating to dryness), 658.2.

Constituents	Parts per million	Percent of total inorganic material in solution
SiO$_2$ (silica).................	19.6	0.18
SO$_4$ (sulphuric acid radicle)..................	5.6	0.05
HCO$_3$ (bicarbonic acid radicle)................	3,585.6	33.02
NO$_3$ (nitric acid radicle).....................	None	
NO$_2$ (nitrous acid radicle)....................	None	
PO$_4$ (phosphoric acid radicle).................	None	
BO$_2$ (metaboric acid radicle).................	(*a*)	
AsO$_4$ (arsenic acid radicle)...................	None	
Cl (chlorin)...............................	3,685.5	33.95
Br (bromin)...............................	26.9	0.25
I (iodin).................................	1.5	0.01
Fe Al } (iron and aluminum).................	9.8	0.00
Ca (calcium).............................	650.7	6.00
Mg (magnesium).........................	228.8	2.11
Ba (barium).............................	5.5	0.05
Sr (strontium)...........................	Trace	
K (potassium)...........................	197.2	1.82
Na (sodium).............................	2,430.7	22.38
Li (lithium).............................	0.5	
NH$_4$ (ammonium).........................	10.7	0.09
Total......................	10,858.6	100.00
Free ammonia.............................	10.1	
Albuminoid ammonia......................	Trace	
Nitrogen as nitrates......................	None	
Nitrogen as nitrites......................	None	

a Small amount.

HYPOTHETICAL FORM OF COMBINATION

Formula and name	Parts per million	Percent of total inorganic material in solution
NH₄Cl (ammonium chlorid)..................	31.8	0.29
LiCl (lithium chlorid)......................	3.0	0.03
KCl (potassium chlorid).....................	350.7	3.23
NaCl (sodium chlorid)......................	5,762.4	53.06
KBr (potassium bromid).....................	40.0	0.37
KI (potassium iodid)........................	1.9	0.02
Na₂SO₄ (sodium sulphate)...................	8.2	0.08
NaBO₂ (soidum metaborate)	(a)	
Na(HCO₃) (sodium bicarbonate)..............	586.4	5.40
Ba(HCO₃)₂ (barium bicarbonate).............	10.4	0.09
Sr(HCO₃)₂ (strontium bicarbonate)...........	Trace	
Mg(HCO₃)₂ (magnesium bicarbonate).........	1,377.8	12.69
Ca(HCO₃)₂ (calcium bicarbonate).............	2,635.3	24.27
Fe(HCO₃)₂ (ferrous bicarbonate).............	31.1	0.29
SiO₂ (silica)................................	19.6	0.18
Total.....................................	10,858.6	100.00

a Small amount.

Vogue of Saratoga Waters.—The Saratoga waters not only have had a great vogue at the Springs themselves, Saratoga having been for many years a fashionable watering place, but immense quantities have also been bottled and shipped to all parts of the country. A few years ago an attempt was made to tap the sources of the gaseous element and condense the carbon dioxid gas for commercial purposes. The waters which accompanied these gases were usually allowed to run to waste. So extensive was this practice that the water level of the Saratoga springs began rapidly to decline and many of them ceased to flow. This, however, did not stop the pumping both of gas and water.

Saving the Springs.—This process was continued with such vigor as to threaten the very existence of the springs. In order that they might be saved, the area in which they existed was acquired by the State of New York. The State put a stop to the waste of water and gas, limited the output of water, which could be taken only under conditions prescribed, converted the area into a state park, beautified and improved it, and thus saved for future generations both the therapeutic and esthetic effects of the springs. Under the wise conservation of the State authorities the water level is rapidly rising. Springs which had ceased to flow are beginning to yield their waters again. The rape of the gas for commercial purposes has been stopped. It is probable that within a few years all of these springs will be restored to their pristine vigor.

Changing Character of the Saratoga Waters.—One of the peculiar features revealed by the study of the Bureau of Chemistry is disclosed in the fact that as compared with the analyses of former years there has been a gradual diminution in the mineral content of the Saratoga waters. This is not surprising when the great drain to which the waters have been subjected is remembered. In earlier times when the springs were in a natural state the waters thereof received a maximum concentration by remaining as long as possible in contact with the soluble minerals from which their mineral content is derived. When the vogue of the waters was fully established and great drains were made upon their volume both at the spring and for shipping purposes, larger volumes of water in contact with the mineral substances for a shorter time were secured. As a result of this the maximum concentration could not be expected to take place. The vigor of the attack upon the water supply is well illustrated by a letter from the owner of the Arondack spring of Saratoga, which is as follows:

We have here wells varying in depth from 150 feet to 600 feet. In drilling the well the veins of water are liable to be struck at many intervals from 40 feet down to the bottom. The well is tubed, generally with a small iron pipe, down to a point that may be decided on as securing the best water. At the bottom of the pipe a seed bag is attached, which very quickly swells up and for a time, if properly adjusted, will shut out all the water above it, so that the yield is only of the water entering below the seed bag. Now in the course of time, which may be a few months and may be two years, the seed bag frequently shrinks or is eaten away a little, and this would allow a certain amount of the upper water to enter. It is also my experience that within about two years there will be at some point of the pipe perforations which again would allow water to enter.

Another point which we have noted is that many of the Saratoga springs are connected with each other. I do not know whether pumping is carried on at all by the mineral-water companies, but it is by all the gas companies here, involving a matter of about 50 wells. We have at our gas works two wells which are connected with a well of a competitor. When he pumps very hard or at all our water levels are more or less affected, as is also our yield of gas. He can pull our level down more than 20 feet. It seems to me that such a condition would, through hydrostatic pressure, affect the amount of water running into the well from the various veins.

By comparing the analytical results with those of earlier chemists Haywood found that the springs contain now only from one-half to one-fifth as much mineral matter as they did in 1871. Not only is the mineral content of the waters at the present time less than in former years, but the analyses show that this mineral content is continually varying, and especially it has been noticed that the rarer elements, such as the compounds of lithium and bromin, vary to a greater extent proportionately to their amount than the other mineral elements which are present.

FIG. 6.

Official Labels.—There is one point to which particular attention should be called, namely, the public ownership of these waters by the State of New York. Not all of the Saratoga waters have yet been acquired by the State, but a great majority of them are now under state control. All waters shipped from Sara-

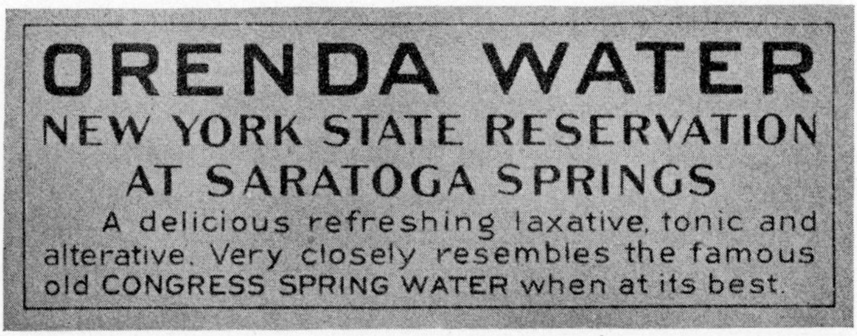

ORENDA WATER
NEW YORK STATE RESERVATION
AT SARATOGA SPRINGS
A delicious refreshing laxative, tonic and
alterative. Very closely resembles the famous
old CONGRESS SPRING WATER when at its best.

FIG. 7.

SOFT SWEET SPRING
NEW YORK STATE RESERVATION
A VERY PURE DRINKING WATER

FIG. 8.

HATHORN NO. 2
NEW YORK STATE RESERVATION
AT SARATOGA SPRINGS
Strong saline tonic cathartic. No discomfort.
No weakening Full of CO_2.

FIG. 9.

COESA SPRING
NEW YORK STATE RESERVATION
AT SARATOGA SPRINGS
Efficient Alkaline-saline laxative, tonic and
effervescing.

FIG. 10.

toga under state control have attached to them labels which indicate that fact. Some of these labels are reproduced above. When these labels are seen on bottled Saratoga waters the purchaser may be certain of the character of the water he is getting. Unfortunately there are still some adulterated or misbranded waters, often of a synthetic character, bottled and shipped from Saratoga, but these of course do not bear the label of the State. If all so-called remedial waters of this country could be under city, state or national control, it would be a great boon to our people. The United States itself has set an example in acquiring and holding the Hot Springs of Arkansas. This example of the United States, later imitated by the State of New York, should be followed through all parts of the country until public ownership of all medicinally advertised waters is established.

Misbranding of Mineral Waters.—One of the most common forms of misbranding of a mineral water of American origin is in the adoption of a foreign name. The name which has been most abused in this connection is Vichy. The Vichy waters of France are owned by the state, and have acquired a fame which is international. Without much reference to the character of the water, however, the name has been adopted both for natural and mineral waters of the United States. There can be no excuse for misbranding of this kind. Its only purpose is to mislead, it is contrary to both the spirit and letter of the law, and is a violation of international ethics. Nearly all nations have signed an agreement to respect each other's trade marks and trade names. There can be no excuse from any point of view for the adoption of the name of a foreign article for one produced in this country. The so-called Vichy spring of Saratoga is one of the most prominent offenders of this rule.

Curative Claims.—There is another form of misbranding which perhaps is even worse than that of the adoption of foreign names, namely, to advertise a water as of great therapeutic virtue by reason of the content of some rare or especially valuable constituent. Among offenders of this kind lithium leads all the rest, though radium is getting to be a close second. Many so-called lithia waters, as has been shown in the courts, contain less lithium than ordinary river waters. The minute trace of lithium which the water contains must of necessity be entirely obscured by the quantities of other mineral ingredients. The remedial effects of lithium as a remedy for rheumatism and gout have been tremendously exaggerated. Modern therapeutics have begun to acknowledge that the vaunted prowess of lithium as a rheumatism or gout remedy is wholly visionary. Nevertheless the presence of lithium, even in minute quantities, in waters has been regarded as a sufficient basis to offer them to the public under the name of lithia waters. In point of fact, the minute quantities of lithium which these waters contain could not possibly be consumed in sufficient volume to exert any therapeutic effect whatever. Unless we accede to the view that a remedy is more potent as it becomes more dilute, we must exclude from the realm of therapeutic remedies the so-called lithia waters of commerce.

Not only have natural waters appropriated names that do not belong to them and assumed virtues which they do not possess, but the same is true of the synthetic waters of commerce. They also have masqueraded under the name of Vichy, Celestins, Carlsbad, etc. The same remarks may be applied to these offenders as in the case of the natural articles.

Adulteration with Carbon Dioxid.—Another form of adulteration or misbranding which is quite common is the representation of a water as being naturally effervescent, in which the effervescence is produced by artificial means. The great difficulty of bottling a naturally effervescent water so as to preserve its qualities is well known. Usually the quantity of gas present is not large enough to give the desired results. On the other hand, the difficulty of introducing it into the bottle without the loss of its natural gas is very great. The practice has grown up in many places of collecting the gas which comes from effervescent waters and after compressing it to use the compressed gas for charging the water after it is bottled. This is a practice to which no objection can be made whatever. Even still waters may be thus carbonated and take on effervescent properties which render them more valuable and more desirable, and no objection can be made to this method of procedure. In all such cases the article must be regarded as either misbranded or adulterated if offered to the consumer as naturally effervescent.

Artificial and Imitation Mineral Waters.—By reason of the fact that the medicinal properties of certain mineral waters are

6

recognized and established by long experience, and by the further reason that it is customary in certain localities to drink synthetic waters, the industry of the manufacture of drinking water has been developed on a very large scale in the United States. There are large industries engaged in the bottling of the natural waters, but perhaps an equally great industry is engaged in the artificial introduction of carbon dioxid in the natural waters and in the manufacture of waters by distillation, and the addition of the necessary or other desirable mineral ingredients. This artificial water is found upon the market at the present time in very large quantities.

Medicinal Properties.—Whether or not such a synthetic has the same medicinal qualities as a natural water is a question of importance. It is quite likely that the materials which nature puts together in the production of natural compounds have properties which cannot be produced by any artificial means. More than that, in natural products there are extremely minute quantities of constituents which escape detection by ordinary analytical methods, and thus fail to be placed in the imitations. For this reason, it may be said that an imitated water cannot be regarded as having the same value in any case as one of natural origin.

Basic Material for Artificial Waters.—The basic material for artificial waters is pure distilled water. This term is not used in the absolute sense, because as has been already intimated the production of absolutely pure water is well-nigh impossible. The term is used in the ordinary sense of practically pure water, free from odor and from appreciable quantity of foreign substances.

Method of Production.—It requires art and experience to prepare a distilled water suitable for the manufacture of an imitation mineral water. To secure the best results requires that apparatus be scrupulously clean and so adjusted as to be easily kept in proper condition. The ordinary pot still is the apparatus usually recommended. In the operation of this still, however, certain precautions must be observed.

Most waters which are used for distilling purposes contain certain odoriferous particles in solution, and these must be removed. The process which is most commonly used requires the proper amount of water in the still for one operation. The distillation is then started and about one-third of the contents evaporated. This first third is allowed to run to waste, because it contains the volatile odoriferous materials above mentioned. The process of distillation is then continued rather gently until a second third of the contents is evaporated. This is received into a condenser where the vapor is converted into water and this water collected in absolutely clean receivers, built of materials on which the distilled water does not appreciably act. The remaining third of the water in the still, which contains concentrated mineral matters, is then allowed to flow in the sewer. Thus, of the whole volume distilled, only half is reserved for the purposes of manufacturing artificial waters, and of the whole quantity of the contents of the still, only one-third.

Properties of the Distillate.—The pure distilled water thus obtained is practically free from dissolved gases, such as nitrogen and oxygen, of which the atmosphere is chiefly composed, and is consequently flat and insipid to the taste. It is of a character sufficiently pure to be used for the manufacture of any imitation waters desired.

Charging with Carbonic Acid.—The simplest form of manufactured water made from this pure distillate is that which is charged with pure carbonic dioxid. The carbon dioxid which is employed is usually collected from natural sources, that is, from springs which give off natural carbonic acid, or is manufactured by chemical means, usually by the action of an acid or marble or limestone.

The carbon dioxid given off naturally from springs is to be preferred, because it is the natural gas which is produced in and found with natural waters.

The carbon dioxid arising from fermentations is not desirable for this purpose, because of its carrying certain odors due to its origin.

The gas produced by chemical means may be purified, and is probably purer carbon dioxid than that derived from either of the other sources mentioned.

It is, of course, the duty of the manufacturer to see that the carbon dioxid which is used is thoroughly washed and purified so that it may enter the water under the most favorable conditions. The water being subjected under pressure to gas of this kind absorbs certain quantities of it, which convert the insipid and dead water into effervescing, pleasant and agreeable drinking water.

Waters thus charged are usually kept in syphons for use. The syphon should be made in a sanitary way, and no lead or zinc pipes or tubes should come in contact with the liquid. The proper material for making the delivery tubes of the syphon is block tin. Thus charged the waters are easily adapted to use, and are kept sparkling until the contents of the syphon are exhausted.

These carbonated waters are also put up in bottles, with corks so secured as not to be blown out by the pressure of the gas. Waters of this kind are known to the trade as carbonated waters, and sometimes simply by the word carbonic.

Kinds of Artificial Waters.—The number of artificial waters which have been offered to the public is almost as great as those of natural origin.

By means of a chemical analysis the composition of a natural mineral water at any given time can be accurately determined. Experience has shown that a natural water does not always have the same quantity of mineral constituents. Not only may the actual contents of mineral matter in natural water vary, but often the proportions of the mineral substances as related to each other may change. This is due to the fact that mineral waters are affected, more or less, by the rainfall, so that the flow of a spring is greater at one season than at another. This increased flow tends, at least in some cases, to diminish the percentage of mineral matter in the water. Then, too, the mineral matters in the water are obtained by solution of mineral substances over which the water passes. It is evident that, in the progress of solution at first, more easily soluble mineral substances are extracted, and thus these will appear in larger quantities. Thus, while it is possible to imitate the natural water by the manufacture of a product containing the same mineral ingredients, it is evident that this imitation would only be true for the analysis of the water at any given time, and not for the variations which might arise at subsequent times.

It is practically impossible to exactly imitate the natural water by any artificial means, but, it is easily possible to make artificially a water which has in general a resemblance to that of natural origin.

Purpose of Manufacture.—In these preparations an attempt is made to simulate the natural mineral waters, the constituents of which are revealed by chemical analysis. If the mineral constituents, as revealed by analysis, are added to pure distilled water in essentially the same proportions as they exist in the natural waters, a product is secured which, in so far as the chemist is concerned, is an exact imitation of the original. As mineral waters are very frequently charged with carbon dioxid naturally, many of the artificial waters are saturated with carbonic acid at a high pressure. When artificial waters are made under strictly sanitary conditions, and many of them, perhaps most of them, are so made, and when they are so labeled as to give to the consumer a correct knowledge of their origin, they are not objectionable.

Merits of Product.—The artificial product has two merits which are not always possessed by the natural water. In the first place, when properly made artificial waters are entirely free of any kind of germ infection. They are clean, sterile, and protected completely from external pollution. They also have a uniform constitution, being made after an exact formula which is not varied. When charged with pure, carefully washed carbon dioxid, they represent, from a sanitary point of view, an excellent product. On the other hand, the natural water is not controlled as to its source, nor as to the character of the soil through which it filters, and hence may at times be contaminated by surface waters or otherwise carry germs of infection which are objectionable. In drinking a natural water the consumer does not, as a rule, have access to so constant a product as when he drinks an artificial water correctly made. On the other hand, the consumer who drinks an artificial water is entirely at the mercy of the compounder. If the manufacturer is ethical and his factory is in a perfectly sanitary condition, the consumer always gets a sound product. If, on the other hand, carelessness is exercised in manufacture, if the premises are not sanitary or the chemical control complete, the consumer may get an article of almost any description.

Adulteration and Misbranding of Artificial Waters.—What has been said regarding the adulteration and misbranding of natural waters is more emphatically to be repeated with reference to artificial waters. It is a very common practice in the manufacture of artificial waters to attach a name which implies that it is a natural product. Even the use of the word "Imitation" or "Artificial" is not a sufficient justification of the use of a name implying a natural origin. An artificial water may be made which resembles in almost every chemical characteristic the natural water, yet it is not one. An artificial water of this kind is best sold under some name of its own, although it is of a potable, pure, and wholesome character, as it often is and always should be. The manufacture and sale of artificial waters of this kind is an industry which deserves encouragement. It is possible to produce artificial waters in which every possibility of contamination or pollution is eliminated. Such waters are sterile. They are made of the purest materials and are commended in a general way as a beverage. This is no excuse for placing upon them any word or phrase or design or device which suggests to the consumer an origin other than that of the laboratory and the factory.

Court Judgments on Misbranded or Adulterated Mineral Waters.—The general practice of misrepresenting synthetic waters as natural and making extravagant claims for the therapeutic properties of such waters has received particular attention at the hands of the Bureau of Chemistry, and the courts have passed upon a great many cases of such alleged adulteration and misbranding. The American Medical Association has collected and published a brochure devoted to a description of mineral waters which are sold under misleading or fraudulent claims. The reader is referred to this pamphlet for detailed descriptions of these waters.[1]

Manufacture of Liquid Carbonic Acid.—Many bottlers of carbonated beverages, instead of using merely the compressed carbonic gas, employ in its stead liquefied carbonic acid. This is obtained as follows: The carbonic acid is derived chiefly by the burning of coke and absorbing the gas in a cold alkaline solution. The gas is then driven out by heat. It is washed, passed through a permanganate of potash solution and charcoal, and then reduced by cold and pressure to a liquid state. It is sold at about five cents per pound, and has been found to be very

[1] Mineral Waters. American Medical Association, 535 North Dearborn Street Chicago.

pure and much cheaper than it can be made on a small scale by the bottler.

Mineral Water Tablets.—As an economy in transportation and a saving of packages, a custom has grown up of making mineral water tablets. There are two methods in which this may be done. The genuine way is to evaporate the natural mineral water to dryness and use the salts thus obtained in tablet form. The second method consists in using these salts in the proportions in which they exist in the mineral water from stock salts which have never been derived from the water at all. Chemically there would be very little prospect of distinguishing between these two sets of tablets, but as far as the effect upon the system is concerned it is not likely that the synthetic tablet would ever be able to take the place of the natural one.

It is probable that the different tablets of this kind on the market are mostly synthetic. We have tablets marked Vichy, Apollinaris, Kissingen, etc. These tablets are accompanied with directions as to the quantity of water in which they should be dissolved in order to restore them to their original dilution in the natural water. As far as economy to the consumer is concerned, there is probably no kind of mineral water that is cheaper than that which is made in this way from prepared tablets. Each tablet is of such a size and composition as to be sufficient for an ordinary glass.

The supposed medicinal effect of the tablets is based upon the observed or supposed medicinal qualities of the waters themselves. It should not be understood that an artificial tablet of this kind would have the same salutary effect as the natural product. Further than this it may be said that the effect of the natural product is not always convincingly established. Regard must also be had to the fact that the continued dosing of the system with laxatives or alteratives, or even with iron or arsenic or lithia, is likely sooner or later to produce harm. The consumption of mineral waters made by artificial tablets should therefore be controlled by the same general principles as pertain to the use of mineral waters in general.

Mineral Salts Used.—The minerals which are used in the manufacture of water should also be pure chemically and manufactured in a sanitary water. The common minerals employed are salt, sulphate of soda, sulphate of magnesia, sulphate of lime, carbonate of lime, carbonate of lithia, phosphates of iron, soda and potash, borax, bromids, iodids, salts of aluminum, salts of manganese and salts of ammonia. These mineral constituents, in order to secure the best results, should be of the finest quality and free from any dirt or contamination of any description.

Magnitude of the Industry.—It is difficult to obtain statistics of the magnitude of the industry in artificial waters. The quantity of bottled mineral water used in the United States is probably over 100,000,000 gallons annually. This includes the quantity which is drunk directly at the mineral springs and bottled for commercial use. The annual value of these waters is probably from 20 to 25 million dollars. The amount of artificial water produced is of course considerably less, but I have no reliable data regarding the entire amount. Plain water without any addition of minerals, both carbonated and still, is bottled and sold in large quantities.

Bacterial Infection of Mineral Waters.—The production by nature of a mineral water entirely free from bacterial infection is not to be expected. Happily most of the bacteria with which we come into contact are either harmless or useful. There are only a comparatively few bacteria commonly met which are harmful. For the purpose of protecting water from bacterial infection, the good as well as the bad organisms are to be excluded. To this end the spring or well producing the mineral water should be as carefully guarded from bacterial infection as possible.

Especially is it necessary that no sources of infection exist in the neighborhood of the spring, nor along the areas of its source. If the source of the spring be contaminated, it is almost impossible to expect a purification even during its course deep under ground and filtration through sand, gravel and soil. This mechanical treatment by nature may serve to remove the greater part of the infection, but it can hardly be expected to remove it all. Particularly the immediate neighborhood of the spring should be free from any infection, and the admixture of infection of any kind from surface waters must be guarded against with care.

Many waters which, as they approach the surface of the earth, are in excellent condition for consumption, become infected at their exit or during the process of bottling and shipping. Thus the exit of the well must be carefully cemented or concreted and so covered as to exclude external infection from surface water and from the air. All utensils used in handling the water must be sterilized, and the bottling must be done under standard conditions of sanitary cleanliness. The same character of problem that faces the safeguarding of ordinary potable waters must be observed with the mineral waters.

The occurrence of large numbers of colon bacillus in water is always a suggestion of pollution. While it is recognized that the colon bacillus may sometimes be innocent and be derived from a source which is unobjectionable, its original habitat in the animal organism, and especially in that part of the digestive apparatus which is known as the colon, makes its presence always a cause of suspicion if an inspection of the sanitary conditions of the premises shows that precautions are not observed to exclude the possibility of such an infection.

Interpretation of the Results of Bacteriological Examination.— Often data which are perfectly reliable concerning a product like water are interpreted in entirely different ways by reason of different points of view. The expert who is employed by the manufacturer will interpret a given set of data in a manner entirely exonerating the manufacturer from any blame either of wilful contamination or criminal negligence. The same data interpreted by an expert representing the consumer may lead to a condemnation of the sample as not suitable for consumption. It is safe, as a rule, to follow the opinions of the expert in the interests of the consumer. To follow the conclusions of the expert for the manufacturer may lead to grave injury. On the other hand, to follow the opinions of the expert for the consumer cannot possibly injure anyone except the producer himself. Very bitter contentions have arisen before the courts regarding this matter, and the courts have interpreted the data differently. Upon the whole the tendency of the judgment of the courts is generally toward the protection of the consumer.

One must take into consideration in this matter various conditions of the environment. The season of the year, the proximity of possible sources of contamination, the magnitude of the flow, the general history of the locality, the dip of the rock strata, the general geological formation, the trend of drainage areas and the character and quality of the soil underlying rocks, are all to be considered.

In general it is safe to follow the rule that all waters intended for drinking purposes which show the presence of colon bacillus in a majority of samples of one cubic centimeter, should be considered as polluted and excluded from consumption. While the application of this rule might exclude some samples which are wholly innocuous, the failure to observe it might admit many others that are dangerously contaminated. The application of science and bacteriology to the study of potable waters and potable mineral waters has been practised only for a few years, and the full benefit which is to come from such examination has not yet been realized. In these circumstances it is not wise to listen to the advice of some few scientific men who speak with some contempt of the efforts which are made to judge the waters by bacterial methods.

HEALING PROPERTIES OF WATERS

Waters as a Medicine.—In addition to the description of water as a beverage attention may be called briefly to its value as a medicine or remedy for disease. The theory of the hydropathic system of medicine is based on the assumption of the curative powers of water. The cardinal principles of hydropathy are accepted by all schools of medicine. The value of external applications of water, both as a means of cleansing and as a control of temperature, are acknowledged by all physicians. The copious drinking of water is also regarded as an efficient remedy in many diseases.

Antiquity of Hydrotherapy.—The use of water as a healing agent dates from the remotest antiquity and is, today, practised by nearly every physician. The school of medicine which depends upon water as the healing agent is called "Hydropathic" and the use of water as a remedy, hydrotherapy. The early records of Asyria, Egypt and China, show that water was a very common healing agent. Among the Greeks and Romans the bath was a public institution, and some of the most remarkable buildings, especially in Rome during the days of its greatest prosperity, were those devoted to the bath. The temperature of the water has a great deal to do with its curative power, and hence cold, warm and hot water and even steam are constantly employed. The so-called Turkish bath depends largely upon the vapor of water at a high temperature for its therapeutic effects. In the writings of Hippocrates a very detailed description of the use of water as a remedial agent is found. In the warfare against the infectious organisms of the body the remedial agent is necessarily carried in solution. Hence water becomes not only the carrier of foods, but also the carrier of remedial agents. The percentage of water in the blood is a dominant factor in the normal activity of that liquid tissue. In certain diseases, such as cholera, the water content of the blood is greatly diminished, and speedy fatality attends the depletion of the water supply.

Water as a Vehicle for Heat and Cold.—Water, on account of its high specific heat, is the most valuable agent for applying heat or cold to the surface of the body or to the internal organs. The cold-water bag or ice bag and the hot-water bag are well-known weapons in the armament of sane therapy. The effect of cold in certain conditions of the brain and in certain high temperatures due to disease is well known. The reduction of the temperature, while only a temporary relief, is necessary to the proper action of the various medicinal healing agents.

The application of cold by means of water or ice is justified on many therapeutic principles. As is well known, a low temperature interferes with the activity of vital forces. In localized inflammation, turgidity and attendant pain, the judicious application of cold may afford at least temporary relief. In cases where the activity of the cells is for the purpose of excreting all poisonous substances cold must be used with careful judgment, or else it may do serious harm. In such cases the application of hot water may result in greater good. The temporary contraction of the blood-vessels after cold water or ice is applied is always followed by a reaction, when it is removed, of greater activity. When one has been out in the cold during the day his cheeks are apt to become very rosy on entering a warm room at night. Thus if increased activity and secretion are desired in the end, it may be justifiable in the beginning to apply the cold compress.

Cold Destroys Life.—Too long exposure to cold is destructive of living beings. Vegetables endure cold much better than animals, but even vegetable life is destroyed when the cold becomes too great. There is no vegetation at the poles or on the tops of high mountains. The effect of cold on the blood is quite marked. The reaction which follows the temporary chilling of the blood is found to increase the number both of white and blood corpuscles, and also the quantity of the coloring matter—hemoglobin. It has long been noticed that people who live in cold climates and out-of-doors for the most part have ruddy complexions, which are not found in people inhabiting warm regions. Thus a judicious use of cold is fruitful of good results.

Water as a Heat Carrier.—The application of heat by means of water is perhaps more commonly practised for therapeutic reasons than even an application of cold. Almost all severe pains in the organs of the body are more or less alleviated by the hot-water bag. Especially is this true of abdominal pains. The alleviation which is secured by the application of the hot-water bag is well marked in many of these painful attacks. The dilatation of the vessels, the increased activity produced by the warmth and the general stimulation of the superficial parts to which the heat is applied, all tend to promote favorable healing action in many congested and pain-racked parts.

It must not be forgotten that there is a very general depressing effect produced by too extensive or too long-continued appliaction of heat. One may remain in the Turkish bath surrounded by steam at a temperature considerably above that of the blood for 10 or 15 or 20 minutes, or even longer in some cases, without being notably depressed. Sooner or later, however, a feeling of oppression and weakness supervenes, which is an index that the application of heat in this way has been carried far enough. It is very probable that if continued for longer periods, even at temperatures only a few degrees above that of blood heat, very serious results might take place.

After being subjected to this excessive heat, applied by means of water vapor, there is of course a tendency to reaction when the source of heat is withdrawn and the surface of the body is rubbed with cold water. The water which is first used for this purpose should be tepid, and gradually made cooler as the reaction takes place. The application of heat in the form of water is well shown, also, in its effects upon the excretory organs. The kidneys, especially, are stimulated and perform their functions more thoroughly when subjected to high temperatures applied in the manner described. Cold also may promote the secretion of urine by closing the pores of the skin.

The Wet Pack.—Perhaps one of the most striking uses of water in diseased conditions is that which is produced by the process known as "packing." In cases of high fever, as for instance in typhoid, the patient experiences much relief and the fever is considerably reduced by wrapping in a wet sheet and keeping the patient therein for a considerable period of time. The sheet may be removed from time to time, washed in tepid water, and replaced. When a medical student I noticed in the wards of the typhoid hospital in Vienna, the general treatment of typhoid patients by giving them large doses of quinine and then immersing them naked in the bath, the temperature of which was high enough to produce no chill, that is, at a temperature of 85, 90 or 95°. Gradually the warmer water would be drawn away from this bath and colder water substituted until the temperature of the bath would sink to 85, 80 or even 75°. Under this treatment the temperature of the patient rapidly fell and he was greatly benefited. It is probable that the therapeutic effect of water in the treatment of fevers of this description has not been properly appreciated by the medical profession.

Value of Bathing.—There is much difference of opinion respecting the value of frequent bathing. The general consensus of opinion is that water is an agent of cleanliness and favorable to the production of health, and that its use should never be omitted even in times of stress when facilities for its application are at hand. It is not recommended that persons suffering from severe diseases should use water in a remedial way on their own volition. The physician should study the condition and apply water in such manner as his judgment may indicate for securing the desired results.

Errors Regarding Healing Virtues of Waters.—Probably there is no other function of waters so generally misunderstood as that which has to do with their medicinal qualities. Nearly all diseases to which flesh is heir yield, according to the claims of the advertisers, to the virtues of this or that water. "Taking the waters" is not only fashionable, but by large numbers of people deemed almost necessary. I have observed the habits of those who go to celebrated watering places to secure the benefits of the waters. That they are benefited by their journey and by their stay, as a rule there can be but little doubt. There are many other factors besides the waters which must be taken into consideration.

Most of those who take the waters are well-to-do people, who have lived on a too generous diet. On reaching the watering place they are placed at once upon a simple and restricted diet, which fact alone helps greatly toward a restoration of health. Their habits of life are also regulated by the rules of the cure, so that they go to bed early and get up early and do not indulge in any excesses of any kind. The complete change of life which they adopt, together with freedom from worry and confinement of business, are all prominent and active factors in the restoration of health. Probably the water itself has little to do with the case.

Specifics in Springs.—In addition to the general therapeutic value ascribed to water there is a widespread belief that certain springs have specific medicinal values. There is perhaps no notion more widely disseminated than this, nor perhaps no one tenet in medical matters which has a greater number of adherents. It is a very common practice to "take the waters" for a great variety of diseases, especially those of a chronic character. Especially efficacious are certain waters supposed to be in rheumatism, gout, dyspepsia, nephritis and many other diseases of a chronic and very persistent character. It will be sufficient to state here that the widespread belief in the medicinal values of waters is not as well founded in fact as it is extensive in belief. That certain waters have medicinal principles may be accepted as beyond contradiction. Many of them are purgative; most all of them are to a certain extent diuretic; and a few contain specific remedial agents such as arsenic and other rare elements such as radium which are highly recommended for certain forms of diseases.

Ferruginous Waters.—Attention has already been called to water containing iron and to the very widespread belief that iron is a tonic and beneficial. This fact cannot be denied. On the whole it must be said that the valuable properties attributed to medicinal waters are as a rule very grossly exaggerated. Mineral waters are not intended at all as a general beverage to be used every day; they are only to be taken at intervals, and the longer the intervals as a rule the better. It is not a normal condition of affairs to be pouring into the alimentary canal a lot of foreign material which is not indicated in a state of perfect health. The value of a remedy as a rule is very greatly enhanced by its intermittent use, and mineral waters are no exception to this rule. It may be that in many cases the drinking of the waters producing more or less cathartic effects and thus freeing the alimentary canal of objectionable matters does tend to help in the good work of restoring health, but if all the factors which are active in this result are analyzed, it is doubtful whether the waters themselves would occupy a very high place in the scale of remedial agents.

Balneology.—The use of baths as a remedial agent (Balneology) is rapidly assuming the position of an exact science. When it is possible to get away from the extravagant and often misleading statements which are published regarding the medicinal properties of privately owned waters, it is evident that the placing of the value of this treatment on a strictly scientific basis is near at hand. It has been very aptly observed by Dr. Ferris, the State physician at Saratoga that:

Hydrotherapy is simply a part of the general practice of medicine, for its agencies may often be used to great advantage in place of vegetable or mineral drugs, and with better results; for, by means of these agencies, recuperation and restoration of vital processes are often secured in cases wherein drugs produce but a temporary condition, under cover of which nature is expected to assert herself. But its principles and practice should be well understood by those who prescribe its measures.

PART III

SOFT DRINKS

Definition.—The term "soft drinks" is applied in the United States to common beverages which contain no alcohol. From the point of view of composition, there is no rule which can be formulated by means of which their composition is controlled, unless it should be by specific legal regulation such as is made by the Bureau of Internal Revenue.

The typical soft drink of the United States is known as soda water. This phrase is, to some extent, a misbranding, since it is so called because originally the gas with which it is charged was derived from carbonate or bicarbonate of soda. The name has become so firmly attached to waters of this kind that it probably will be accepted in the future as distinctive. In this country it is mixed, before drinking, with a syrup, which in addition to the sugar it contains, has a flavoring material to give it character. The syrup is made of pure sugar, or at least should be and the flavoring material is a vegetable juice or extract of some kind, such as vanilla, orange, coffee, raspberry, strawberry, chocolate, pineapple, lemon, banana, cherry. The above are all natural flavors and wholly unobjectionable from an ethical point of view.

Other drinks of this kind are legion. Some of them are of a composition to which the term soft drink is probably not applicable because this appellation should be reserved solely for beverages which contain no added injurious ingredient. The moment a substance is placed in a soft drink which is open to objection on account of its injury to health, the article ought to lose the right to be known by that appellation.

The industry of soft drinks has grown to an enormous extent in the United States, and much can be said in its favor and very little against it. There is an undoubted tendency in man to drink other things beside water and milk and if this tendency and desire can be controlled and directed into the consumption of harmless beverages containing no habit-forming drug, it may aid greatly in the cause of temperance and in the restriction of the alcoholic habit.

I think the great majority of the parents of this country would prefer to see their children patronizing the soda fountain rather than the saloon, and if a choice must be made between the two, the soda fountain must certainly be preferred.

A soft beverage or a soft drink may be regarded as an institution particularly American, because the industry in soft drinks is much more extensive in this country, in proportion to its population, than it is in any other country of the world.

A concoction which is pleasant to the taste from any cause, provided it be not due to an injurious substance or one which may threaten health, and which at the same time is refreshing and satisfying and made from wholesome articles in a sanitary manner, cannot be open to any serious objection. The common occurrence of sugar in the soft drink is a matter of regret in respect of its use by children.

There is another matter which ought to be mentioned in this connection and that is the danger which may come from the drinking of ice-cold liquids in extremely hot weather, such as prevails in most parts of the United States during the summer months. The sudden chilling of the alimentary canal, or at least the upper portion of it, by the ice-cold liquid on a hot day cannot be looked upon as wholly innocuous. The danger is especially great if at the same time of the consumption of the cold beverage the body is exceptionally warm from exercise as well as from the temperature of the ambient atmosphere. One of the great attractions of the soft drink counter in summer is the coldness of its products, and they usually are swallowed at a temperature not much above the freezing point of water. This consumption would be robbed of most of its objectionable character if the beverage could be slowly sipped instead of being swallowed almost instantly, as is often the case. The dangers to the mucous membrane from constant chilling in cases of this kind are not inconsiderable and must not be neglected.

Character of Soft Drinks.—It would be difficult to give any typical description of the general character of the soft beverage. Nearly all of these beverages contain sugar and this is a point

7

against them. Sugar is not a good ingredient of a "cooling" drink. Many of them contain carbon dioxid, aromatics, spices, and flavors. Others contain bitter principles which are supposed to be stimulants and tonics, and unfortunately many of them contain drugs which are objectionable from every point of view. Some of these products are decidedly of the habit-forming family, such as cocain and its preparations, while others belong to less objectionable classes, such as caffein, theobromin, etc. The principle may be broadly stated that the addition of any drug of this kind or character to a soft beverage which is consumed so extensively by the children of the country is highly objectionable and if it cannot be prevented in any other way, should be prohibited by law. The remarkable fact is exhibited in this country of parents who refuse to allow their children to drink coffee or tea at home, on the assumption that it may prove injurious to them, unwittingly permitting them to drink the active principle of these two bodies, namely, caffein, under the guise of coca cola, pepsi cola, tokola, cola coke, and other similar beverages at the soft beverage fountain. It is perfectly certain that if these parents were informed of the nature of beverages of this kind they would forbid their children to use them.

The extent of the soft beverage industry, its popular qualities, and the popularity of its products in this country, merit a more particular description of some of the more important members of the series.

SUBSTANCES USUALLY EMPLOYED IN THE PREPARATION OF SOFT BEVERAGES

Water.—All soft beverages consist very largely of water. The water which is used in the manufacture ought to be potable and pure. No suspected waters or those contaminated in any possible manner should be employed in the manufacture of these drinks. It would not perhaps be advisable to say that the water used in the manufacture of these soft drinks should be pure distilled water, because it is evident that the natural state of water used as a beverage by man is not of that description. If a pure spring or well water of good potable quality should be secured for the manufacture of beverages of this kind, it would probably, in most instances, be as acceptable and unobjectionable as pure distilled water. At the same time, such water would contain the natural but minute quantities of mineral substances, which not only improve the taste of water but add much to its nutritive qualities. From the chapters on water in this book it is easily seen what kind of waters should be avoided.

It is particularly important that the water should be free of organic matters, as well as mineral matters, which may render it either unwholesome or unpalatable.

Carbon Dioxid.—The carbon dioxid which is used for charging the waters of soft drinks should be a pure gas, free of all injurious ingredients or contamination of any kind. The source of the gas is of no consequence, provided it is pure. The methods of producing and handling carbon dioxid are described under mineral waters, p. 83.

Artificial Flavors.—The natural flavors used in soda waters have been described. In addition to these, soda waters and other drinks often contain artificial flavors, that is synthetic flavoring matters produced by the chemist and which resemble more or less closely the natural flavors of fruits. These artificial flavors, although used in small quantity, are highly objectionable both upon ethical grounds and because of their possible injury to health. Even the sugar formerly was substituted in many cases by the artificial sweetener, saccharin, a very little of which would give an intensely sweet taste to a large volume of beverage. Since the inhibition of the use of saccharin in food products such as soft drinks, very little of it is now found and it may be regarded as on the road to speedy exclusion.

The increase in price of artificial colors and synthetic ethers which are largely used in making flavors has been of a considerable magnitude since the beginning of the European War. The manufacturers, it is stated, have endeavored to sell at the old prices, but this is difficult unless some sacrifice of the quality is made. It would be a difficult task to give any idea of the nature of any of these particular flavors. They váry according to the wish and the skill of the manufacturer and the taste of the consumer. A very good rule would be to demand only natural flavors at the soda fountain. There would then be no danger of securing an artificial article which might contain a color not certified as being pure or some ingredient which possibly might be injurious to health. In a country like ours, where there is such an abundance of natural flavors of an unobjectionable character and where sugar is reasonably abundant and cheap, except in time of war, there can be no possible excuse for the substitution of artificial bodies for these natural products.

Purity of Soda Water.—The general principles of sanitation which are applied to the preparation and transportation of potable and mineral waters should be applied also to the so-called soda fountain. The carbonated water employed should be pure potable water or distilled water, prepared with every precaution to insure freedom from contamination. It should not be kept in tanks lined with zinc or lead, nor should it pass through lead pipes. Contamination with lead, a very dangerous poison, may ensue by keeping soft water in contact with lead for any length of time.

The carbon dioxid which is used for charging should be thoroughly washed and pure, free from all contamination of every kind. The glasses used for drinking purposes should be thoroughly sterilized before using a second time. Dipping them in a tank of water, so commonly practised, which has served the same purpose for many hundreds of times without changing its contents is worse than no treatment. The best precaution is to use paper cups and then throw them away. Rigid inspection should be exercised over all fountains of this kind, in order that the public may be guarded against injury and the purity of the product preserved.

Adulterations of Soda Water.—Adulterations are happily very little practised. They consist chiefly in the substitution of the artificial flavors and sweeteners for the natural. As soda waters are usually sold at five cents a glass, there can be no excuse for using these cheap substitutes. Even at the highest price of genuine articles of flavors and of sugar and the cost of procuring clean water, the profit on a glass of soda water ought to be satisfactory to the most grasping.

Look Behind the Counters.—It would be a means of securing sanitary conditions of this kind if the public had free admission behind the counter. I have often turned away from the soda fountain without patronizing it when I have seen, by accident, the method of washing the glasses and the insanitary conditions with which the dispenser of the beverage was surrounded. As kitchens of public restaurants should occupy the front space next to the street, so should the counter of the soda fountain be opened to the public. Publicity is the best sanitary activity which is known.

Synthetic Lemonade.—It is sometimes customary to substitute for lemon juice a yellow solution of citric acid. This of course is simply an adulteration, and a very undesirable one at that. Anyone who has ever compared the flavor of a real lemonade, either straight or at a soda fountain with carbonated water, with an artificial preparation of a yellow solution of citric acid, will understand exactly the force of this admonition. The same is true of orange, where a little orange peel or the flavor thereof and citric acid colored yellow is employed. All of these practices are properly catalogued under the head of adulterations.

Other Soft Drinks.—In addition to the drinks which are served at the soda fountain, similar drinks are bottled and sold generally to the consumer. There is a vast array of ginger ales, pops and materials of this kind, on the market. To this group may be added, also, the so-called root beers, which are sweetened water flavored with various extracts of medicinal and other herbs and sometimes slightly fermented. The same precautions in regard to the preparation of these drinks as were made in regard to soda waters should be observed. The basic material, that is, the water, should be pure and free of infection. The flavoring materials should be natural products. The sweetening material should be sugar only, and no alkaloidal body or active medicinal agent should be employed in their manufacture.

Magnitude of Sale.—These preparations are sold extensively at all outdoor gatherings, and especially those of summer sports, as tennis, baseball, and golf. The strictest precautions should be used in regard to sterilizing the bottles in which they are contained, and in using only high-grade flavoring materials in their preparation.

Ginger Ale.—Ginger ale is probably the most popular of these soft beverages. In the manufacture of ginger ale it has been a very common custom to use capsicum instead of ginger as the flavoring re-agent; or if ginger be used, only a small quantity is employed, while the capsicum gives the bite and snap which is

expected. Capsicum is not objectionable from a health point
of view, except when used in excess. Its use in a product labeled
"Ginger Ale" can only be regarded as fraudulent. According to
the regulations under the Food and Drugs Act, if capsicum be
used in a ginger ale product and the product be called "ginger
ale," it should be so labeled as to show that at least a part of its
spicing is due to capsicum. Since the food law has required the
proper labeling of these beverages, many manufacturers have
gradually come over to the making of the pure ginger ale product,
that is, one in which the ginger is the dominant or perhaps the
sole flavoring re-agent.

Official Definitions.[1]—1. *Ginger ale* is the carbonated or artificially carbonated
beverage prepared with potable water, acidulated sugar (sucrose) sirup, and
ginger ale flavor.

2. *Ginger ale with capsicum* is the carbonated or artificially carbonated beverage
prepared with potable water, acidulated sugar (sucrose) sirup, and ginger ale with
capsicum flavor.

3. *Sarsaparilla* is the carbonated or artificially carbonated beverage prepared
with potable water, sugar (sucrose) sirup, and sarsaparilla flavor. It may or may
not be acidulated.

Root Beer.—There is a class of soft drinks known as root
beers, which are composed with water as a base and with materials
which are subjected to more or less fermentation, producing
alcohol. The root beers are probably so called by reason of the
fact that they are flavored with certain vegetable, medicinal
substances, some of which may be derived from roots. The "beer"
part of the name relates, of course, to the fact that they have undergone incipient fermentation.

The quantity of alcohol which is produced in these beers is
not very large and yet I think it inadvisable to recommend even
these small quantities for children. A root beer, because it contains
a trace of alcohol, is hardly suitable for classification under the
general heading of "soft drinks," inasmuch as this term is applied
only to beverages which contain no alcohol.

Alcohol in Home-brewed Root Beers.—La Wall has found
in some recent investigations[2] that many so-called root beers have
a large percentage of alcohol. Such beverages as these should
be classed under alcoholic drinks. He says:

[1] Food Inspection Decision 177 issued Aug. 20, 1918.
[2] American Journal of Pharmacy, August, 1916, page 355.

The conditions under which the beverage is made are very favorable for the development of appreciable amounts of alcohol. Yeast, sugar, water and a flavoring extract which usually contains some inorganic salts for the stimulation and nutrition of the yeast, are combined under conditions favorable to the rapid growth of the yeast, and the mixture is then bottled and the bottles are directed to be tightly closed.

When the pressure of carbon dioxid, evolved by the fermenting mixture, reaches a certain point, the fermentation automatically ceases. It may easily be seen that if the mixture is allowed to stand for a short time before bottling, or if the bottles are not entirely filled so that a comparatively large air space remains, fermentation may proceed for some time, and the alcohol content is correspondingly varied or increased.

I accordingly made some experiments to ascertain just how high the alcohol would go under the most favorable conditions, and also to see what the average alcohol content of a product made strictly according to directions would be. The following results were obtained:

After standing 2 days.............0.25 percent alcohol.
After standing 3 days.............0.32 percent alcohol.
After standing 4 days.............0.35 percent alcohol.
After standing 5 days.............0.53 percent alcohol.
After standing 6 days.............0.64 percent alcohol.
After standing 7 days.............0.81 percent alcohol.
After standing 9 days.............1.29 percent alcohol.
After standing 10 days.............1.36 percent alcohol.
After standing 11 days.............1.52 percent alcohol.

No higher alcohol content was observed in this series even after standing for 10 days longer.

Later some additional experiments were made, allowing the fermenting liquid to stand for three hours before bottling and also only partially filling the bottles, and while, of course, the alcohol content rose more rapidly in each case, the highest amount noted under the most favorable circumstances was 1.77 percent.

There is a very delicate question involved as to whether a beverage made in this manner could be sold as a non-alcoholic drink, using the expression "non-alcoholic" in its popular significance as equivalent to non-intoxicating. Under the internal revenue rulings we find No. 804, June 29, 1904, J. W. Yerkes, Commissioner, the following:

"The question whether fermented malt liquor is intoxicating or non-intoxicating is immaterial under the internal revenue laws, although it may be a material question under the prohibitory laws of a State or under local ordinances."

No ruling, so far as I can find, has ever been made with reference to root beer, nor can I find any literature on its alcohol content when made as above described. The soda-fountain root beer, of course, is made by diluting a flavored syrup with carbonated water and therefore contains no more alcohol than the minute amount contributed by the extract used to flavor the syrup, which would not exceed 0.05 percent, and is not to be confused with the home-brewed or fermented product, which is the subject of this article.

It is recorded in literature that koumiss, which is made from milk fermented under somewhat similar conditions, sometimes contains over 2 percent of alcohol.

The foregoing facts may come as a surprise to many who have looked upon home-brewed root beer as a strictly temperance drink. With beer averaging 4 percent alcohol, the mathematical ratio becomes apparent that three bottles of home-brewed root beer, which have been allowed to stand for 10 days or over, are equivalent to one bottle of ordinary beer.

Flavors Employed in the Making of Root Beer.—The so-called root beers are aqueous or alcoholic extracts of certain aromatic bitter principles, some of which have medicinal properties and are included in the Pharmacopœia of the United States. Among the materials which are used for flavoring in this way are juniper berries, pipsissewa, sarsaparilla, spikenard, wintergreen, vanilla, various barks and roots, bitter principles such as tannin, and other allied bodies. These materials are sometimes furnished together with the necessary quantity of yeast so that the fermentation of the beers may be made at the home.

Near Beer.—Near beer is held to be a malt beverage containing less than ½ percent of alcohol. In many states where prohibition is established the sale of a malt beverage containing less than ½ percent of alcohol is permitted. A Minnesota court lately decided that a temperance beverage known as "Malta" is not a violation of the Hanson law of that state, which forbids the sale of malted liquor in any other place except licensed saloons. It was testified that the beverage was not an alcoholic or fermented liquor, and not made from malted grains.

Apparently the case was one of misbranding, as the sale of a beverage under the name "Malta" is clearly an indication that it is made of malt.

The federal revenue agents in northern Illinois in examining what was sold as "near beer" discovered that it contained more than one-half of 1 percent alcohol, which would require that the maker, bottler and retailer of this beverage must all secure government licenses.

Pop.—The term "pop" is applied to a carbonated beverage, and doubtless the name has been derived from the slight explosion which takes place when the stopper is removed. There is no rule nor regulation followed by the manufacturers of so-called pops. Generally the water which is used is colored, usually with caramel, and also flavored with different flavoring substances, either of a natural or synthetic character. If the flavoring substances are of natural origin they can hardly be objected to. The use of synthetic flavors, that is, ethereal compounds derived from alcohols, is not to be commended in any case. While many of these flavors are practically harmless in small quantities, they belong to a class of bodies which must be used with great discrimination, and it is best to have them eliminated altogether. There is one component of these waters which the consumer should be aware of, namely, that they are generally sweetened. Pops are sold largely in hot weather, and especially at great outdoor gatherings, like baseball games. A sweetened water has very little, if any, value as a remedy for thirst. In fact, one of the best ways to induce thirst is to eat large quantities of sugar or drink large quantities of sweetened water. Nature requires a lot of dilution in order to make this excess of sweets tolerable. Therefore, there is little if any benefit, as far as quenching thirst is concerned, derived from the consumption of these articles.

Lemonade.—Lemonade is a beverage almost as widely consumed as soda water throughout all parts of the United States. Lemonade is a beverage made from the expressed juice of lemons, sweetened to suit the taste with sugar. The water which is employed may be either plain, potable water or sometimes water impregnated with carbon dioxid. The lemon juice which is used in lemonade contains not only the acid citrate of potash as its acidifying agent, but also the aromatic flavors, consisting of oil, ethers, etc., characteristic of the lemon itself. The rinds are frequently added and give an additional bitter taste due to a glucoside of some kind, probably of a tannic nature, residing particularly in the skins.

Orangeade.—Orangeade is a beverage similar to that described under lemonade, in which the orange juice is employed instead of lemon juice. The same descriptive matter is applied to it as to lemonade. Inasmuch as orange juice is less acid than lemon juice, a smaller quantity of sugar is employed in the manu-facture of this beverage.

Limeade.—The term "limeade" is applied to a beverage made from the juice of limes and in the same manner as described for lemonade. As lime juice is considerably more acid than lemon juice, a larger quantity of sugar is employed than for lemonade. All of these beverages are usually served very cold, in fact, colder than is required for health and flavor. Instead of having them near the temperature of ice, they would be much better in every respect if drunk at a temperature of about 50°.

Soft Drinks Containing Caffein.—The laws controlling the traffic in cocain and its sales have become so strict that the use of this body in so-called soft drinks is practically prohibited. The only alkaloid which is employed at the present time in the prepa-ration of these beverages is caffein. Caffein is an alkaloid. of somewhat restricted occurrence, being found chiefly in coffee, tea, cola nuts, yerba maté, and guarana. In the cylindrical sticks of the plant guarana prepared for the market, as high as 5 percent of caffein is sometimes found.

The caffein which is used commercially is extracted chiefly from tea sweepings, although it may be obtained from coffee which is used in making Kaffee Hag and similar preparations. It is a white crystalline powder, colorless when pure, and at high temperatures is sublimed unchanged. It has a somewhat bitter taste and is soluble in water. It is a substance known in chemis-try as a base, that is, it unites with acids to form salts.

The effects of caffein are well known, and in large quantities it produces sleeplessness and other injurious effects. Even in small quantities it acts strongly on certain individuals who are more or less idiosyncratic to its effects. It is consumed in large quantities by the tea and coffee drinkers of the world. Chocolate and cocoa also contain small quantities of caffein, but the principal alkaloid in cocoa is theobromin.

The chief soft drink which contains caffein is coca cola. In addition to the content of caffein, coca cola also contains the extract from coca leaves from which the cocain has been pre-viously obtained. In addition to this, also, a very small quantity of cola nut is employed, together with sugar, caramel and acid and aromatic substances. The composition of coca cola syrup as disclosed before the federal court in Chattanooga is as follows:

Caffein (grains per fluid ounce)................. 0.92–1.30
Phosphoric acid (H₃PO₄) (percent)............... 0.26–0.30
Sugar, total (percent)........................ 48.86–58.00
Alcohol (percent by volume).................... 0.90–1.27
Caramel, glycerin, lime juice, essential oils, and
    plant extractive........................... Present
Water (percent)............................ 34.00–41.00

From the above composition it is seen that coca cola, that is, the syrup from which the drink is made, consists of about half sugar, one-third water, and the remainder the substances mentioned in the analysis above. In preparing the beverage about one ounce of the syrup, as indicted above, is used per glass. The active principle on which its vogue mainly depends is the caffein. When caffein is taken in from one- to two-grain doses, as is the case with those who drink tea, coffee and coca cola, the effects are prompt and, as a rule, agreeable. There is a sense of gentle exhilaration, a relief from the oppression of fatigue, and a decided wakefulness or stimulation of the nerve centers. This condition continues for several hours, so that the effect produced is never by any means transitory.

Authorities competent to judge of the effects of caffein from a scientific point of view, differ widely in their estimates of its effect upon the human system. All admit that in excessive quantities it is harmful. Some maintain that even in small quantities, when its use is continued, harm is done the organism. Others maintain that in small quantities in most persons, that is, in quantities not exceeding from 5 to 10 grains of caffein per day, no evil effects are produced, and even beneficial effects may be expected. The layman is of necessity somewhat confused in this multitude of conflicting opinions. The safe attitude to take on a question of this kind, it seems to me, is that which should be assumed in reference to alcoholic beverages and tobacco, namely, that inasmuch as there is a difference of opinion among experts, the layman will do well to follow the rule of avoiding in his diet anything which is not necessary and which may possibly produce injury. These various opinions are well illustrated by the testimony of experts before the federal court. Dr. John Witherspoon, former

President of the American Medical Association, testified that he had treated twenty or thirty patients afflicted with coca cola habit in four or five years. When they gave up the habit their health improved. He said:

I regard coca cola as habit-forming. One glass creates a demand for another, because it stimulates the user and makes him feel better. Then when the effect wears off the feeling is one of depression, and he gets very nervous and seemingly cannot do without it very well.

The above may be regarded as a type of the testimony given by physicians who believe that coca cola is harmful.

Dr. Victor C. Vaughn, former President of the American Medical Association, testified:

I am of the opinion that coca cola syrup, taken in the form of a beverage in proportion of one ounce of syrup to 6 or 7 ounces of carbonated water, taken five or six times in the course of a day, would not produce injurious effects. I have no doubt it would be stimulating to the brain and muscles and to some extent possibly to the kidneys slightly, but such stimulation would be normal . . . . I should say that caffein should not be given to children under 7 years of age . . . . Because a certain drug does not produce an observably harmful effect, does not at all prove that it is not deleterious. Even one or one and a half grains of caffein may prove harmful to many persons, and I have no doubt there are many people who should not take caffein at all. I would prohibit caffein altogether to children under 7 years of age, and even above that age there may be some, and no doubt there are many, to whom it should not be given.

The above-contrasted opinions of two eminent medical men, each of whom has been President of the American Medical Association, show the wide variance of view of competent experts in the case, and yet a quite agreeing opinion that at times and in some individuals caffein is harmful.

When coca cola was first made the unextracted leaves of the coca were employed, and thus a small amount of cocain was found in the beverage. Subsequently the use of the fresh leaves of coca was abandoned and the exhausted leaves employed in their place. Many attempts have been made to introduce other caffeinated beverages to the people of the country, at first by giving them a name similar in some respects to coca cola. The courts have uniformly decided that such imitations are a violation of the copyrights of the Coca Cola Company. These imitations of coca cola, both in name and composition, have therefore not secured any lasting vogue. It is of importance that the people of the country should understand the nature of a beverage of this kind, in order that they may intelligently assume a proper position in respect of its use.

The Cola or Ola Drinks.—The remarkable success of the Coca Cola Company of Atlanta, Georgia, in building up a very great and lucrative business in the sale of coca cola has induced many other manufacturers to imitate, as nearly as possible, the name adopted by the Coca Cola Company. Aside from the unethical principle involved, it is gratifying to know that the public is not slow to distinguish the genuine from the imitation. In other words, if anyone wishes to drink a cola or an ola he prefers to take the original article rather than its substitutes. Whether or not all the various beverages which use "cola" or "ola" in some form in their names are also imitations of coca cola in composition, I am unable to say. At least, however, an informed person, knowing the nature of coca cola, would reasonably expect that any "cola" or "ola" drink of any kind would not be content with imitating only the name of the original article, but also to some extent its composition. It is fair to presume, therefore, that the "colas" or "olas" as a class contain caffein as the active and valuable ingredient.

The family of colas or olas is increasingly great. The Report of the President's Homes Commission, appointed by President Roosevelt to study the condition existing in the homes of the country, contained a list of cola or ola drinks, so-called, which is found on pages 268 and following of the Report. The names of these beverages, as printed in this Report, are as follows:

Afri Cola, The Afri Cola Co., Atlanta, Ga.

Ala Cola, Ala Bottling Works, Bessemer, Ala.

Carre Cola, E. Carre Co., Mobile, Ala.

Celery Cola, The Celery Cola Co., Birmingham, Ala.; Dallas, Texas; Nashville, Tenn., and St. Louis, Mo.

Chan Ola, L. M. Channell, New Orleans, La.

Chera Cola, Union Bottling Works, Columbus, Ga.

Coca Cola, Coca Cola Co., Atlanta, Ga.

Cola Coke, Lehman-Rosenfeld Co., Cincinnati, Ohio. (This preparation was formerly sold under the name of Rocco Cola.)

Cream Cola, Jebeles & Calias Co., Birmingham, Ala.

Four Kola, Big Four Bottling Works, Waco, Texas.

Hayo Kola, Hayo Kola Co., Norfolk, Va.
Heck's Cola, Heck & Co., Nashville, Tenn.
Kaye Ola, A. W. Kaye, Meridian, Miss.
Kola Ade, Wiley Manufacturing Co., Atlanta, Ga.
Kola Kola, W. J. Stange Co., Chicago, Ill.
Kola Phos, John Wyeth and Bro., Philadelphia, Pa.
Kos Kola, Sethness Co.,Chicago, Ill.
Lime Cola, Alabama Grocery Co., Birmingham, Ala.
Lima Ola, Wine Brew Co., Macon, Ga.
Nerv Ola, Henry K. Wampole & Co., Philadelphia, Pa.
Revive Ola, O. L. Gregory Vinegar Co., Birmingham, Ala.
Rocola, American Manufacturing Co., Savannah, Ga.
Rye Ola, Rye Ola Co., Birmingham, Ala.
Standard Cola, The Standard Bottling Co., Denver, Colo.
Tokola, Samuel Smith & Co., Chicago, Ill.
Vani Kola, Vani Kola Company, Canton, Ohio.
Wise Ola, The Wise Ola Co., Birmingham, Ala.
Citro Cola, Miners Fruit Nectar Co., Boston, Mass.
Koke Ola, Eagle Bottling Co., Frankfort, Ky.
Lon Kola, Lon Kola Co., Danville, Ky.
Mexicola, Celiko Bottling Works, Raleigh, N. C.
Pau Pau Cola, Pau Pau Cola Co., Detroit, Mich.
Pepsi Cola, C. D. Bradham, New Bern, N. C.
Charcola, H. C. Metzger, Meridian, Miss.
Cherry Kola, Williamsport, Pa.
Cola Soda, Jacob House and Sons, Buffalo, N. Y.
Field's Cola, H. C. Field, High Point, N. C.
Imported French Cola, Alabama Grocery Co., Birmingham, Ala.
(Claimed to be carbonated Wiseola.)
Jacob's Kola, Tampa, Florida.
Kola Cream, The Henzerling Co., Baltimore, Md.
Kola Pepsin Celery Wine Tonic, W. J. Miller, Cleveland, Ohio.
Kola Vena.
Loco Kola, Norton, Virginia.
Mintola, Davis Kelley Co., Louisville, Ky.
Ro-Cola, Savannah, Ga.
Schelhorns Cola, Evansville Bottling Co., Evansville, Ind.
Vine Cola, California Commerical Co., Los Angeles, Cal.

The Report of the Homes Commission says:

During the past decade soda fountain specialties containing caffein, extract of kola nut and extract of coca leaf, the active principle of which is cocain, have been offered in considerable quantities and, due to extensive and attractive advertising, both as beverages and as headache remedies and nerve tonics, their sale has assumed large proportions.

Judging from the names of most of these products it would appear that extract of kola nut is one of the chief ingredients, and, while in certain instances this drug is undoubtedly present, in most cases the caffein has been added as the alkaloid caffein obtained from refuse tea sweepings.

The use of the coca leaf, by reason of the fact that it introduces cocain into the drinks, has been now generally discontinued. In the case of coca cola the decocainized leaf, the refuse product discarded in the manufacture of cocain, is employed. So perfectly is the cocain extracted in the process of manufacture that the extract used in coca cola does not apparently contain any of the cocain alkaloid. The Report of the President's Homes Commission was issued in 1908, and since that time it is certain that many more drinks of the ola or cola type have been placed upon the market. While I have intimated that one acquainted with the composition of the original beverage would infer that the substitutes contained caffein and are also made partly from the kola nut which contains considerable quantities of caffein, it would not be just to intimate that all of them are of that composition. It is possible that many of them are merely imitations of name, and not imitations of composition.

In order to determine whether or not the manufacture and sale of ola or cola beverages is still practised, I have consulted the National Bottlers' Gazette, the issue of January 5, 1916. In that issue it is evident that the practice of the cola or ola habit is still in vogue. It is said, on page 64:

Bottlers interested in what's what in cola drinks should turn to the two-color advertisement of the Sher-A-Coca Co. of Pittsburgh, Pa., on which is advertised its two high class drinks, Sher-A-Coca and Sher-A-Coca Punch. An interesting little booklet has recently been issued by the company, describing the merits of these new beverages, and our readers will do well to get a copy and read it.

Thus it is seen that even in drinks in which the word "ola" or "cola" does not occur, but where "coca" does occur, it is recognized that they belong to the cola or ola type. On the same page attention is called to a new cola which is attracting considerable attention, namely, Penn-Cola, manufactured in Pittsburgh, Pennsylvania. On the same page attention is called to the fact that the Gay-Ola Company of Memphis, Tennessee, is now known as the Gay-Ola Syrup Company. And still on the same page it is stated that the proprietor of the spicy specialty, Roxa-Kola, of Winchester, Ky., recently gave testimony before the Kentucky Railroad Commission in the interest of the Kentucky Bottlers' Association. Thus it is seen that five new cola drinks have started out with the first month of 1916. To this list may now be added another viz., Christo-Cola which I first noticed in the early summer of 1916.

These competitors with Coca Cola's name are of course endeavoring to fight shy of infringing the trade mark. A correspondent from Missouri, where they still want to be shown, says in the National Bottlers' Gazette:

We are in need of a little information, and ask you to give your opinion on the following: We are preparing an advertising card, and are in doubt if the words "A genuine Coca and Cola flavored beverage," would be infringing on the "Coca Cola" trademark. The Louisville Carbonating Syrup Co., who manufacture a Coca and Cola flavored syrup, put the following upon their crowns: "A genuine Coca and Cola Flavor."

To this the editor replies:

"A Genuine Coca and Cola flavored beverage" would not infringe on the "Coca Cola" trade mark. This is providing, however, that the word "and" between the words "Coca" and "Cola" is as plain and distinct as the rest of the lettering.

A formula for cola compound syrup is given, also, on page 100 of the publication just cited, as follows:

Two centigrammes of quinine hydrochloride, 0.04 gramme of citric acid dissolved in sufficient water to obtain 10 grammes of solution, to which are added 195 grammes of simple sirup, 10 grammes of sodium glycerophosphate, 10 grammes of fluid extract of kola, 15 grammes of saccharated iron oxide and 4 drops of oil of orange peel.

Legislation and Court Decisions in re Coca Cola.—In the so-called Harrison "Narcotic Act" which regulates among other things, the distribution of cocain, a provision was inserted exempting spent coca leaves and products made therefrom from the provisions of the law. This provision authorizes the re-extraction of the exhausted coca leaves after the cocain has been removed and the employment of this extract in beverages without registration.

The case of the United States against 40 barrels and 20 kegs of coca cola before the District Court of the United States,

Southern Division, Eastern District of Tennessee, is one justly celebrated. The Libel of the United States alleged that this merchandise as advertised was misbranded in that it did not contain as essential ingredients any coca or cola and further that it was adulterated, in that it contained an added ingredient, caffein, which was deleterious to health. After a trial lasting more than a month the Court took the case from the jury and decided its principal points in favor of the Coca Cola Company, namely that it was not misbranded and that the caffein therein was not an added substance. An Abstract of this case is published by the Department of Agriculture as Notice of Judgment No. 1455, May 27, 1912.

The case was appealed to the Circuit Court of the United States, Sixth Circuit, before Judges Warrington, Knappen and Denison who by a unanimous opinion supported the rulings of the lower court. The number of the case is 2415 and an abstract of the case was issued by the Department of Agriculture, from the Office of the Solicitor, June 30, 1914, as Circular No. 80.

The case was appealed by the Government to the Supreme Court. This Court by a unanimous opinion No. 562, May 23, 1916, overruled the courts below, both on the decision of misbranding and adulteration and remanded the case for a new trial.

The case was called before Judge Sanford at Chattanooga on the 12th day of November, 1917. The Attorney for the United States agreed with the attorney for the Coca Cola Company to forego a new trial and in view of the statements made by the Company that the formula had been changed the case was dismissed without prejudice to the Coca Cola Company under the following order of the Court:

"AND IT IS FURTHER ORDERED that the said goods, wares or merchandise seized herein, to wit, the forty barrels and twenty kegs of Coca Cola, shall be released to the claimant upon said claimant paying the costs above adjudged and giving sufficient bond, conditioned that the product shall not be sold or otherwise disposed of contrary to the provisions of the Federal Food and Drugs Act, or the laws of any State, Territory, district or Insular possession of the United States."

"In open Court, this 12th day of November, 1917."

(Signed) Edward T. Sanford,
United States Judge.

8

The nature of the changes proposed by the Company in the composition of coca cola does not appear in the published proceedings of the court to which I have had access. In a recent advertisement of the Coca Cola Company it was stated:

This Company regards it a privilege to comply with the Government's request, made similarly to all manufacturers employing sugar in quantity, to reduce our output fifty percent.

To the end of conservation we pledge our further efforts in every direction that opportunity may disclose, in manufacture as well as beyond the scope of our immediate interests; and in this effort generally we bespeak the co-operation of dealers and consumers everywhere.

The Greater Danger of Free Caffein.—Dr. John Uri Lloyd, in The Eclectic Medical Gleaner, emphasizes the increased dangerous activity of certain drugs when used alone as compared with their mild and favorable action when used in their native environment. He says, under the head of "Dangerous Products:"

Remedies that in their natural condition were kindly and beneficial proved, when elaborated from surrounding structures, to be somewhat as is nitro-glycerin if contrasted with gunpowder. The explosion is likely to burst the gun if enough be used to blow out the ball.

It is the custom of some who make compounds to be given to the people indiscriminately and in large quantities, to liberally use caffein as a constituent of their headache cures. Some people argue that because caffein is obtained from coffee and tea, it is the one "active principle" of coffee and of tea, whilst others imagine that because coffee and tea can be used without stint in the making of a structural decoction which carries the caffein in assimilable combination, caffein itself may be likened to the drugs from which destructive chemistry produces caffein. Dr. Claisse, in La Clinique, France, has reported several accidents from solutions of caffein, and even from caffein in its usual dose.

In a brief communication to the same journal, Dr. Triboulet indorses his colleagues warning, and declares that caffein, although a valued therapeutical agent, is a brutal medicament, the action of which is extremely difficult to regulate. In the first place, when given hypodermically, it often excites local inflammation when the drug is deposited too superficially under the skin. Secondly, it is a powerful cerebral excitant, and capable of causing maniacal delirium, especially in aged persons.

Soft Drinks in the Far North.—There is a soft drink used by the Russians which is composed of honey, pepper, hot water and boiling milk. A soft drink used in Lapland is made of hot water and meal, strongly flavored with tallow, or reindeer blood if obtainable.

Adulterations and Misbranding of the Above Beverages.—The principal adulterations of lemonade, orangeade and limeade, are made by using citric acid and sugar instead of the juices of the fruits, and adding a dash of lemon flavor or orange flavor or lime flavor as the case may be. These flavors may be the natural oils of the fruits dissolved in alcohol or of an artificial character. The use of citric acid instead of the natural juice of the fruits is objectionable because the citric acid of the fruits is combined more or less completely with potash and other bases. Instead of adding merely an acid to the contents of the stomach, therefore, there is added a salt which upon digestion produces alkalinity instead of acidity. The natural or artificial flavors are not by any means comparable in character to the natural flavor of the fruit itself.

Once at a summer resort I saw a booth where orangeade was being sold. The beverage was presumably made while you waited. A few oranges were constantly rolling up an inclined plane and dropping into a machine and disappearing, and the juice was going out of a side exit. The food authorities having examined this machine, found that the oranges were doing an endless stunt. It was Sisyphus translated into a citrous fruit. They escaped destruction entirely, while the reputed juice was wholly of artificial character. The application of the term "lemonade," "orangeade" or "limeade," to any preparation of this kind is misbranding, and the articles themselves are adulterated.

PART IV

FRUIT JUICES

Cider.—Cider is a beverage which is universally known wherever apples grow. Unfortunately around most orchards the best apples are not used for cider making. They are reserved for sale in boxes or barrels, or sent to the factory to be worked up into jellies, apple butter and dried apples. As a result of this common practice a great part of the cider which is made in this country is from imperfect apples, either undergrown, overripe or decayed. Cider made from imperfect material cannot be expected to measure up to the standard of excellence required. When cider is made of select material, under sanitary conditions, it is a beverage more desirable than grape juice, because it contains a much less quantity of sugar and is not so apt to unbalance the ration. Even cider made from culls may be of excellent quality. Especially apples that are blown down during a storm and immediately gathered and manufactured into cider are quite as acceptable as they would have been if picked from the limbs in the usual careful way. If overripe or underripe or decayed apples are used, the character of the beverage produced is very inferior.

Almost any kind of a small press will make small quantities of cider, and it is manufactured in many different varieties of press. The modern cider press combines an effective rasping machine and hydraulic press of high resistance, by means of which the yield of cider from a given number of apples may be greatly increased. Cider is often made during the earlier part of the harvest when the weather is still at times excessively warm, and fermentation sets in within 24 hours. In such circumstances cider is not a soft drink for a longer period than a day or two at most.

Preservation of Cider.—Extensive investigations have been made in the preservation of cider, especially in the Department of

116

Agriculture, where under my supervision Mr. Gore conducted experiments to keep the cider from fermenting and to exclude all added preservative substances. At the same time we tried to preserve the taste and flavor of the cider, which often are impaired by heating to a sterilizing temperature. The most effective methods employed were as follows: The cider, as soon as it was expressed, was carried to a clarifier, something like an ordinary cream separator. It was passed through this instrument at a high rate of speed, some 6 or 7 thousand revolutions a minute. The cider, treated in this way, deposited its gummy and other solid substances upon the walls of the separator and discharged a limpid, brilliant liquid from which practically the yeasts had been removed together with other solid matter. This mechanical treatment vastly improved the character of the product. Enough successful work was done to show that this method of clarification, even on a large scale, would be entirely feasible.

After the cider was clarified in the way above described it was conducted to a pasteurizing apparatus and kept at a temperature of from 140 to 145° for 30 minutes. It was then drawn off into a barrel which had been thoroughly sterilized by steaming, first passing through a cooling apparatus without coming into contact with the air, and passing down into the barrel through a tube surrounded at the bung hole with sterilized cotton. In this way the pasteurized cider was conducted without coming into contact with the air into a sterilized barrel, which, as it was filled, drove the air out through the cotton at the bung. When the barrel was full the delivery tube and the sterilized cotton were removed from the bung and the barrel stoppered with a sterilized bung. Cider prepared in this way and kept in a cool place, not necessarily in a refrigerator, keeps indefinitely and has no bad or objectionable taste. Instead of conveying the cider into barrels for keeping, it can also be placed in bottles, which, if exposed in any way, should be re-sterilized after the corks are inserted. If cider is made late in November, after the weather has grown cool, and placed in well-sterilized barrels it will keep even if not sterilized well into the winter if the weather does not become too warm, but will eventually set up a slow fermentation and become hard.

I feel sure that the objection which is usually urged against sterilized cider, namely, that its taste is impaired, could be removed were cider treated as described above. In this way I feel certain that a cider could be obtained which, in every respect as regards excellence of flavor, would be quite comparable with the cider of Normandy.

Adulterations of Cider.—The principal adulteration of cider is one already referred to, namely, the use of unfit apples in its manufacture. The most common adulteration of cider at the present day, aside from that arising from unsatisfactory raw materials, is the preservation thereof by means of benzoate of soda. The use of benzoate of soda being permitted by the Department of Agriculture, without restriction of quantity or material preserved, has led to the very extensive use in the preservation of cider of this objectionable material. Although under the rulings of the Department no prosecutions can be brought against those who sell preserved cider, there may be a public opinion excited of such weight as to practically eliminate it from this class of beverages. The makers of grape juice are to be congratulated, in so far as I know, on the fact that they never have taken advantage of the permission to use benzoate of soda in the preservation of grape juice. So much, however, cannot be said of the makers of cider. It is so simple an operation, and before the European War was so cheap, as to have allurements for all who care more for the mercenary part of business than they do for the health of consumers.

Dr. Lucas, of New York, in a very interesting experiment secured a lot of pure cider and divided it into two parts. To one part he added the usual amount of benzoate of soda required to preserve the cider from fermenting. The other was used without manipulation. The cider was given to an equal number of healthy young men. All those who received the portions containing benzoate of soda were rendered more or less ill; while the others who drank the pure cider were unaffected. There is no doubt in my mind of the fact that the addition of benzoate of soda to a substance like cider, or in fact to any substance, is highly deleterious and is forbidden by the Food and Drugs Act. It is only a wrongful interpretation of this Act which permits such abuses to continue. In so far as I know cider is no longer preserved by the fumes of burning sulphur. The ease with which it can be preserved with benzoate of soda does not warrant experiments with bodies more difficult to handle.

Perry.—"Perry" is the name applied to the expressed juice of pears, a product in every way, except the character of the fruit, which is identical with cider. The perry industry has not assumed any large proportions in this country. I do not remember ever to have seen perry offered commercially in the markets, though of course it doubtless has been an object of merchandise. Other beverages may be found in the home from time to time, expressed from cherries, peaches, blackberries and other small fruits, but the quantities of these are minute and they do not demand any further description.

Hard Cider.—When cider is left standing for a short time, the length of which depends largely upon the temperature, fermentation sets in and the cider becomes hard, that is, the sugar which it contains is converted largely into alcohol and carbon dioxid. The taste of the cider changes, it loses its sweetness, it acquires a sharp, penetrating taste due to a combination of alcohol and carbon dioxid, and becomes intoxicating. Hard cider is often used as a beverage, and its activity apparently is increased by the raw products of fermentation which it contains. If hard cider could be stored in wood in such a way as to avoid acetic fermentation its taste and flavor and general character would improve, just as beer is improved by being placed in lager or distilled spirits by being kept in wood. The character of the fermentation depends upon the character of the natural yeasts which are distributed everywhere and which are attached to the fruits before they enter the press. These yeasts are usually of a character to produce a good product. They may be, however, of a bad character. Thus all kinds of flavors may occur in self-fermented ciders.

In localities where the cider industry is carried on as a profession, as, for instance, in Normandy, in certain parts of Germany, and in certain localities in England, steps have been taken to control the character of the fermentation so as to produce always a finished product of an agreeable taste and flavor. Professor W. B. Alwood[1] has studied the effects of different kinds of

[1] Bureau of Chemistry, Department of Agriculture, Bulletin No. 129.

yeast upon the character and flavor of the product. These yeasts
he secured by visiting the localities in Europe where pure cultures
are used in the manufacture of cider. These living yeasts were
brought to this country and cultivated in the laboratory at
Charlottesville, which is under the immediate direction of Prof.
Alwood. In the preparation of apple juice for the introduction
of pure cultures it is first pasteurized so that all the yeasts
which it contains are killed. It is then seeded with the pure
culture and left to fermentation in the usual way.

The content of alcohol developed in hard-cider making depends, of course, upon the amount of fermentable matter present.
In low-grade apples, grown in Nebraska, Gore[1] found the following
average composition of several varieties, including those which
are commonly grown: York Imperial, Black Twig, Winesap,
Rhode Island Greening, Grimes Golden, Jonathan and Ben
Davis. These apples were all unmerchantable and therefore may
be regarded as the waste material of the orchard. The average
composition of the juices of these varieties is as follows:

Total solids.............................. 13.24 percent
Sugar-free solids......................... 1.48 percent
Invert sugar.............................. 7.87 percent
Sugar..................................... 3.35 percent
Malic acid................................ 0.42 percent

These data may be taken to represent very accurately the ordinary
materials, culls, specked apples, windfalls, etc., which are used
in the United States in the making of cider.

The cider was made under the supervision of Prof. Alwood in
the most approved manner, and a very fair article of bottled cider
was secured. The cider was made during hot September weather,
and after racking and storing in sterilized barrels and kegs
analyses were made in the following April. The different lots
contained the following percentages of alcohol: Lot 1, kept in
kegs, 5 percent; stored in barrels, 4.84 percent. From the above
data it is seen that a cider properly fermented and protected from
acetic fermentation by proper storage produces almost the
theoretical percentage of alcohol which could be expected.

[1] Journal of American Chemical Society, 1907, Volume 29, page 1112.

Theoretically a sugar will produce one-half of its weight in alcohol. An 11 percent sugar would be expected to produce 5½ percent of alcohol. In point of fact all of the sugar is not converted into alcohol. During fermentation some of it is converted into organic acids and other products, and a little of it always escapes fermentation.

In order to produce a sparkling cider the final fermentation is allowed to take place in a stoppered bottle in the same way that champagne is made, though of course the pressure is not usually so great. Hard ciders made in this way produce a fine foam, and their potable qualities are greatly benefited by finishing the fermentation in the bottle. Ciders which are fermented in the manner described, so as to be clean and free of germ life except that necessary to produce the final fermentation, keep perfectly in the bottle. Alwood found that the samples thus made were in excellent condition at the end of two years. Ciders made from winter apples and allowed to ferment, either through their own yeasts or by the addition of yeasts, were found to keep perfectly, although in many instances the fermentation was stopped before all the sugar was consumed. The apples from which the winter juices were made contained more sugar than those of the low-grade apples, namely, a little over 12 percent. The percentage of alcohol obtained was generally about 5 percent, though in one instance it fell to 3.64 percent by weight and rose as high as 5.12 percent by weight in another.

These valuable researches of Alwood show beyond any doubt that it would be possible in this country to make a cider, especially a bottled cider, containing approximately 5 percent of alcohol, which would be quite equal to the best sparkling ciders made in Normandy.

Production of Vinegar.—Cider vinegar, which is recognized in this country as the type of vinegars and which the law regards as being the substance which should be furnished when vinegar is asked for, is usually made without any artificial fermentation or treatment whatever except to leave hard cider for a year or two in a barrel, trusting to the natural ferments to produce the acetification. The formation of acetic acid goes forward usually to a degree inversely proportional to the temperature to which it is subjected. In winter acetification goes forward slowly, while in summer it goes forward more rapidly. The acetic ferment gradually increases in quantity, and by agglutination forms ropy deposits which are known as "mother of vinegar." The quantity of acid formed when acetification is complete is slightly greater than that of the original alcohol. In Alwood's experiments as high as 7.1 percent of acidity was secured when acetification was complete. The sample above cited, which contained a little over 7 percent of acetic acid, was completely fermented at ordinary cellar temperatures in a little over a year.

Grape Juice.—A very large industry is now established in this country in the manufacture of unfermented juice of the grape. The ease with which this unfermented juice may be pasteurized or sterilized and kept indefinitely has caused the business to grow rapidly in magnitude. Localities in which grape juice is manufactured are widespread, but especially is the business conducted on a large scale in the State of New York. There are many areas in this state suited to grape culture, and especially a narrow strip of land extending along Lake Erie from near Buffalo to **Erie.** This locality is peculiarly suited to the growth of grapes and without doubt is the best grape region on the East side of the Rocky Mountains. In Ohio there is a like region of excellent grape land, especially near Sandusky. Other important areas for the manufacture of grape juice are California and, to a limited extent, New Jersey. The variety of grapes employed is rather extensive, but the principal varieties are Catawba, Norton, Clinton and Concord.

Composition of Grape Juice.—It is important, in selecting grapes for making an unfermented beverage, to have them reasonably low in acid and with a comparatively high content of sugar. Well-ripened grapes, grown in the regions mentioned, have approximately the following average composition: total solids in the expressed juice, 22 percent; solids not sugar, 2 percent; invert sugar, 19 percent; total acidity, reckoned as tartaric, 0.65 percent. There are many wide differences from these mean data, according to the season of the year and the character of the grape. As the grapes grow riper the acidity tends to decrease and the sugar content rises. The mean composition of different varieties of grapes, as reported by Professor W. B. Alwood, Bureau of Chemistry, is as follows:

TABLE I.—AVERAGE COMPOSITION OF 8 IMPORTANT NATIVE AMERICAN GRAPES
Results in grams per 100 cubic centimeters.)

Variety	No. of samples	Specific gravity of juice	Total solids calculated	Sugar-free solids calculated	Total sugar as reducing	Total acid as tartaric	Acid-sugar ratio.
Catawba.........	221	1.0858	22.30	2.42	19.88	0.928	1:22.5
Clinton..........	25	1.0896	23.31	3.51	19.80	1.464	1:14.4
Concord.........	127	1.0754	19.57	2.48	17.12	0.650	1:27.6
Cynthiana........	5	1.0942	24.51	4.21	20.30	1.108	1:19.8
Delaware........	53	1.0993	25.85	2.39	23.37	0.646	1:37.7
Iona.............	61	1.0877	22.79	2.30	20.50	0.753	1:27.7
Ives.............	100	1.0704	18.27	2.79	15.45	0.693	1:22.7
Norton..........	50	1.0987	22.50	3.24	19.24	1.080	1:17.8

The above table is made up of examination of a large number of samples of the leading varieties of grapes, extending over a period of time from 1908 to 1911 inclusive. The districts represented are almost entirely those of Northern Ohio and Virginia. The data, therefore, are truly representative of the normal composition of the varieties mentioned. Inasmuch as the Concord grape is the one which is chiefly used for making grape juice, it will be interesting to compare commercial grape juices with the grapes from which they are chiefly made.

COMPOSITION OF SAMPLES OF GRAPE JUICE BOUGHT ON THE OPEN MARKET

	Total solids	Sugars	Solids not sugar	Cane sugar (sucrose)
Royal Purple grape juice........	16.22	14.48		
Red Wing grape juice...........	19.27	16.70	2.57	0.76
Grape juice....................	18.83	17.78	1.05	1.35
Monticello grape juice..........	20.64	18.26		
Grape juice....................	17.79	14.31		

In the examination of commercial samples of grape juice I have not found any glucose, saccharin, salicylic acid, benzoic acid, borax, sulphurous acid, or any of their compounds. This fact shows that the use of preservatives in commercial grape juices is no longer practised. Occasionally we find added cane sugar, usually not to exceed about 2 percent. The most striking fact, however, is that the total sugars in the grape juices of commerce are not as high as are found in the grapes of good quality, as determined by the analyses of Alwood. This discrepancy may be due to the fact that grapes of less desirable quality are used. I should hate to hint that water is added in the process of manufacture. There is no evidence that I know of this practice, save the variation between the amount of sugar in the grapes and the amount which is found in the commercial samples.

Method of Manufacture.—For red grape juice the grapes are crushed, but not pressed. The crushed grapes are taken into large heating caldrons, usually made of aluminum, and furnished with a steam jacket. The temperature of the crushed grape is carried to about 175°F. for a period of 15 minutes, the mass being well stirred during this period of heating. The purpose of heating the grape juice is first to pasteurize it, but especially to extract the red coloring matter from the skins. After extracting the color and pasteurization the juice is separated from the pulp, either by gravity or generally afterward by pressure, and conducted into the containers which are used for storage. Sometimes these containers are barrels, but more often glass carboys. The containers are properly sterilized before the pasteurized juice is admitted. The bungs are also sterilized and coated with wax or some other preparation so as to make them air-tight. The juices are left in these storage containers until complete sedimentation take place and the liquid is clear. In some instances the pulp is not pressed, but additional sugar is added to it, and a low-grade sugared wine is made therefrom. When the sedimentation is complete the clear juice is siphoned off carefully, so as not to disturb the sediment, into bottles, which are then corked and sterilized as is ordinarily done for bottled goods. When white juice is to be made the crushed grapes are first pressed so as to secure the juice apart from the skins, and the subsequent treatment is the same as that just described. Bulletin 656 of the Bureau of Chemistry, issued May 8, 1918, gives detailed instructions for the manufacture of grape juice.

PART V

COFFEE

Historical.—In connection with the active use of coffee as a beverage a short history of this substance will not be out of place. In the following historical sketch I am indebted chiefly to Mr. John Uri Lloyd.[1]

It appears from the researches of Lloyd that coffee was not known to the ancients. It is not mentioned in the history of the Jews, nor do the early Greek and Latin writers make any reference to it. No description of it, moreover, is found in the older writings of Confucius, and other Eastern Philosophers. Nevertheless, it is evident that, in its native country, the coffee must have been used from times immemorial. The coffee tree appears to be originally of African origin.

There is a belief that coffee was introduced in Abyssinia by Robert Manns about the year 875, and from Abyssinia it reached Arabia in the thirteenth century, and from Arabia it was introduced into Syria, Persia, Turkey, Asia and in other countries.

The first introduction of coffee into Great Britain appears to be about the year 1640, just two years after the foundation of Harvard University in the United States. It is thus seen that the European and American use of coffee is not of very ancient origin. At the time of its first introduction into London (1651) it commanded a very high price, ranging from 20 to 25 dollars per pound. In 1657 the use of coffee was introduced into Paris, but it met with some disfavor on the part of Louis XIV, and afterward Louis XV, and neither of these monarchs started drinking coffee, and it did not become popular in France before 1669. The popularity of coffee in Paris was due largely to the acts of the Ambassador of Mahomet, who gave magnificent entertainments, and introduced the use of Turkish coffee. Black

[1] The Western Druggist.

slaves of the Ambassador, arrayed in most gorgeous oriental costumes, presented the very choicest Mocha in very small cups of fine porcelain, made with the art which seems known alone in oriental countries. This magnificent presentation of a beverage, which, in itself, is calculated to win its way on account of its taste and its influence upon the nerves, succeeded in making it popular with those who were able to afford it. The use of coffee in the French Capitol became one of the fads of the aristocracy. They took much interest in it, and the daughters of Louis XIV, notwithstanding the position of their father relative to the use of the beverage, became addicted to the use of it, and finally it became a common beverage of the Royal household. It is related that the cost of the berry at that time was so high that, although not drunk in very great quantities, it cost the Royal family as much as $15,000 per year.

In 1664 the use of coffee was introduced in Germany, and from Germany it spread rapidly to neighboring countries.

Lloyd attributes the rapid spread of the use of coffee to the fact that man naturally craves a stimulant at times, and is never willing to be wholly without it; and the stimulating effect of the caffein, the active principle of coffee, modified as it is by its form of combination and its association with the other ingredients of the coffee, presented to man a beverage which was almost ideal and satisfied his craving for a stimulant.

The Effect of the Introduction of the Use of Coffee and the Cultivation of the Coffee Trade.—As soon as the nations of Europe began the use of coffee a great stimulus was exerted upon the cultivation of the tree. The countries which possessed the tree became jealous of its privilege, and, in Arabia, the penalty of death was threatened against anyone who would permit the exportation of a single coffee shrub to a foreign country. The only place where European vessels were permitted to secure Arabia coffee for transportation was at the Port of Mocha, and this is the reason that "Mocha" is a representative term for Arabian coffees. As is well known, no coffee trees grow in the vicinity of Mocha, and hence the merchants who produced the coffee berry had no opportunity to get hold of the coffee trees for the purpose of transporting them to other countries where the climate and soil were suitable for their grow h.

The Dutch were the first to get hold of the plants, and, as is well known, established its culture in the Dutch East Indies where it has thrived ever since. Today the coffee shrub is developed throughout the world wherever the climate, soil and rainfall are favorable to its growth.

Methods of Using Coffee.—It is believed that in early days and in the countries where the coffee tree grew indigenously the berry was not used itself as a beverage. On the contrary it appears to have been roasted as in the case of today, ground or pounded to a powder, mixed with lard, butter or other oily substances, and employed in this state as a food.

Bruce, who traveled toward the source of the Nile in 1678 published the following statement:

The Galae is a wandering nation of Africa, who, in their incursions into Abyssinia, are obliged to travel over immense deserts, and being desirous of falling on the towns and villages of that country without warning, carry nothing to eat with them but the berries of the coffee tree roasted and pulverized, which they mix with grease to a certain consistency that will permit of its being rolled into masses about the size of billiard balls, and then put in leathern bags until required for use. One of these balls they claim will support them for a whole day, when on a marauding incursion or in active war, better than a loaf of bread or a meal of meat, because it cheers their spirits as well as feeds them.

Lloyd's belief that the coffee berry is a good food as is easily seen from the quantity of *nitrogenous* and *carbohydrate* substances, as well of oils which it contains, is not shared by other dietitians. Its chief value, when on forced marches of the kind described above, is in its stimulating effect upon the nerve centers and through them upon the muscular activities.

Inhibition By Mahomet.—It is well known that the Koran prohibits absolutely the use of alcoholic beverages and also coffee but the Mohammedan Church and the Turkish Government have been unable to prevent the use of coffee from becoming quite general among the followers of the Prophet. Thus at the present day the believers in the Koran are not only the most enthusiastic drinkers of coffee but also apparently the most skilled in its preparation. It is drunk by the Musselmen at all hours of the day and night and by all classes of people from the Sultan to the Mufti, merchant, artizan and peasant. Nevertheless there are many devout followers of Mahomet who even today refuse to drink coffee.

The use of coffee in England was also at first bitterly resisted. When the first coffee houses opened in London, followed soon by others, they were called by their opponents "seminaries of sedition." Charles II attempted to suppress the coffee houses in 1675, under the pretense that they served as places of meeting for the disaffected, who invented and circulated false and calumnious reports to defame the government of the King and to disturb the repose of the nation. There are among others a woman's petition against coffee, and a man's also. Charles II, however, failed in his campaign against coffee, and it was taken up by Oliver Cromwell, who as Lord High Protector, attempted to prohibit the use of coffee in London. A heavy tax was placed upon the drink, which had the usual effect of producing smuggling and afterward the coffee habit conquered in England, also both the Church and State. The Proclamation of King Charles denouncing coffee is as follows:

Charles R.

Whereas it is most apparent that the multitude of coffee houses of late years set up and kept within this kingdom, the Dominion of Wales, and the town of Berwick-upon-Tweed, and the great resort of idle and disaffected persons to them have produced very evil and dangerous effects, as well for that many tradesmen and others do herein mis-spend much of their time which might and probably would be employed in and about their lawful calling and affairs, but also for that in such houses, divers false, malitious and scandalous reports are devised and spread abroad to the defamation of His Majestie's Government, and to the disturbance of the peace and quiet of this realm, His Majesty hath thought fit and necessary that the said coffee houses be (for the future) put down and suppressed and doth strictly charge and command all manner of persons, that they or any of them do not presume from and after the tenth day of January next ensuing, to keep any public coffee house, or to utter or sell by retail, in his, or her or their house or houses (to be spent or consumed within the same), any coffee, chocolate, sherbett or tea as they will answer the contrary at their utmost peril.

Given under our court at Whitehall this third and twentieth day of December, 1675, in the seven and twentieth year of our reign.

"God Save the King."

The Royal decree probably did not have much to do toward suppressing the coffee habit, but it did call attention to those engaged in it so that they were regarded as quite important personages.   On account of the opposition which these personages made to the terms of the Proclamation, His Majesty, by reason of what he called his "princely consideration and royal compassion," caused it to be withdrawn before it was put into effect.

The Coffee Habit.—Lloyd says:

This stimulant that does not intoxicate, may become man's enemy. A fair friend is it, but a dogmatic master.   Fortunately it does not inebriate. So long as it is used in judicious moderation, most persons find it a cheering, exhilarating companion, but when coffee obtains the upper hand of man or woman it shatters the nerves, and demoralizes the victim's disposition. In our opinion, the soporific action of an alcoholic drink is to be preferred to the wakefulness of the person who tosses the night through, a victim to coffee dissipation.   But it is not safe to suggest, even to a nerve-wrecked coffee slave, that he is to be classed with the man who abuses himself by means of alcohol, the indignation of such people is most pronounced.   Nor is it discreet, with some people, to include coffee with the stimulants that enslave.   But yet, when one views the condition of nervous America, the question arises as to why the pronounced prohibitionists who object to even the temperate use of malt or other alcoholic beverages that by their abuse are harmful, are overlooking the tyranny of coffee and tea.   But this is a problem in itself, and somewhat foreign to this historical article.

Coffee Houses.—After the introduction of coffee a place for drinking coffee was soon established.   In French the term "café," equivalent to the English "coffee house" not only came into use but has passed into a different signification, being applied to a kind of restaurant where drinks are chiefly served with only light refreshment, coffee being only one of the many beverages dispensed.   The term "café" has broadened still further in being Anglo-Saxonized so that in this country it is now practically synonymous with restaurant or eating house (cafeteria), the food being the chief thing dispensed instead of liquids.

In this connection it is interesting to read what is said by McCullough, concerning the first of the coffee houses of London. He says:

The first of the establishments to which the name of "coffee house" was given was opened in London in 1652.   A merchant, named Edwards, who traded with Turkey, having brought some bags of coffee from the Levant, and brought with him a Greek slave accustomed to make it, soon saw his house besieged by a crowd of people who, under the pretense of visiting him, came to taste this new liquor.   To satisfy his friends, who became more numerous, and at the same time to free himself from the embarrassment

9

that they caused him, he allowed his servant, to establish himself where he pleased to make a coffee and sell it to the public. In consequence of this permission the Greek opened an establishment at the very spot where now is the Virginia Coffee House.

An account of the establishment of the café in Paris relates that:

Toward the close of the year 1669 an Armenian, named Pascal, set up a shop in the market of St. Germain, where he publicly sold coffee; the knights of Malta, who had dwelt in the East, travelers and officers gave so great a reputation to his establishment that it was sold very dear. He had numerous successors who all did an excellent business. One named Stephen of Alleppo finally built in the Rue St. Abdre des Ares, a room magnificent for the time, which he decorated with glasses and marble tables. Voltaire immortalized the Procope Coffee House, where a table of J. J. Rousseau and the encyclopedists is still shown.

Rousseau says in his memoires that he besotted himself in the society of the widow Laurent's Coffee House, and there, it is said, he composed his celebrated couplets, criticizing the other visitors; these satires being the cause of his banishment by an act of the French Parliament.

Introduction of Coffee Houses into Germany.—The first coffee house in Germany was said to have arisen in this way.

In 1683 the Turks besieged Vienna with an enormous army, but were completely routed. In their camp large quantities of coffee berries were found and presented to a valiant soldier, who established the first Vienna coffe house. His drink sprang at once into favor, and it has ever since remained in favor.

Journalism became associated with the coffee house in 1709, when Richard Steele brought out the Tatler in the advertisement of which it was announced:

All accounts of gallantry, pleasure and entertainment shall be under the article of White's coffee house; poetry under that of Will's coffee house; learning under the title of Grecian; foreign and domestic news you will have from St. James' coffee house, and what else I shall on any other subject offer shall be dated from my own apartment.

The Great Insurance Company of Lloyd's was founded in the coffee house, for it is stated:

"There can be no doubt," says Frederick Martin, the historian of Lloyd's, "that the Coffee House in Tower St. here referred to formed the small seed from which sprang the greatest marine insurance corporation in the world, and that Mr. Edward Lloyd, owner of the coffee house, was the founder and name-giving god-father of Lloyd's."

In Drake's "History of Boston" mention is made of the "London Coffee House" which existed there in 1689. In the "Annals of Philadelphia" Watson mentioned a coffee house in the neighborhood of Front and Walnut Streets, at which a Common Council of the city was held in 1704. A coffee house was founded in New York probably about the same date. In 1729 the New York Gazette contained its first reference to a coffee house, and in 1730 advertised a sale of land by public auction at the Exchange Coffee House, which was situated at the foot of Broad Street.

Derivation of the Term "Coffee."—It is, perhaps, not generally known that the word "coffee" is of Abyssinian origin. Its name is said to be derived from the city of Kaffa, near which tradition places the origin of the coffee tree.[1] It is stated on the contrary by Macfarlane[2] that "coffee" is derived from the Arabic word "Kahwah," meaning wine, and that it was the name given by the Arabians to the decoction obtained from the pulp of the coffee berries. The botanical name *Coffea arabica*—indicates an Arabian origin.

The coffee tree on the American continent, where it grows over wide areas from Mexico southward to the southern limit of toleration, is described as a shrub rather than a tree, growing to a height of from 14 to 18 feet. It has a rather long and slender trunk, without branches except near the top. The roots of the plant are thin and very numerous; they grow deeply into the earth, and there is usually one central root, a counterpart of the stem of the tree, running straight down, if not interfered with by obstructions. If the soil be poor on which the coffee tree is cultivated, it does not reach the dimensions above stated. According to the richness of the soil it is pruned so as to vary from 6 feet in height in poor soil, to 8 or 10 feet in soil of better quality. One of the objects of thus preventing the growth of the tree to the extreme height, is to facilitate the cultivating of the tree and the harvesting of the berries.

The leaves, when they first grow, are a bright green color,

[1] Bulletin of the International Bureau of the American Republics.
[2] Tea and Coffee Trade Journal, January, 1910.

turning, however, to a deeper shade of the olive tint at full maturity. In the spring, in the axilla of each leaf of the growing tree appear a number of buds, from 12 to 16 in number, which, when in bloom, have an exquisite perfume. The bloom of the tree is evanescent, and after even a few days the petals begin to fall and the glory of the floral display is at an end. It is not until the leaves fall that the coffee berries; that is, the seeds of the tree, begin to assume their full significance. The seeds at first on the tree are yellow, and they begin to redden as they develop into the genuine coffee berries. The bean of the coffee seed varies greatly in shape and size, according to the variety of the coffee and the country in which it grows. The size and shape of the seeds are relied upon by experts, to a large extent, in determining the origin of the coffee and even the country in which it is produced. There are two seeds in each berry, lying side by side. Each of the seeds is covered by a delicate silver-colored skin; next to that is a tougher membrane of somewhat rougher constitution, and afterward, the pulp of the seed, which contains some saccharin substance of a mucilaginous nature, a material which is well suited to fermentation.

Relative Number.—It has been estimated that a vessel of 50 cubic centimeters in volume will hold 187 berries of fine Java; 203 berries of Costa Rica; 207 of Guatemala; 210 of Caracas; 213 of Brazilian Santos; 217 of Mocha; 236 of Rio; 248 of Manila or Philippine coffee, and 313 of Western Africa. It is thus seen that the Java beans are the largest of those usually met with in commerce, and the West African beans the smallest.

Localities Suited to the Growth of Coffee.—Any productive soil which will grow ordinary trees may be devoted to coffee culture. The distribution of the rainfall of course is important, and those localities are preferred where the natural rainfall is sufficient, without irrigation, to produce a crop. One of the most important considerations is the temperature. The temperature should never be very low, nor should it be very high. In tropical countries, therefore, the coffee trees are best grown at elevations a considerable distance above the sea, securing a temperature which is not too high. The coffee tree does not grow well by itself in the open, but does best when protected by the shade of other trees, especially during its early growth. There are some localities where the temperature is so modified by altitude and other climatic conditions as to require no shading. The coffee tree is hardly a tree, but rather a bush, both by its nature and by the fact that it is purposely kept rather low to favor picking. The ordinary care which would be given to the apple or orange orchard is necessary for the proper growth and fruiting of the coffee tree. The coffee bush begins to bear at about the same age as the apple tree, from 7 to 10 years after planting. The production per acre of the coffee orchard does not reach in value that of the ordinary orange or lemon crop, but the cost of care and cultivation is usually less.

Classification of Coffees.—Coffees are known in commerce both by geographical and special names. The principal coffee-producing countries are Brazil, the Dutch East Indies, Abyssinia, Arabia, the Central American States, Hawaiian Islands, Philippine Islands, the East Indian Islands, and Porto Rico and Cuba. On the continent of North America coffee is produced only in Mexico. In the countries named there are many varieties, with local appellations and for which claims of different kinds of quality are strenuously made. All coffees have a common property in this, that they contain caffein as an active stimulating agent and all of them develop aroma and flavor by roasting. The expert, by inspection of the beans, roasted or unroasted, and especially the unroasted, is usually able to distinguish the origin of the coffee. The expert tasters, also, have so trained their olfactory and gustatory nerves that they are able, on brewing coffee in a definite manner, to distinguish by the flavor, with more or less certainty, its origin. The highly trained expert can even distinguish the different components of a blend by this method, which is known as "testing the coffee in the cup."

Among the principal kinds of coffee which are most generally known may be mentioned the following: Mocha, grown in the Yemen district of Arabia; Java, grown in the Dutch East Indies; not, however, as a rule upon the Island of Java. It is held by the Bureau of Chemistry that the term "Java" should be confined in its use not to the coffees produced in the Dutch East Indies, but only to those produced upon the Island of Java. Brazilian coffees are those chiefly consumed in the United States, and are known and divided in general into two large classes, named from the ports of export, viz., Santos and Rio.

Delimitation of Coffee Areas.—It is of the utmost importance in the proper naming of coffees from different countries that the exact area which is covered by a given name should be known.

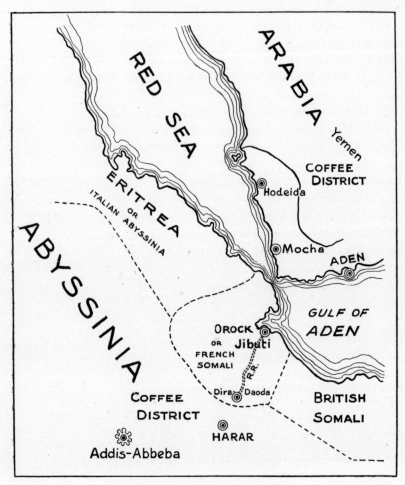

FIG. 11.—Map of the coffee districts producing Mocha and Abyssinian coffees. (*Consular Service, State Department.*)

For the purposes of the enactment of the Food and Drugs Act information has been sought from competent authorities respecting the application of the terms Mocha and Java. Many high-grade coffees not coming from the distinct region in Arabia or the Island of Java, in fact often coming from other countries, have been called by these names. In order that a proper restriction could be placed upon these names, the Bureau of Chemistry after consulting the best authorities accessible, has made the following rules respecting the area in which coffee entitled to the names of Mocha and Java may be produced:

Java Coffee.—There seems to be in the trade much uncertainty respecting the requirements of the Department of Agriculture as to the labeling of coffee produced in the Dutch East Indies. This question has been under advisement by the Department for some time, and, with the coöperation of the Department of State, important information has been secured.

The Department of State was asked to communicate with the Government of the Netherlands and ascertain to what extent in the opinion of that government the term "Java" should be used in harmony with the provisions of the law as applicable to coffees produced in the Dutch East Indies.

A communication has been received through the Department of State from the American Minister at the Hague, who has consulted the Netherlands Government on this subject. In this communication the American Minister states:

The term "Java Coffee" has been abused for many years, hence it arises that of both roasted and unroasted coffee, perchance ten times as much coffee is sold to the consumer, under the name of "Java Coffee," as is grown in Java.

In conformance with the provisions of the "Pure Food Act," all coffee coming from the island of Java might be called "Java Coffee," that from the Padang districts "Padang Coffee," that from Celebes "Celebes Coffee," and all other sorts from the Netherlands Indies "Dutch East Indies Coffee."

In the Netherlands what is known as "Java Coffee" is only the Coffee arabica produced in Java, so that the Coffea liberica coming from that island under the name of "Java Coffee" falls as little under that term as all the coffee from the rest of the islands of the Indian Archipelago.

The Department is of the opinion that the suggestions which are incorporated in this quotation from the American Minister at the Hague indicate a proper method of labeling coffee coming from the Dutch East Indies.

Coffee grown on the island of Sumatra would also be properly labeled if called "Sumatra Coffee," or, if desired, the label may state specifically and correctly the particular location in which the coffee in question was really grown.

In Food Inspection Decision 82 the Department of Agriculture has indicated its views with respect to the restrictions which it is necessary to place upon the sale of coffees coming from the Dutch East Indies, particularly in respect of such coffees as are known under the name of "Java" coffee.

Mocha Coffee.—Among the coffees largely sold upon the American market are those which go by the name of "Mocha." Because of the commercial value of the true Mocha bean, it becomes necessary to indicate the restrictions which must be placed upon the coffees put upon the market and sold under the name of "Mocha."

This matter has been fully investigated and valuable information obtained through the Department of State and from the consul and consular agent in those districts where the true Mocha coffee is grown and whence it is shipped to America and other parts of the world.

The following quotations are taken from the report submitted to the Department of State from the consular agent at Aden under date of January 3, 1908:

The Mocha coffee is produced in that district of southern Arabia known as "Yemen." The latter is a strip of territory commencing at a point on the Red Sea a little north of the port of Hodeidah and extending first southeast to the Strait of Bab-el-Mandeb and then east nearly to Aden. Yemen is, with the exception of a narrow fringe of land along the Red Sea and the Gulf of Aden, rugged and mountainous, embracing innumerable small, elevated valleys of high fertility which are irrigated by waters from the melting snows. This is the coffee district of Arabia.

The term "Mocha" was bestowed upon "Yemen" coffee early in the last century, when Mocha was the port from which all Arabian coffee was shipped. The formation of huge sandbars in the Red Sea off Mocha, practically barring out all shipping, caused the port to be abandoned, and its trade went to Hodeidah and Aden, the bulk of it going to the latter place.

As all of the coffee raised in Yemen may properly be called "Mocha" coffee, all coffee shipped from the port of Hodeidah comes within such classification. With regard to that exported from Aden, however, the case is somewhat different. There is a coffee grown in the upland regions of Abyssinia, in the vicinity of Harrar, which is known locally and to the coffee trade of the world as "Longberry" or "Harrar" in contrast with that of Mocha, which is sometimes called the "Shortberry." The colors of both coffees are practically the same, but the Abyssinian product has a raw, rank, leathery odor, while that of the berry grown in Arabia is delicate and agreeable. The Harrar berry is much longer than the Mocha one, besides being much less regular in form.

While a considerable quantity of Abyssinian coffee is brought to Aden for shipment to Europe and to the United States, it is doubtful whether very little of it, if any, is exported as being Mocha coffee, the local merchants as a rule dealing in both grades of coffee and being very careful of the reputation of their houses. In Aden the only way in which a dishonest dealer might adulterate Mocha coffee would be by mixing it with the Abyssinian article. Such a proceeding would be at best but a clumsy fraud and would be readily and rapidly detected. It is safe to say that practically all of the coffees shipped directly from Hodeidah or Aden to the United States and labeled "Mocha" are pure and unadulterated.

The consensus of opinion is that the term "Mocha" as applied to coffee, should be restricted as indicated in the above communication from the consular agent at Aden, that is, to coffee grown in that part of Arabia to the north and east of Hodeidah, known as Yemen.

Coffee Extracts.—A genuine coffee extract is obtained by securing the extract in the usual way, then evaporating the water, preferably in a vacuum, until the residue is dry. These extracts are ready for use when dissolved in the proper quantity of water. The quality of the coffee does not seem to suffer very greatly by the evaporation of the extract to make the dry powder, although it is undoubtedly true that some volatile, aromatic substances must escape in this process. The soluble coffee thus prepared is useful for emergencies, as, for instance, picnic parties, early breakfasts, and in other circumstances where it is not convenient to make coffee in the usual way. All that is necessary to produce a cup of coffee with the extract is to heat the water and dissolve the proper amount of extract therein. A very considerable trade has been established in this country in extracts of this kind.

Coffee Essence.—There are other substances on the market known as coffee essence, which are not of as high a grade, as a rule, as the dry extracted coffee mentioned above. The coffee essences often contain materials other than coffee. In other words they are not true essences, which would mean a concentrated extract of coffee, but such an extract mixed with other bodies. The consumer sometimes is brought into contact with this kind of coffee on buffet cars or even in regular dining cars or in hotels or restaurants. There is not much to be said in favor of such essences. They are not comparable in quality with the genuine extracts.

Composition of Coffee.—Many conflicting data are given by various authorities respecting the composition of coffee. Many attempts have been made to ascribe the properties which are valuable in coffee to particular chemical characteristics. In only one particular are these statements reliable. The properties of caffein, the alkaloid which gives coffee its principal stimulating value, are well known, and the effects which coffee and tea produce upon the nerve centers are correctly ascribed to this ingredient. Caffein, however, has so little taste or flavor that it plays no important rôle whatever in giving to coffee the characteristics of aroma, taste and flavor, which commend it to the palates of so many people.

Spencer, in his studies in the Bureau of Chemistry under my direction (Bulletin No. 13, Part Seventh, pages 902–903), gives the following data for the average composition of coffee:

Green or unroasted coffee.
Moisture............................... 11.23 percent
Fat or oil............................. 13.27 percent
Ash................................... 3.92 percent
Caffein................................ 1.21 percent
Sugar.................................. 8.55 percent
Tannin................................ 8.84 percent
Roasted coffees.
Moisture............................... 1.15 percent
Fat or oil............................. 14.48 percent
Ash................................... 4.75 percent
Caffein................................ 1.24 percent
Sugar.................................. 0.66 percent

Unfortunately the samples representing the roasted coffees were not the same as those representing the green coffees, so the data are not strictly comparable. The principal changes, however, which are produced in roasting coffee are: first, reduction of the moisture; second, the practical elimination of sugar. The percentage of fat and the percentage of caffein do not seem to be materially influenced by the roasting process. The percentage of tannin as reported by Spencer is not nearly so great as that of other analyses which he quotes in the same bulletin. For instance, in the analyses of 7 samples the caffetannic and tannic acid combined, which represent tannin, varied from a maximum of 23.10 percent to a minimum of 19.50 percent, and in these same analyses the caffein varied from a maximum of 1.53 percent to a minimum of 0.64 percent.

One of the important constituents of green coffee is sugar. Approximately 8 percent of sugar are found in green coffees. In the work conducted by Spencer considerable quantities of this sugar were extracted and obtained in a crystallized state and identified as pure sucrose. The presence of sugar in green coffees has not been given sufficient emphasis in the explanations which have been made concerning the development of the aroma of roasted coffee. The analytical data show that the sugar is practically eliminated, that is, converted largely into caramel. Caramel is a flavoring substance of high value, and there is no question of the fact that much of the prized aroma and flavor of roasted coffee are due to the presence of caramel.

Although it may reflect upon the skill and ability of the chemist, the truth should be told respecting the nature of the changes produced on roasting. The delicacy of flavor, taste and aroma, which is produced, while doubtless correlated with chemical changes, has not yet been accurately connected with any particular change that occurs. It is quite customary to state that the pronounced flavor of roasted coffee is due to certain changes in the oils. The data show very little change of any kind in the oils, and the greater quantity of them is fixed and non-volatile. Of all the problems of chemistry there is none more difficult than the determination by chemical means of a flavoring or odoriferous substance. Minute quantities of these substances are sufficient to give character to a product, and yet they elude the most delicate balance of the chemist.

For historical purposes rather than for any expression of definite composition references may be made to the other constituents of coffees which some analysts have claimed to discover. The data on which these discoveries rest are not of sufficient accuracy to warrant any definite statements.

The Ash of Coffee.—Coffee has a high content of mineral substances, and as these are largely soluble they give to the extracted coffee an additional value. Spencer found as the average

of the composition of ash of four representative coffees, consisting
of Mocha, Maracaibo, Java and Rio, the following data:

Sand	1.06	percent
Silica	0.84	percent
Ferric Oxid	1.18	percent
Lime	5.51	percent
Magnesia	10.98	percent
Potash	61.84	percent
Soda	0.36	percent
Phosphoric acid	12.94	percent
Sulphuric acid	4.63	percent
Chlorin	0.76	percent

Coffee Making.—Some confusion may exist at times in the
use of the word "Coffee," which is applied alike both to the bean
roasted or unroasted and also to the decoction thereof. Both
green and roasted coffee contain large quantities of soluble
material, the green bean, as has already been stated, yielding a
larger percentage of soluble material than the roasted article.
This is probably due to the fact that the sugar is converted not
only into caramel but into lower forms of decomposition which are
more or less insoluble. The extracted matter from the roasted
coffee is determined in two ways: one, a short extractive period
of two or three minutes; and second, a complete extractive period.
From 3 to 4 percent more extracted matter will be obtained by
complete extraction than is obtained by the ordinary method of
making coffee.

For a general description it may be assumed that fully 20 percent of the coffee used will appear in the extract when made
in the ordinary way. There are two ordinary methods of making
coffee decoction. One is boiling for a short period with the
proper amount of water, and the other consists in passing the boiling water over the finely powdered coffee, thus securing its soluble
constitutents by percolation. This latter method makes a finer
flavored coffee to most tastes but not so strong for the amount of
coffee used as the three- or four-minute boiling. In the United
States, at least, most people prefer a perfectly limpid, clear
coffee without any sediment. This is obtained either by mixing
white of egg with the coffee before extraction, or by enclosing the
coffee in a linen bag or a very fine gauze which prevents the fine

particles of coffee from passing into the decoction. There is little to say regarding the preference between the two methods of extraction. Where economy is desired and at the same time a coffee which is of fine quality, the boiling method is preferred. Where larger quantities of coffee can be used for a given amount, the percolation method is to be preferred.

Relative Proportions of Caffein and Tannin.—The injurious effect which excessive coffee drinking produces has been attributed both to the caffein and tannin therein. It is a common opinion that a coffee made of partial percolation has a less relative content of tannin and is therefore of better flavor and less injurious. Investigations made by the laboratory of Good Housekeeping show that these opinions are not well founded.[1]

The data obtained show that the more thorough the extraction the less the proportion of tannin. In a typical test it was shown that in one percolation the percentage of tannic acid in the solids of the extract was 68. When the extract was prepared from the same sample by boiling for some time the percentage of tannin in the solids of the extract was only 54.7. It appears that the tannin in the roasted coffee bean is more soluble than the sum of the other soluble constituents. Moreover, it is interesting to note that more than half of the soluble constitutents of the roasted bean consist of tannin.

There is still another method for making coffee, which is practised largely in the Orient, affording a product known as Turkish coffee. In this method the coffee is ground into as fine a powder as possible. It is then added to the cup, the hot water poured in, and the wholes stirred around until the water is saturated as far as possible. The grounds are left in the cup. This is a more reasonable and economical way of using coffee. All the berry including the oil, sugar and nitrogenous materials are thus utilized for food. Those who are accustomed to drinking Turkish coffee like it even better than that made by the extraction or percolation method.

Extraction of Caffein from Coffee.—One of the principal objections to coffee drinking is the tendency which is produced in many instances to create caffein dyspepsia or poisoning. It cannot be doubted that the excessive use of coffee, or its too

[1] Good Housekeeping, October, 1917, page 144.

frequent use, or its use in infancy or childhood, are all attended with some danger and frequently with serious and permanent injury. The chief objectionable ingredient of the coffee bean is the alkaloid, caffein. The average quantity of caffein in the raw bean is a little over 1 percent. The percentage quantity in the roasted bean is somewhat increased, because of the loss of moisture and other volatile substances during the process of roasting. The roasted coffee will contain perhaps an average of 1.2 percent of caffein, whereas the unroasted will be a little less than 1.1 percent.

Method of Extraction.—Many attempts have been made, and by various agents, to remove practically the whole of the caffein from the coffee bean. The coffee bean, as is well known, is an extremely resistant substance; it is very tough; it is almost impossible to powder it or disintegrate it by pounding, and hence it is quite impracticable to attempt to grind or reduce to a powder the green coffee bean before extraction. If this could take place, the actual labor of extracting the caffein would be very much decreased, and the degree of extraction increased. Since this, however, is not practicable in commercial transactions, other methods of proceeding have been studied and adopted to a greater or less extent. The more common one is to soften the bean first by subjecting it to the action of steam under pressure. This method of procedure softens the bean to a greater or less extent and opens up its pores so that an extracting medium can penetrate into the interior of the bean. Any of the common solvents of caffein may be used, but benzol, or trichlorethylene are generally employed. These bodies have very little effect upon the other constituents of the bean. After the extraction has been completed, any residue of the solvent remaining in the bean may be expelled by a re-application of the superheated steam, or by distilling in a current of steam.

Degree of Extraction.—The quantity of caffein which is left in coffee after it is thoroughly treated in the manner above described, may be placed at about 0.2 of 1 percent. In some instances, almost complete extraction is secured, as for instance 0.02 of 1 percent, while in others a comparatively large quantity, amounting to over 0.25 of 1 percent may be left. It is evident that the phrase caffeinless cannot be properly applied to a body which has been treated in this way, because to be caffeinless all of the caffein must have been abstracted.

Taste and Aroma.—If a coffee be subjected to treatment for the extraction of its caffein, and subsequently roasted and placed upon the market the person not an expert drinking it, without comparison with any other coffee, would discover nothing more in regard to it than possibly a peculiar taste, not unpleasant, which would be still quite like the coffee to which he had been accustomed. The beverage which is made has a fragrance and aroma and taste which remind the drinker very forcibly of coffee, even of a good grade, when the ordinary coffee used is of that variety. If coffee before and after extraction be roasted and otherwise treated exactly alike, the superiority in flavor and character of the untreated coffee becomes at once apparent on the comparison of the two samples. It cannot be said with verity, that the treatment with the extracting material does not to a certain extent deteriorate the product. When it is remembered that the point to be kept in view is the protection of the health of those who are particularly susceptible to the influence of caffein, it is advisable to sacrifice to a certain extent the taste and flavor if one can secure thereby a product which will no longer have a deleterious effect even upon those most susceptible to the caffein poisoning.

Commercial Importance.—It cannot be said that there is any great commercial importance at the present time to be attached to this partially extracted coffee. It is true that factories have been built, especially in Germany, and operate on a scale more or less large, and this extracted coffee is to be had upon the markets in many localities. It will require a number of years of experimental work and instruction on the part of the makers of this product, to bring it into as general use as that of the untreated coffee. At the present time (1918) importations from Germany are not possible and factories for making the decaffeinated product have been erected in the United States.

Blending Coffees.—A great deal of the coffee which is offered for sale is represented as being blended for the purpose of securing a uniform flavor and aroma.

While it is evident that in many cases the blending claims are not justified by the facts, it is still known that it is possible for experts to mix certain strains of coffee in such a way as to secure a mixture giving practically uniform taste and flavor. This, of course, takes away the individualities, and hence, it must be supposed that it will be confined chiefly to the medium or lower grades of coffee. This, however, is not the case, since dealers frequently use the highest grades of coffee in making these mixtures. Why high-grade coffee, which has a character of its own, should thus be reduced to a common level so that its peculiar flavor and aroma are lost in those of other coffees, it is difficult to determine. A person who would blend the high-grade vintages to make a drink of uniform character, would be deemed almost guilty of sacrilege. The mixing of medium or lower grades of coffee might be justified, but from the point of view of the connoisseur there can be little justification for the mixing of high grades. The suspicion is with some warrant, although it may not always be just, that the blending of the high-grade coffees has for its chief purpose the giving of a good flavor to the low-grade coffee, so that the mixture may sell for a higher price than could be obtained for the coffees entering into it when sold alone. This may be a justifiable practice. In other words, a low-grade coffee which would sell at a low price, may be mixed with enough high-grade flavor to give the coffee an additional flavor, so that the mixture can be sold at a higher profit than both of these constituents would bring if sold separately.

It is evident, of course, without argument, that to claim that the coffee is mixed or blended without it actually is so should not be done, and no misstatement should be made respecting the character of the coffees entering into the blend.

Blending may be practised both on washed and unwashed coffees, and the same principle is followed in both, namely, the mixing of an acid and a bitter coffee with a sweet one, namely, to quote the language of an expert, "Blend an acid and a bitter coffee with a sweet one to tone down the qualities of the others if necessary, so that neither quality will be too pronounced."

Washed Bogota and washed Bourbon Santos may be taken as the base for blends, but each must be first tested for its particular quality.

It is the work of an expert to test all these coffees to determine their qualities, then to mix small quantities in varying proportions, drawing the coffee and determining its flavor. He finally secures, by mixing, the flavor which he wishes to establish for his grade or brand, and in subsequent processes he varies the quantities of the different constituents of the blend as those constituents themselves vary in quality, so as to maintain as nearly as possible uniform character throughout.

It is only when an expert of high character blends a coffee that any really good results can be expected.

Food Value of Coffee.—The food value of coffee and of tea may be regarded as of no importance. The only food of any consequence in coffee is sugar, and that is largely caramelized in roasting. The effects of coffee are due solely to its stimulating powers. Apparently coffee and tea diminish, or (perhaps a better expression) relieve fatigue. They do not seem to do this in any material way, but by so influencing the nerve centers as to counteract or destroy the sense of fatigue.

Inasmuch as coffee has no tissue-building power at all, and no food value except that which is given to it by added milk, cream or sugar, it cannot be regarded as a food in the true sense, though legally, as a beverage, it is classified in the Food and Drugs Act as a food. Coffee contains a considerable quantity of oils, but as these are insoluble in water they do not appear in the infusion. A little bit of protein substance is extracted from the coffee by hot water, but so small in quantity as to have no significance from a dietetic point of view.

Adulteration of Coffee.—Quite extensive adulterations of coffee have been practised, but since stringent laws, national and state, have been enacted the evil is rapidly decreasing. One of the crude forms of adulteration which was found on the market many years ago was the fabrication of artificial beans, molded in the shape of coffee beans and colored to represent either the green bean or the roasted bean as the case might be. There was no difficulty in detecting these artificial beans when the mass was looked at carefully, as there were always differences which would be discriminatory. A person not expecting anything of the kind, however, might easily buy, without suspicion or without detection, coffee beans containing as high as 25 percent of these artificial bodies. I believe that this method of adulteration has entirely disappeared.

Other forms of adulteration recall at once the use of chicory.

Of all the roasted vegetable substances which have been used in coffee, or as substitutes for coffee, there is none that has had the vogue of chicory. This plant is of wide distribution and the United States is cultivating it particularly in the Middle West, though it grows in many localities. The tops of chicory are used as spinach or greens. The roots, when roasted, are used extensively for mixture with coffee. In some countries, as in France, where the customers require a coffee of good body and very dark color, chicory is commonly employed in the household and in the café or restaurant to give these characteristics to the coffee. It is stated by Windmuller[1] that the motto of the French in so far as coffee is concerned, is "Chaud comme l'enfer, noir comme le diable." The chicory, however, is often mixed with the coffee before it is sold, and extensive frauds have been practised in this way. While it cannot be said that unlimited quantities of chicory are altogether wholesome, it is certain that in the quantities used in mixing with coffee it has no properties which may definitely be called injurious. To those who knowingly use coffee with chicory and like it there can be no objection to its addition to coffee, as such an addition would not be an adulteration. Its addition to coffee, on the contrary, for sale to the unwary or innocent purchaser is purely a commercial fraud. Those who advocate the use of chicory say that connoisseurs prefer the taste of coffee to which chicory has been added and that famous restaurateurs buy it for the sake of its flavor to mix with the coffee of these fastidious customers. Inquiry among the chefs of the best restaurants has failed to confirm this statement.

It is true, however, as already noticed, that many consumers do like a blacker color and a heavier body. Roasted chicory is in no sense a substitute for coffee, as it is deficient both in the peculiar flavor, taste and aroma of coffee, and in its alkaloidal constituent. It contains a little tannin, as do most vegetables, but not sufficient to give any particular character to its decoction. Chicory contains scarcely any fat, and less sugar than coffee itself. It contains no alkaloidal substance and about as much ash as coffee. The starch which exists in many root crops in chicory is represented by another body, called inulin. When starch is hydrolized to form a sugar by the action of an enzyme or by an

[1] The Forum, January–March, 1908, page 429.

acid the final result is the formation of dextrose, a right-handed optical body. When inulin is hydrolized in the same way a sugar is formed with a left-handed rotation. This sugar is known as levulose. There is nothing, however, in the constitution of inulin or levulose which in any degree threatens the health of the consumer.

Misbranding of Coffees.—One of the principal faults of the coffee trade has been misrepresentation of origin. The coffees known as Mocha and Java, on account of having excellent quality and great popularity, are seized upon as a convenient method of representing other coffees, excellent in themselves but not entitled to these names, as having a like origin.

Until the enactment of the Food Law coffees of various characters, especially Brazilian coffees, were commonly called in this country by the terms Mocha and Java, although the records of the Treasury show that the importations of so-called Java and Mocha coffees annually are very small, yet at retail, very large quantities of the total coffee offered for consumption were known as Mocha and Java. The use of these terms evidently was purely fraudulent. The real coffees bearing these names have exceptionally high qualities and command exceptionally high prices. It therefore became extremely profitable to offer cheaper coffees under the names of the more expensive. When it is recalled that at the present time raw coffees in this country of very good quality can be had at from 10 to 15 cents per pound, it is not surprising that it is a very tempting opportunity to offer them under other names at a very largely increased price.

Consumption of Coffee.—For the years, 1916 and 1917, the latter being the latest of the published data available, the total consumption of coffee in the United States is given in the following table:

IMPORTS, EXPORTS AND CONSUMPTION OF COFFEE, 1916–1917

	1916—Pounds	Value $	1917—Pounds	Value $
Imported....	1,166,888,327	118,813,421	1,286,524,073	122,607,254
Exported	35,293,688	4,550,522	67,162,222	8,300,943
Consumed ...	1,131,594,639	114,262,899	1,219,361,851	114,306,311

Assuming the population of the United States to be 110,000,000 the consumption of coffee amounts to 11 pounds per person per year.

PART VI

TEA (THEA SINENSIS, L.)

Common Beverages.—Among the beverages in common use in all civilized countries tea is next to coffee in importance. This beverage is used principally at the table, but tea is often served, especially in England and Russia, as a beverage to be taken between meals.

Tea Drinking Customs.—The American traveler in Russia would be surprised to know that he may enter Russian business houses at certain hours of the day, for instance at ten o'clock in the morning, and four o'clock in the afternoon, and find business entirely suspended while tea is served to the employees throughout the establishment.

Use of Samovar in Russia.—If one were traveling in Russia in *ante bellum* days and desired a very early breakfast before the usual hour of serving, he would find a samovar placed in his room ready for use, and a sufficient amount of bread and butter, with sugar and milk, to prepare a simple breakfast of tea and bread and butter.

The samovars of Russia are particularly prized for their artistic merits. One of the great industries of Russia is the manufacture of household utensils in brass, nickel, bronze, copper and copper alloys. Among these one of the most important is the samovar, or hot-water urn. Every home, however humble, has one or more of them and they are in constant use. The samovar is not an old utensil in Russia, but has been introduced within the last 80 years. The most valued, of course, are the old-fashioned ones which have been hammered by hand. Modern machinery has replaced this style by the machine-made samovar which is doubtless as beautiful at a distance and quite as useful, but does not appeal quite as much to the taste of the connoisseur as the hand-made article. The samovar is heated

148

with charcoal, burned in a cylindrical chimney which runs up through the center of the vessel.

The exhibition of the long lines of samovars which are found in the parks on holidays are most interesting. Especially in Moscow will the traveler find on the holidays in pleasing array in the parks, long lines of samovars, each attended by a peasant woman in the attractive garb of that class. A striking illustration of the holiday samovars I saw on a pleasant Sunday afternoon in the park on the hill beyond the Moskova river, from which Napoleon first looked upon the Russian capital.

Tea Serving in England.—In England the serving of tea in the afternoon about four o'clock is almost a universal custom among

FIG. 12.—Tea samovars, Sparrow Hill, Moscow. (*Photo by Author.*)

the well-to-do people. The custom of drinking afternoon tea is also rapidly spreading in the United States. But after all the most important use of the beverages which have been mentioned is at the table.

Beginning of Tea Drinking.—The use of tea as a beverage probably dates to the remotest antiquity, but not among nations (with the exception of China) whose literature is available. Confucius speaks of tea drinking 500 years before Christ. It is not mentioned by any of the classical writers. It was not until the organization of the East Indian Company that tea was brought into England. In regard to its use in America the only thing that seems to be certain is that it was not brought over in the Mayflower. It is stated by authority that the first merchant who put tea on sale in England was named Garway.

By 1660 tea was pretty well known in England among the wealthy and fashionable. By 1664 it was on sale at the coffee houses. Even in 1664 the cost was excessive, sixty shillings a pound being the price. While the first use of this leaf was as a medicine, a German named Olearius recognized its value as a beverage as early as 1633. But many there were who vilified it, calling it an "impertinent novelty and the sellers of it immoral and mercenary persons." In Boston, tea was on sale by 1690 and in 1691 there were two houses besides those kept by Daniel Vernon and Benjamin Harris. By 1712 it was advertised in the Boston News Letter and you could buy it from Zabdiel Bolton at his apothecary shop. The favorite variety was green, but the advertisement reads "green and ordinary." Bohea was the favorite and by 1725 it could be purchased in apothecary, tobacco and drygoods shops, as well as those devoted to "small wares."

From a commercial point of view the China tea trade began about 1678 by the East Indian Company carrying to England about 5,000 pounds of tea as a speculation. This quantity supplied London for many years before it was finally disposed of. The exports of tea from China reached a maximum about 1886, when a total of about 300,000,000 pounds were exported. From 1886 there has been a gradual decline in the amount exported, largely due to the growing popularity of India teas.

Tea-producing Countries.—The principal tea-producing countries are China, Japan, Java, India, Ceylon, Sumatra and Formosa, now a possession of Japan. Attempts to grow tea have been made in the United States, as will be mentioned further on, but our dear labor will long prevent us from being competitors with the nations of the Orient.

In the United States tea has been produced only experimentally in gardens near Charleston, S. C.

The Tea Plant.—China teas, according to the "Tea and Coffee Trade Journal," are produced chiefly from the variety of the plant known as *Thea bohea*.[1] The leaves of this tree-shrub are of a

---

[1] Botanists now recognize only one true tea plant, *Thea Sinensis*.

dark green color and are from one and a half to three inches in length. This variety is a small shrub compared with the variety of the tea plant which grows in India and which is a large, strong, growing plant, which becomes almost a tree, and the leaves of which range from three to five inches in length.

Generally the tea is grown by small farmers, who cultivate only a few bushes around the house in which they live, although others have more extensive gardens, but, as a rule the tea farms are very small.

After picking, the tea is dried in the sun, rolled and again semi-dried. The tea is then sold to merchants at Tea Hong, who dry and grade it before exporting. The product made by this process is black tea.

Time of Harvesting.—The first crop, in China, is gathered in April and May; the second in June and July, and the third in the Autumn.

The chief centers for collection for export of the product are Hankow and Shanghai in the north, Foochow in the center and Canton in the south. There are 30 or 40 different varieties of tea on the market.

Blending.—There is a great deal of blending of tea among the China varieties. The practice of blending is a scientific art. The expert blender will take a low-priced tea and mix it with medium and fine qualities, in such a way, as to make a mixture which produces a good cup of tea. There are no rules that can be laid down to get this, but only the expert blender, by selecting different quantities and varieties can get the desired taste and flavor.

Indian Tea.—We learn from the Tea and Coffee Trade Journal that the term "Indian Tea" is one which is made to cover a lot of territory. It naturally would be applied to the whole area covered by the geographical territory known as India. The character of the teas which are included in Indian Tea, therefore, must vary as much as the different localities which cover that vast territory. Among Indian teas are those which come from Assam, Darjeeling, Dooars, Kangra, Nilgiris and Travancore. Of these teas, experts claim that the finest are produced in Darjeeling, but often in that locality teas produced on different estates vary largely in quality.

Darjeeling is situated at the foothills of the Himalayas, and the climate in that region is favorable to production of a tea with fine flavor and aroma and general character. The yield of the trees, however, is somewhat limited in quantity, hence inasmuch as the farmers are not able to produce a large quantity of tea they use every means possible to produce teas of a high grade. The crop in Assam yields, it is stated, twice as much as in Darjeeling. The climate is much warmer, and the large crop and high temperatures result in a tea of less desirable flavor in character.

A large part of the Indian tea goes to England, and, as the English are great tea drinkers, and competent judges of it, it must be acknowledged that their preference for Indian tea is not wholly of a patriotic kind. It has not been very many years ago since China furnished Great Britain with the principal part of its tea. The great quantity of tea drunk in Great Britain, amounting to about 400,000,000 pounds, comes mostly from India and Ceylon. It is stated that this change in the character of the tea consumed in England is due solely to its quality, and not, of course, to any protection by import duties on teas of other countries, nor by any preference of the English for the tea simply because it is Indian.

One of the reasons for the rapid development of the use of Indian tea is found in the action of the Indian Government establishing a commission for the scientific testing of teas. Probably it is better to say, not the action of the Government, but of the action of the Indian Tea Association. It, however, receives quite considerable aid from the government, and hence may be regarded in that sense as a government body. This scientific commission called attention to the fact that the fermentation which takes place in the tea leaf when curing is not due to the bacterial change, but wholly to enzymic action.

Formosa Tea.—Since Japan took possession of the Island of Formosa as related by Tea and Coffee Trade Journal the tea industry has been greatly stimulated. The quantity of tea imported to the United States from Formosa amounts to about 18,000,000 pounds annually. The quantity has not varied greatly in many years. The proportion of Formosa tea received is not keeping pace with the increase of population. The Formosa teas are of high value, especially those belonging to the Oolong class.

The process of manufacturing Oolong Formosa tea is described as follows:[1]

"The green freshly picked leaves are placed loosely in an open place for ½ hour in the sun. They are allowed to lie there long enough to undergo a certain degree of fermentation, during which a special flavor is developed. After the edges of the leaves become slightly brown, they are stirred in a hot pan over a charcoal fire, thus checking the fermentation and rapidly bringing it to a close. The leaves are then rolled with the hands or feet and dried over a charcoal fire in baskets.

The method of making green or black teas is quite different from the above. In preparing green tea, the freshly plucked leaves immediately after being picked are steamed in order to stop fermentation. They are then dried over a fire, and are twisted while drying. Formerly if the green color was not sufficiently marked, an artificial green was added. No green teas are now imported from Formosa.

Japan Teas.—The principal teas produced in Japan are described by Mr. Geo. P. Mitchell, Chief Tea Examiner, as follows:

"Sencha, ordinary tea.
Gyokuro, dew drop tea, very high grade.
Tencha, ceremony tea before grinding.
Motcha, ceremony tea ground into the powder as used.
Bancha, poor peoples tea, flat leaf tea made from prunings after picking.
The Japan teas imported are either pan fired—Kama-cha or basket fired—Kago-cha.
Regular schools are supported where the proper use of the ceremonial teas is taught in great detail. Particularly are the pupils instructed to venerate and praise the bowls in which the tea is served and to so order the drinking that no two persons touch the same part of its circumference to their lips. This is apparently the Japanese loving cup.
Different kinds of tea are also designated sometimes by the localities in which they grow, or the ports from which they are shipped."

Tea is cultivated in Japan by local farmers who have all the way from 1 to 20 acres under their control. The farmer either rents or owns the ground. If big enough, he has a small firing-place where the tea is put through a semifiring process. The local farmers sell to the collectors or country merchants, who bring it down to Kobe, Yokohama and Shizuoka, where the different toyas (tea merchants) are located. It is usually shipped in 100-pound boxes by rail or boat to these ports. The semifiring process followed by the local farmers consists of steaming the tea leaves in a basket,

[1] Adapted with certain additions from Tea and Coffee Trade Journal.

after which it is dried in trays and then given a light firing over a charcoal fire in wooden trays having paper bottoms, the tea being rolled by hand, it being then allowed to dry in a similar firing-tray but at a lower temperature. The rolling and firing process usually takes about 12 hours. The toyas sell to the Japanese or foreign exporters, who subject the tea to a further process of manufacture and then ship the tea to America.

The firing process followed in the tea-firing godowns at the ports named consists of stirring the leaves in hot firing-pans from 30 to 50 minutes and then in cold pans from 15 to 20 minutes. When finished the tea is taken to the packing godown where it is sifted to remove the dust and packed while still warm into the half chests, lined with lead, which are to convey it to the grocers and tea drinkers of America.

No coloring matters are now permitted in Japan teas. When our government decided to exclude all artificially colored teas, the Japanese government and the Central Tea Association quickly accepted the new ruling. It is highly desirable that all tea growing countries should take the same action.

Basket firing consists in simply re-firing the tea without any of the stirring process as practised in the pans. A bamboo basket, shaped like a dice-box, but open at both ends, is placed over a large iron brazier containing lighted charcoal (well covered with ashes) and the tea is strewed, about an inch in thickness, on a close-woven bamboo tray which fits the neck of the dice-box The baskets are occasionally removed from the brazier and the tea turned over by hand in order that all may be equally fired; they are carefully replaced on the brazier, without allowing any dust or leaves to fall through the tray on the charcoal, and in the course of 40 to 60 minutes the tea is ready for packing.

The foregoing description applies to the preparation of ordinary Japan tea, during which no fermentation of the leaf has been allowed to take place. But in the preparation of black tea (congou), of which a considerable quantity was made some seasons ago, fermentation has to occur. In this case the tea is withered in the sun and then rolled on tables by coolies for about 30 minutes, after which it is packed tightly in large round baskets and covered with a cloth for an hour or so and allowed to ferment. The leaf is then tipped out onto the rolling tables, well shaken out and rolled for 15 minutes more and fired on iron gauze sieves or in drum baskets over charcoal fires. The first process takes 45 minutes, and the drum baskets 70 minutes.[1]

[1] The Tea and Coffee Trade Journal, July, 1907, page 20.

Java Tea.—Accustomed as we are to think of Java as the home of coffee, it is important also to note that it has acquired some reputation as a producer of tea. The magnitude of the industry at the present time amounts to about 80,000,000 pounds a year, and the method of cultivation, which is practised, as well as the method of manufacture, may be said in many respects to be superior to those of Ceylon and India. So important is the trade becoming that there are stated intervals in Amsterdam when tea brought from Java is sold at public auction. Most of the teas produced in Java are of good quality, and many are extremely fine. According to credible historical evidence, tea was first introduced into Java from China about 1826. The Chinese varieties have been, however, of late years replaced by the seed from the Assam tea country, which is found to be very productive and more profitable. Before the war little Java tea has come to the United States, most of it going to Holland, to Australia, and especially to Great Britain. Russia takes a little over a million pounds, and smaller quantities go to other countries. In the opinion of some experts, Java teas have a quality which is distinctly superior to those of China and Japan, judging at least by the English standard of quality.

The tea factories in Java are said to be the most modern and complete of any in the world.

Many of the tea plants are driven by electric power, and have the fermenting rooms laid in white China tile.

The tea areas are subjected to heavy rainfalls, the rainy season extending from September to April. While the average rainfall is 72 inches, in some cases it reaches as much as 124 inches.

The price of the Java teas upon the London market is about the same as that of the Assam teas.

Soudanese labor is employed mostly in Java, the men being paid from 10 to 15 cents per day, and the women from 6 to 12 cents per day.

Java teas are divided into the following grades:

Orange Pekoe	Pekoe	Souchong
Broken Orange Pekoe	Pekoe Souchong	Dust

In addition to the above there is also produced in Java a grade known as Silver Pekoe, or white point pekoe. This is a high-grade tea, made from the top or first leaves, and is rolled by hand usually in the sun. It is not fermented previously to rolling. This brand of tea often sells as high as $1.25 per pound.

American Tea.—It is evident from a study of the climate and soil conditions connected with the growth of tea, that the United States is not excluded from the possible areas in which it can be cultivated. For many years, the late Dr. Shepard, of South Carolina, has promoted the cultivation of tea in the vicinity of Charleston, Summerville, and the United States Department of Agriculture has collaborated in this work from time to time during the past 30 years. The tea gardens at Summerville have gained a deserved reputation for producing a tea of excellent quality. Unfortunately, the conditions of labor and wages in this country render it difficult to compete with the remarkably cheap labor of the countries most famous for tea production. Dr. Shepard has devoted the leisure of many years of his active life to the promotion of this industry, but it does not appear to have assumed any vogue outside of the immediate vicinity of his own activities. He is of the opinion that in certain sections of the United States, American grown tea may hold its own against the imported article.

The adaptability of the soil and climate of the vicinity mentioned, to tea growing has been demonstrated, and the very fact that the tea plants thrive there answers all doubts respecting the possibility of their growth. The tea which is produced in South Carolina is not of sufficient quantity to have affected in any way commerce in that article, nor to have established throughout a very wide range of consumers, its own standard of excellence. It is undoubtedly true that tea grown in the United States will have a character, flavor and quality of its own, and hence, at first might not be relished by those accustomed to drinking the teas of established grade and quality. It is only fair to presume that by use this prejudice against American tea would disappear, and assuming that it were produced in quantity to warrant the expectation, it would assume a just place among the tea drinkers of the world.

Typical American Tea Bush.—The character of the growth, size, shape and appearance of the tea bush is well illustrated by the figure herein, showing one of the typical tea orchards grown at Summerville, South Carolina.

FIG. 13.—Tea bush at "Pinehurst," Summerville, S. C.
(*Courtesy of the late Dr. Shepard.*)

FIG. 14.—Colored children plucking tea at "Pinehurst," Summerville, S. C.
(*Courtesy of the late Dr. Shepard.*)

The tea bush requires a large amount of nourishment, and also intensive cultivation, as well as fertilization, and, therefore, tea gardens require soil as rich as can be secured. The greater the fertility of the soil originally, the less the expense of fertilization and the greater the economy of growth. In this country a rich loam, according to Dr. Shepard, containing a large amount of humus, affords the best promise for both quantity and quality. Recent investigations have shown that an abundance of available phosphoric acid in the soil contributes largely to the flavor of the tea made thereon.

The tea plant, according to the same authority, has generally a long tap root, hence a rich, readily penetrable sub-soil has its advantages, but it must be free from stagnant water, which the tea bush cannot tolerate. Slightly sloping land, sufficient to secure natural drainage, but not steep enough for heavy rains to carry off the most valuable top-soil, is preferable. Where the surface is too sloping, well-placed banks and ditches can be employed to prevent denudation, but all such devices are costly and interfere with the cultivation of the land. In the present condition of cultivation, Dr. Shepard advises that only the cleared land be utilized, as the cost of destroying a forest or of eradicating a jungle, and converting the soil therein to a proper condition for the growth of the bush, requires an amount of labor which is prohibitive to successful culture. Importance is also attached to the shipping facilities, both because of the convenience in bringing in supplies, as well as for the transportation of the products of the garden to the markets. It is realized that in the United States the fulfilment of all the conditions necessary to the successful and profitable cultivation of the tea plant is yet far in the future. It may be that the ingenuity of the inventor, the care and intelligence of the supervisor, and the skill of the laborer may in the future establish, preserve and extend the tea industry in the United States to such an extent as shall be worthy of mention among the commercial centers of tea production of the world.

Dr. Shepard says:

The establishment of a tea garden on a commercial scale in any country involves the expenditure of a considerable amount of money and a rather long wait for a pecuniary return. It demands very careful attention in the field and factory, but especially in the marketing of a new product. Such was peculiarly the case with Ceylon, where the planters have been forced to roundly tax themselves to successfully introduce their new teas on the foreign markets. As the scope of supervision embraces not only agricultural operations, but also the working of a factory filled with machinery regulated by the watch and thermometer, it follows that the services of a capable and, therefore, expensive superintendent are required.

The substitution of machinery for all the manual operations except plucking and final culling has become imperative from the motive of economy and the necessity of observing every precaution against the contamination of the tea from human handling, as well as for greater ease in the maintenance of uniformity of production.

As has been already stated the aim of the American grower should be the production of the higher grades of tea by a wise selection of seed, by limiting the plucking to the youngest and smallest, hence tenderest and most valuable leaf, and by the careful use of the latest and best machinery. Repeated chemical analysis has proven that American tea is comparatively rich in the nerve stimulant and food, thein, and low in tannin, the astringent and deleterious principle; it is not subject to the wily Oriental's passion for adulteration with artificial coloring and make-weight matter; it escapes, for home consumption, the injurious effect of a long ocean transportation, whereby chests are broken and dampness and all manner of germs admitted. Indeed, because relieved from this last danger, American tea has not to be "fired" at the high temperatures requisite for resisting long voyages, but which are at the sacrifice of both fragrance and taste.

While we cannot agree with Dr. Shepard that thein or caffein, is a food nor that tannin is its injurious principle; nor would many agree with his ideas that transportation injured the value of the tea, we must admit that his enthusiasm, his intelligence, his industry and his patience in working out this problem are deserving of all commendation.

Picking.—The proper condition of the leaf for plucking must be learned by experience, and is determined by sight and by touch. The harvest begins at Pinehurst about May 1st and continues until October. The pickers must learn to discriminate, so as not to mix the poor and worthless with those that are suitable for subsequent manufacture for drinking purposes. He should pick for quality rather than quantity. For making good tea, only the bud and the first two leaves which must be tender, are taken; otherwise the leaf will crumble or flatten out in the process of manufacture, instead of producing the familiar twisted or curved appearance.   The harvest is repeated every two weeks. Each period of growth is known as a "fluh."

Care is required in the handling of the leaf so that it may not be bruised.   For the same reason it must not be packed too tightly in the baskets; otherwise it would become pressed, heated and subject to fermentation.   At Pinehurst the leaf gathering is done by colored children, who when skilled will harvest enough leaves each day to make 10 pounds of dry tea.

In general the tea is divided into two principal classes, green and black tea.   These two classes are determined by the degree of oxidation to which the chlorophyll or coloring matter of the leaf has been subjected.

Black Tea.—In the manufacture of black tea there are four principal processes, withering, rolling, oxidizing and firing. During this time the leaf, of course, is continually losing moisture. When, in the process of withering, the leaf has obtained the proper degree of pliability, it is rolled either by hand or by a rolling machine.   The rolling breaks up the oily cells in the leaf and gives to the tea leaf that twisted appearance by which it is so universally distinguished.   At Pinehurst all the rolling is done by machines and the process requires about three quarters of an hour.   After the leaf has been sufficiently rolled to break up the cells it is spread on table trays, in layers about two inches thick, for two or three hours, during which an additional or final oxidation takes place. The oxidizing room should be kept damp and as cool as possible. It is then taken into the drying or firing machines, and subjected to an even heat until the leaf has lost all of its moisture, and all the exudations which have come to the surface as a result of drying and rolling combined remain deposited as dry matter upon the surface of the leaf, which should then be crisp and fragrant.   The best method of firing is to drive hot air over the tea by means of a fan.

The quality of the tea is largely dependent upon the care observed in its firing, and the temperature should not be so high as to dissipate all the volatile oils.   On the other hand, the drying must be at a temperature sufficiently high and continued sufficiently long, to remove the moisture; otherwise rapid deterioration will take place.   The treated tea is placed in air-tight cans or bins and allowed to remain until ready for shipping.

Green Tea.—In the manufacture of green tea the principal point to keep in view is to prevent any oxidation or darkening of the leaf. As this oxidation starts practically as soon as the leaf is plucked, it requires great skill and celerity to produce a high-grade tea. The production of green tea, insofar as at present is known, requires hand labor exclusively. Inasmuch as the oxidizing effects are produced by an enzyme it is important in making green tea to kill this ferment as soon as possible.

FIG. 15.—Loading end of sterilizer to destroy enzymes.
(*Bureau of Plant Industry, U. S. Dept. of Agriculture.*)

This is accomplished by subjecting the fresh picked leaf to a high temperature. Dr. Shepard invented a rotary steam heated sterilizer for this purpose. This cylinder is so constructed that it may be fed continuously at one end and the cured leaves are constantly being delivered from the other end. It requires about 7 minutes for the complete process. After the enzymes are destroyed by this process the leaves are ready for rolling and firing.

Tea made in this way is necessarily very expensive and so the art of artificially coloring a tea green to resemble tea made as above has been devised. It is certain that very little green tea is now imported into the United States which has been adulterated by coloring. It is claimed, and it is doubtless true, that the green teas at Pinehurst gardens near Summerville are not "faced" and are genuine unadulterated teas.

Composition of Tea.—Attention has already been called to the chief constituents of tea as far as their active principles are concerned. Of the active ingredients, tannin is the most abundant and caffein next. The quantity of moisture contained in ordinary commercial tea, which has been kept dry, rarely goes above 10 percent, and is usually about 5 percent.

The analysis of 63 samples of tea by Spencer in the Bureau of Chemistry, purchased on the open market, gave the following averages:

Moisture................................ 6.77 percent
Total Ash.............................. 6.25 percent
Total extract.......................... 36.99 percent
Tannin................................. 10.25 percent
Caffein................................ 2.02 percent

The chemical analysis of a tea, while it gives a correct notion of its composition, does not necessarily reveal anything respecting its character and comparative merits. All teas contain essentially the substances which have been mentioned in the table of analysis, but they vary greatly among themselves in palatability and character. It is difficult to determine the chemical characteristics of a tea on which its value depends. The terms flavor, strength and aroma are not associated with any definite ideas of chemical composition. In many cases, in fact, in all cases, the qualities of the tea do depend to a certain extent upon chemical composition. For instance, the astringency of a tea is due to its tannin; its exhilarating effect to its caffein; and its aroma to certain *volatile* principles, the nature of which is not well known. Usually the teas which have the larger quantities of these subtle matters are found to be the most valuable, but it is not true in all cases.

Making Tea.—The infusion of the dry tea leaves in hot water is an extremely simple process. First of all the tea should not be boiled. The best method of infusion is to place the tea leaves loose in the pot, fill the tea pot with boiling water, and allow to remain three minutes. The first infusion is not highly colored but possesses a fine aroma and taste. After several minutes the tea takes on a deeper amber color. A second extraction from the same tea gives a deeper colored but less fragrant product. The tea leaves may be dropped directly into the hot water, but in this case the infusion should be poured through a strainer.

Flavor of Tea.—Only the expert tea taster can distinguish the blended flavoring of different teas. The rest of us can only say, we like, or dislike. There is much in habit. We are likely to prefer that to which we have become accustomed. The higher flavored teas bring higher prices on the market, but the price is not an unerring gauge of excellence. If the tea does not please we know it without knowing why.

> "I do not like thee Dr. Fell,
> The reason why I cannot tell."

Praise and Blame. Pro.—In the Critic and Guide for February, 1916, page 68, is quoted, an extract from an article published by Thomas Garway in 1660, which is entitled "An Exact Description of the Growth, Quality, and Virtues of the Leaf Tea." According to this author the particular virtues of tea are these:

It maketh the body active and lusty.

It helpeth the headache, giddiness and heaviness thereof.

It removeth the obstructions of the spleen.

It taketh away the difficulty of breathing, opening obstruction.

It is good against turpitude, distillations, and cleareth the sight.

It removeth lassitude, and cleanseth and purifieth acrid humors, and a hot liver.

It is good against crudities, strengthening the weakness of the ventricle or stomach, causing good appetite and digestion, and particularly for men of corpulent body, and such as are great eaters of flesh.

It vanquisheth heavy dreams, easeth the frame, and strengtheneth the memory.

It overcometh superfluous sleep, and prevents sleepiness in general, a draught of the infusion being taken; so that, without trouble, whose nights may be spent in study without hurt to the body, in that it moderately healeth and bindeth the mouth of the stomach.

It prevents and cures agues, surfets and fevers, by infusing a fit quantity of the leaf, thereby provoking a most gentle vomit and breathing of the pores, and hath been given with wonderful success.

It strengtheneth the inward parts, and prevents consumption; and powerfully assuageth the pains of the bowels, or griping of the guts, and looseness.

Con.—Windmüller, in the Circle calls particular attention to the abuses of tea drinking and its baneful effects. He cites the case of one of his friends, a busy lawyer who studies frequently all night in the preparation of the briefs he has to bring into court the next day. During this time he constantly drinks tea. When finally he is overmastered by sleep the caffein still keeps his mind active and carries him in his dreams into court, where, after a brilliant delivery of his convincing argument in his dream he is horrified to hear the judge decide against him. In the morning it is necessary to drag him from his bed and then he discovers that under the spell of night-mare his mind has been wandering. He calls for more tea, staggers into court, still dominated by the influence of the tea under which he delivers his plea. Although fully aware that exertions such as this, made possible by unnatural stimulants, will end in the destruction of his mental faculties, he repeats the experiment whenever he must snatch from the hours needed for rest the time to work. These repetitions have resulted, sometimes in success, sometimes in failure. But the strain has already begun to tell on his system. The minds of all literary persons who burn the midnight oil and who acquire this tea habit become greatly enfeebled, it matters not how they prepare the drink.

Paraguayan Tea.—A kind of tea is grown in Paraguay and Brazil which is now finding a market in some parts of the world, known as Yerba Maté. The botanical name of the plant is *Ilex Paraguayensis*. The merchantable leaves are also known under the trade name Juno. Tea drinkers, as a rule, are not greatly impressed with the virtues of Yerbe Maté as a substitute for tea. It has certain characteristics of tea, which may render it a beverage of some importance in the future. Quite a quantity of the tea is imported into the United States at the present time and a vigorous propaganda for its extended use has been inaugurated. The principal part of the trade comes through Brazil. The trade is already great and is growing rapidly, and is to be taken into consideration in all calculations regarding the commerce of Brazil, Argentina, Uruguay, and Paraguay. In fact, contrary to the general understanding of the matter, it is true that far greater quantities of Yerba Maté are grown in Brazil than are grown in Paraguay itself.

The quantity of exports from Brazil, at the present time, is considerably over 50,000 tons annually, and its value ranges from $8,000,000 to $10,000,000. Of European countries Italy appears to have used the most; Germany also used a considerable quantity of it before the war, and France and Portugal small amounts.

There is practically no limit to the amount of this leaf which can be produced in Brazil. As part of Brazil is practically an undeveloped country, if this variety of tea should become popular, it may be that Brazil will become the great tea as well as the great coffee-producing country of the world.

Caffein Content of Maté.—From the few analyses accessible it appears that the caffein content of maté is about that of tea. One analysis made under my supervision showed a content of alkaloid of 1.35 percent. It has been known to the inhabitants of South America for hundreds of years. Wonderful claims are made for its sustaining powers and fatigue-relieving properties. One enthusiastic writer says: "Alone and independent of all other foods the infusion of maté sustains the strength and vigor during entire days. In the practice of medicine maté is recommended in all cases where it is necessary to feed without exciting the nervous system and without loading the stomach." This is quite equal to the claims made for the virtues of the caffein containing drinks in sustaining muscular exertion and relieving fatigue. Any apparent relief in such cases is illusionary.

Adulterations of Tea.—Teas are subjected to many forms of sophistication. Chief among these once were reckoned artificial coloring and facing, that is, the attachment of heavy bodies to the surface of the tea to increase weight. Facing and coloring may be practised together, since the facing material may be a coloring body, or the coloring itself be a heavy material to add weight. In general, facing is practiced by treating the prepared leaves with a mixture containing Prussian blue, turmeric or indigo. In general, it may be assumed that this coloring is intended to conceal damage or inferiority, but perhaps, in some cases, it is not. Very often, however, tea leaves which have been damaged in the manufacture, or which from their age or imperfection are found to be of an inferior grade, are faced solely to improve their appearance. The addition of heavy bodies to increase weight is no longer practised.

It may be stated, as a general rule, that in tea-growing countries the teas which are used for home consumption are not subjected to the treatment as described above. It is very common, however, in preparation of teas for export to subject them to this treatment.

Whatever may be said in extenuation of the coloring matters of tea on the ground that they are not injurious to health in the quantities used, it must not be forgotten that they do not in any case ever add anything to the value of the product.

Before facing to increase weight ceased, the addition of chemicals that cost only a few cents per pound to tea which is sold at prices ranging from 50 cents to $5 per pound, could not fail to be extremely remunerative to those who practise the process. The materials which were used cannot be wholly regarded as free from injurious effects; especially is this the case with Prussian blue, which is one of the most common of the materials used for facing. Sometimes very large quantities of these facing materials, such as Prussian blue, soapstone, and other substances, have been found, the quantities having been reported from 1 to 3 percent. This is very much above the average, which may be regarded as not exceeding one-half of 1 percent of the total weight of the treated tea.

Whether or not a tea has been faced can be determined by chemical and microscopical methods. The microscope is usually sufficient for the purpose. Following are methods of detecting ordinary facing materials:[1]

Prussian Blue.—This substance is easily detected by means of the microscope. Shake the leaves in a glass cylinder with water and examine the detached particles with the microscope. If the coloring matter sought is present, transparent particles of a brilliant blue may be seen. Prussian blue may often be identified by the microscope on the leaf mounted as an opaque object. The particles detached as above may be examined chemically as follows: Treat with hot sodium hydroxid solution, acidulate with acetic acid, and add ferric chlorid. If Prussian blue is present in the facing the characteristic blue precipitate will be formed. The powdered tea leaf may be examined by the chemical method, but it is advisable to remove the tannin by precipitation with gelatin solution and filtration through powdered kaolin, after

[1] Spencer, Bulletin No. 13, Division Chemistry, Part 7, page 881.

acidulating with acetic acid. The color of Prussian blue is discharged by sodium or potassium hydroxid.

Indigo.—Under the microscope indigo appears of a greenish blue. Its color is not discharged by sodium hydroxid, a distinction from Prussian blue. Indigo forms a deep blue solution with sulphuric acid.

Turmeric.—Turmeric is identified by means of the microscope. According to Hassell turmeric consists of characteristic yellow cells of a rounded form which are filled with peculiar shaped starch granules. On the addition of an alkali the cells turn brown, swell up, and the outlines of the starch granules become visible.

Plumbago.—The microscope is employed in the detection of plumbago. A thin slice of the tea leaf will exhibit numerous bright particles if plumbago facing has been used.

Gypsum, Soapstone, Etc.—These substances, employed with the coloring matter in facing teas, may be separated by shaking the leaves in a cylinder with water. The sediment is examined by the usual qualitative methods for these substances.

Special Tea Inspection.—Laws regulating the importation of adulterated teas are the oldest restrictions of that kind in the legislation of this country. Several years before the civil war, laws on this subject were enacted. These laws were amended and extended by the Act of March 2, 1883. In 1897 the Act of 1883 was repealed and the law which is at present in force was approved March 2d of that year. This Act set up an entirely new and more rigid inspection of imports. It provided for a Tea Board to be appointed annually by the Secretary of the Treasury and to consist of 7 members. This Board was empowered to establish standards of the leading types of teas and prepare these standards of actual teas by which all imports had to be tested. The standards now in vogue are those established in 1918 and are as follows:[1]

1. Formosa Oolong.
2. Foochow Oolong.
3. Congou.
4. India (used for Ceylon).
5. Gunpowder, green.
6. Young Hyson, green.
7. Japan, pan fired.
8. Japan, basket fired.
9. Japan, dust.
10. Scented Orange Pekoe (used for capers).
11. Scented Canton.
12. Canton Oolong.

[1] T. D. 37566, March 29, 1918. Regulation 19.

Comparison with Standards.—In comparing with standards, examiners are to test all the teas on these points, namely— for quality, for artificial coloring or facing matter, and other impurity, and for quality of infused leaf. Quality shall be ascertained by drawing, according to the custom of the tea trade, with the weight of a silver half dime to the cup. The quality must be equal to standard, but the flavor may be that of a different district, as long as it is equally fit for consumption. As an illustration, a Teenkai may be equal to a Moyune, but a distinctly smoky or rank Fychow or Wenchow of sour character is not considered equal to the first two mentioned.

In examining Japans and all other green (unfermented) teas, while limiting the comparisons in the matter of infused leaf to the specific standard called for, examiners are advised to admit teas upon the question of quality, in the two kinds above cited, provided that they are equal in the case of Japans to either the pan-fired or the basket-fired standard; in all other greens to either the Gun-powder or the Young Hyson standard.

In order to test the quality of the infused leaf in comparison with the standard, a second drawing should be made of double weight. After pouring off the water, the infused leaf should be taken out so as to exhibit the lower side which rested against the cup. Should the mass show a larger quantity of exhausted or decayed leaf than the standard it affords sufficient evidence to be judged inferior in quality and consequently to be rejected.

To examine for artificial coloring or facing matter, it is advised that the following-described method be used in comparison with the standard, viz.:

Read Color Method.[1]—Place two ounces of tea in a sieve 5 or 6 inches in diameter, having 60 meshes to the inch and provided with a top. Sift a small quantity of the dust onto a semiglazed white paper about 8 by 10 inches. The amount of dust placed on the paper should be approximately one grain. To get the requisite amount of dust it is sometimes necessary to rub the leaf gently against the bottom of the sieve, but this must not be done until the sieve has been well shaken over the test paper. The dust thus collected should be poured from the paper into the scales, and after weighing the amount of one grain it is returned to

[1] Modified in harmony with Supreme Court decision T. D. 37637, May 16, 1918.

the same paper, and should be well distributed over the surface of the paper. The paper should then be placed on a plain, firm surface, preferably glass or marble, and the dust crushed by pushing over it, with considerable pressure, a flat steel spatula about five inches long. This is done repeatedly, the tea dust being ground almost to a powder and the particles of coloring matter, if any, being thus spread or streaked on the paper, so as to become more apparent. The loose dust should then be brushed off and the paper examined by means of a simple lens magnifying $7\frac{1}{2}$ diameters. In distinguishing these particles and streaks bright light is essential.

The crushed leaf in either black or green tea appears in such quantity that there is no chance of mistaking the leaf for artificial coloring or facing material.

This test is performed in comparison with the standard, and if the tea is clearly equal to the standard as regards artificial coloring or facing matter, the operation need not be repeated. If particles of artificial coloring or facing are found in the sample under comparison with the standard, this operation should be repeated a sufficient number of times for the examiner to satisfy himself whether or not the tea contains impurities consisting of artificial coloring or facing matter in excess of the standard. If found to contain artificial coloring or facing matter in excess of the standard, samples should be drawn from packages representing at least 5 percent of the line in question and subjected to the above test to see if a majority of these samples contain artificial coloring or facing matter in excess of the standard.

The above test may be applied to all varieties of tea.

In the case of Japans and all other green (unfermented) teas, in addition to the above white-paper test, repeat the operation in comparison with the respective standard on semiglazed black paper for facings, and if it is not equal to the standard, additional samples should be drawn and tested as provided above in the test on white paper. Should the examination of the sample by the "cup test" double weight, for scum, sediment, etc., or the "Read Test," or both, disclose more impurities than the standard, then a pound sample should be sent to the local appraiser's chemist or to the nearest pure food laboratory of the Department of Agriculture and an analysis made in comparison with the standard to determine whether it contains more impurities than the standard. If the tea in question is found to contain more impurities than the standard, it would properly be rejected as not being equal to the standard in purity.

All extraneous substances are impurities, and the presence of any may be detected in any way found efficient. This black-paper test detects all facings like talc, gypsum, barium sulphate, clay, etc.

Should a tea prove on examination to be inferior to the standard in any one of the requisites—namely, quality, quality of infused leaf, or purity—it would justly be rejected, notwithstanding that it be superior to the standard in some of the qualifications. No consideration shall be given to the appearance or so-called style of the dry leaf.

Legality of the Read Test.—Importers of teas protested against the application of the Read Color Test on the ground that it was extra legal. The Treasury Department won the case in the United States District Court. The case was appealed to the United States Circuit Court of Appeals which reversed the decision of the lower court. The Treasury Department then took the case to the Supreme Court. This court issued its decree on April 22, 1918, affirming the decision of the Circuit Court of Appeals. In substance, the Supreme Court ruled that the Treasury Department could not reject tea offered for import into the United States on the ground of artificial color alone, unless it could be shown that the artificial color is harmful. The court thus renewed its decision in the bleached flour case. Mr. George F. Mitchell, Supervising Tea Examiner of the Treasury Department, has given me the following opinion respecting the effect of the Court decision on the importation of colored teas:

My personal opinion is that the Read Test as amended will continue to keep all colored teas out of the United States, and at the same time meet the decision of the Supreme Court; but in case our regulations fail to do this, I plan in coöperation with the tea trade, who are unanimously against colored teas, to ask Congress to change the law so that we might reject teas for artificial coloring matter alone.

Efficiency of Tea Examination.—The efficiency of the Tea Examiners as shown in the month of February 1918 is indicated by the data which show that there was only one rejection for color. The fact that only one of the teas was rejected for color, although the test was applied in every case, shows that the Read test has practically broken up the practice of importing colored teas into the United States.

Comparison of teas imported during fiscal yar ending June 30, 1917, with those imported during fiscal year ending June 30, 1918.[1]

KINDS OF TEA EXAMINED

Kind	Pounds, 1917	Percent	Pounds, 1918	Percent
Green.....................	45,333,340	42.77	46,665,737	31.39
Oolong....................	21,317,813	20.11	19,062,635	12.81
Black.....................	39,330,005	37.11	82,956,012	55.79
Total....................	105,981,158		148,684,384	

COUNTRIES

	Pounds, 1917	Percent	Pounds, 1918	Percent
Ceylon and India............	28,632,215	27.02	44,395,552	29.86
China.......................	20,375,569	19.23	23,693,932	15.94
Japan.......................	54,833,093	51.74	52,316,298	35.19
Dutch East Indies...........	2,140,281	2.02	28,278,602	19.02
Total....................	105,981,158		148,684,384	

VARIETIES OF TEA EXAMINED

Variety	Pounds, 1917	Percent	Pounds, 1918	Percent
Formosa Oolong.............	19,379,755	18.28	17,957,489	12.75
Foochow Oolong.............	1,092,252	1.03	350,367	0.24
Congou.....................	8,676,484	8.19	10,438,905	7.02
India and Ceylon............	28,482,564	26.88	44,217,331	29.74
Java and Sumatra...........	2,140,281	2.02	28,278,602	19.02
Ceylon Green................	149,651	0.14	178,221	0.12
Ping Suey Green.............	8,192,796	7.73	8,381,763	5.37
Country Green...............	1,537,555	1.45	3,746,944	2.52
Japan.......................	31,754,960	29.96	31,218,651	21.00
Japan Dust..................	3,698,378	3.49	3,140,158	2.11
Capers......................	4,449	0.0042	876	0.0006
Scented Orange Pekoe........	26,227	0.024	19,649	0.13
Scented Canton..............	497,484	0.47	427,502	0.29
Canton Oolong...............	348,322	0.33	327,277	0.22
Flowery Pekoe...............	....		649	0.0004
Total....................	105,981,158		148,684,384	

Detection of Spent or Exhausted Leaves.—Attention has already been called to the fact that one of the methods of adulterating tea is the use of spent or exhausted leaves. These may be skilfully treated, faced, and brought into a condition which re-

[1] Report of George F. Mitchell. Fiscal year ended June 30, 1918.

sembles very much unexhausted leaves. The microscope is not of much value for detecting exhausted leaves, because the microscopic appearance of the exhausted leaf is usually made to resemble very closely that of the natural leaf.

The best method of detecting the adulteration is, by chemical examination. A considerable addition of exhausted leaves to a tea will result in a diminution of the chemical constituents which are characteristic of teas. Happily this form of adulteration is not often found. Also the use of foreign leaves, a form of adulteration easily detected, has lost its vogue.

There is a difficulty attending the chemical method of detection to which attention should be called. While the average constitution of a tea is very well known, there are many very wide variations from that average. For instance, the tannin itself may sometimes reach as much as 15 or 20 percent and sometimes fall below 7 or 8. It follows, therefore, that a tea from a leaf which originally contained 20 percent of tannin may be well exhausted, and still contain as much tannin as an unexhausted leaf of another kind. Generally, however, the best way is to rely upon the total extracted matter. If that is far below the average for tea, the use of exhausted leaves may be suspected, and a more careful examination made.

An important thing also in this matter is the careful study of the leaves themselves. There are certain physical changes which take place in an exhausted leaf which, with a little experience, would indicate to the expert whether or not this form of adulteration had been practised. If the leaves are very much broken, frayed, or partially unrolled, it is evident that it has once been extracted. The percentage of ash is also another means of aiding in the detection.

Amount of Extract in Tea.—If tea be dried until its content of moisture be only about 5 percent, it will yield by boiling water almost half of its weight in extract. For instance, it was found that:

Indian tea yielded an average of..45.58 percent of extract.
Oolong tea an average of........46.03 percent of extract.
Congo tea an average of.........37.48 percent of extract.

PLATE I

Genuine Tea Leaves and Possible Adulterants. (*U. S. Dept. of Agriculture, Bureau of Chemistry Bulletin 13, Part VII.*)

1. Tea.   2. Maté or Paraguay tea (Ilex Paraguayaensis.)   3. Camellia (Camellia Japonica).   4. Hawthorn.   5. Box Elder.   6. Horse Chestnut.   7. Sycamore.   8. Rose. 9. Plum.   10. Elm.

PLATE II

Genuine Tea Leaves and Possible Adulterants. (*U. S. Dept. of Agriculture,*
Bureau of Chemistry Bulletin 13, *Part VII.*)

11. Ash. 12 and 13. Willow. 14. Beech. 15. Oak. 16. Missouri, or Golden Currant. 17. Ash. 18. Common Red Currant. 19. Birch. 20. Poplar. 21. Raspberry. 22. Jersey Tea (Ceanothus Americanus).

Adulteration with Foreign Leaves.—This is a form of adulteration which at one time was much practised. It is so easy of detection, however, that it has largely gone out of vogue. There are no other leaves which resemble tea leaves sufficiently well to deceive an expert in their examination. The additional fact that almost every form of leaf has some peculiar flavor renders it difficult to use foreign leaves to any great extent without affecting the taste of the tea to such a degree as to excite suspicion. The microscope, however, is by far the best means by which to detect added foreign matter. The leaf of the tea plant, as already stated, is quite characteristic, as shown in the appended plates.[1]

Description of The Plates.—"These illustrations were prepared from photographic prints made by the following simple method. The natural leaf was used in making an ordinary silver print, precisely as copied by a photo-engraving process. Many of these illustrations show even the delicate veins of the leaves; the tea leaf, however, is quite fleshy, and did not yield a photographic print as distinct as those from the other plants. The lower epidermis of the leaf contains most of the stomata, which are surrounded by curved cells. There are few stomata in the upper epidermis. The stomata are useful indices. Hairs are very numerous on the younger tea leaves; they always contain them. They are sometimes entirely wanting in old leaves. Dr. Thomas Taylor, in a report to the Department, mentions the presence of stone cells in tea leaves and states that his observations confirm those of Blyth in regard to the absence of these formations in certain leaves, viz., those of the willow, sloe, beech, Paraguay tea, ash, black currants, two species of hawthorn, and raspberry. After the boiling a fragment of the leaf is placed on a slide under a cover-glass and the latter is pressed down firmly with a sliding motion until the specimen is thin enough for microscopic examination. The stone cells are valuable aids in the microscopical study.

In the general study of the serration and venation of a tea leaf the specimen should be steeped in hot water, and, after softening, the leaves should be unrolled and spread upon a glass plate for examination by transmitted light. Even small fragments of tea leaves will usually show some distinctive characteristic.

[1] Bulletin No. 13, Division of Chemistry, Part 7, Plates 39 and 40.

In general it may be stated that a microscopic examination is only necessary in exceptional cases. In doubtful samples the stomata should be examined, and a search should be made for stone cells; the epidermis of both the upper and lower leaf should be examined. Even in the case of dust the microscope will furnish conclusive evidence as to whether it is from tea or some other plant."

Adulteration with Other Substances.—Many other substances have been detected in tea from time to time, but not with sufficient frequency to render it necessary to go into a very full account. It is said by some authorities that catechu has been sometimes used to give character and strength to the tea. Besides the soapstone, gypsum, etc., already mentioned, barytes, and other heavy mineral matters have been used for facing to increase weight. Even metallic iron has been found added to teas in some cases to increase weight; also the magnetic oxid of iron, particles of sand, pieces of brick, etc., are sometimes found, but probably accidentally occurring.

It is a mistake to suppose, however, that the green color of tea is due to copper, although this has often been charged. It is doubtful if copper has ever been employed for coloring tea.

Iron salts have been added to tea for the purpose of deepening the color of the infusion.

Lie Tea.—This substance is an imitation of tea, usually containing fragments, or dust of the genuine leaves, certain foreign leaves, and certain mineral matters, held together by means of paste, usually made from dextrin or starch, usually colored by some one of the facing preparations already noted. This form of adulteration, however, is very seldom made use of. The lie teas are particularly distinguished by their high content of ash, which has been found in some cases to be over 50 percent of their entire weight. Lie tea is easily detected by treating the suspected sample with boiling water. This process disintegrates the aggregation and each of the constituent portions breaks up into separate parts. Lie teas have followed foreign leaves into oblivion.

PART VII

COCOA AND CHOCOLATE

Confusion of Terms.—When one hears the word "koko" he does not always form the correct mental image of what it means.

Coco is the name of a tree, *Cocos nucifera*.

Coca is the name of a shrub, *Erythoxylon Coca*.

Cacao is the name of a tree, *Theobroma Cacao*.

Cocoa is the name of a product, viz., the powdered fruit of the Theobroma Cacao, from which a part of the oil has been extracted.

Cocoa is a beverage prepared from the above with milk and sugar.

Chocolate is a beverage prepared from the crushed unexpressed fruit of the theobroma.

Chocolate is the term applied to a confection made from the above with sugar.

These words with the exception of chocolate are all usually pronounced "koko". Thompson, the poet of "The Seasons" was misled by the similarity of sound when he wrote:

> "Amid these orchards of the sun,
> Give me to drain the cocoa's bowl
> And from the palm to draw its freshening wine."

He had in mind the coconut palm, but the illustrations are quite confusing.

Definitions.—Cocoa is a corruption of the botanical name of the plant *cacao*. It is applied to the roasted and crushed fruit of the Theobroma from which about one-half of its oil has been expressed. The term cocoa nibs is applied to the roasted whole seeds of the Theobroma.

Chocolate is the name of the roasted and crushed cocoa nibs before any of the oil has been extracted. The term chocolate is also applied to the beverage made from the crushed cocoa nibs.

175

Unless the above facts are clearly comprehended much confusion of ideas will ensue. It is unfortunate that the term cocoa was not applied to chocolate and *vice versa*. The matter is further complicated by applying the term chocolate to a highly sweetened confection containing chocolate as a base. The sugar amounts to as much as 60 or 70 percent of the whole mass. The plain chocolate which has not been manipulated at all is often called "bitter chocolate," in distinction from the confection which is called "sweet chocolate."

Production of Cocoa.—The small seeds which make cocoa have quite a peculiar manner of growth. Most fruits and nuts are

FIG. 16.—Branch of tree with pods attached.

borne upon the new growths of the tree. The cocoa seeds grow in large pods and are attached to the older parts of the trunk or the larger limbs of the tree. This is shown in a striking way by a branch of the tree to which the pods are attached, as indicated in the figure.[1] The relation of the pods to the leaf is also best shown by an illustration.[2] The cocoa tree like coffee requires, in very hot climates, a shelter to promote especially its early growth. This shelter is usually provided by planting it under another

[1] Courtesy of Walter Baker & Co., "Cocoa and Chocolate," page 14.
[2] Courtesy of Walter Baker & Co., "Cocoa and Chocolate," page 13.

tree of larger size. The tree begins to bear at about the seventh
year, and it reaches its full growth about the tenth year. The
yield of cocoa per tree is very small. The average for trees of
10 years is only about one pound each. The cocoa tree grows to
a height of 20 or 30 feet. It is an evergreen, and usually the
diameter of the trunk near the ground on a full grown tree is
from 6 to 12 inches. It is a continual bearer, though the heavier
crop comes in the spring. The blossom is a very small one,
almost white but with a pinkish tint, and looks very much like

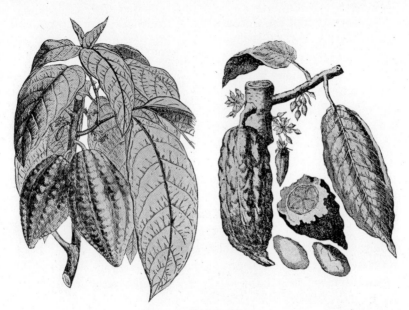

FIG. 17.—Pods and leaves.

an artificial petal made of wax. As already stated, instead of
growing on the new wood it grows directly out of the trunk or
larger branches of the tree and the fruit is harvested by long
poles something like the long handled knives which are used for
pruning apple and other fruit trees. The fruit ripens in from three
to four months or at least grows to such maturity that it can be
harvested at that time. As soon as the fruit is harvested the
pods are split open and the seeds are spread out to dry. It is
important that the drying should be as prompt and as complete
as possible to prevent the pods from molding.

12

Introduction of Cocoa.—Cocoa was unknown to the European world until after the voyages of Columbus. Its use in Central and South America and in Mexico was fully established at the time of the discovery of this Continent. In fact, the name "chocolate" which has become identified with the drink which has already been described, is said to be of Mexican origin— Chocolatl. After the conquest of Mexico by Cortez it was not long until this drink was introduced into Spain. Finck, in his remarkable work on "Food and Flavor" recommends cocoa as a substitute for coffee.[1]

Composition of the Cocoa Bean.—The cocoa bean is composed chiefly of fat, this ingredient amounting to fully 50 percent and often a larger percentage of the total weight. The next most important ingredient is the protein or nitrogenous elements, including the theobromin and caffein, which exist in small quantities in the bean. Starches and sugars together form from 20 to 25 percent of the weight of the bean. In round numbers the ash amounts to 4 percent and the theobromin to about 1 percent. Various analysts, however, give very wide differences in the content of theobromin, some rising to as much as 2½ percent and others falling to as little as five-tenths of 1 percent. From the above data it will be seen that chocolate has a real food value.

The chief food value of the cocoa bean resides of course in the fat. Owing to the high calorific power of fat, cocoa furnishes a very large amount of heat and energy. The digestible protein and the starch and sugar add no mean value to cocoa as a food. Inasmuch as the whole mass is mixed with the drink, all of the nutrient portions become available for the sustenance of the body. If in the preparation of the beverage milk and sugar are added, the food value, of course, is increased to that degree. Theobromin, which is the principal active alkaloid of the cacao bean, is much less objectionable in foods than its first cousin, caffein. Theobromin and its isomer, theophyllin, do not exert so powerful a stimulating influence as caffein, and hence for this reason cocoa is preferable to tea or coffee.

Consumption of Cocoa and Chocolate.—The data relating to consumption of cocoa and chocolate in the United States

[1] "Food and Flavor," by Henry T. Finck, page 577, The Century Company, New York.

are secured by taking the whole imports of these bodies and subtracting therefrom the exports. The difference will give the consumption and the amount remaining in stores. As this amount may be assumed to be practically constant from year to year, the difference may be regarded as measuring the consumption. For the two years, 1916 and 1917, the latter being the latest of the published data available, the total consumption of cocoa, including chocolate made therefrom, is given in the following table:

Description	1916	
	Pounds	Values
Imports, crude..................................	242,617,054	33,926,029
Imports, prepared...............................	2,293,733	676,439
Total Imported.................................	244,910,787	34,602,465
Exported.......................................	10,295,040	1,773,554
Consumed......................................	234,605,747	32,828,912

Description	1917	
	Pounds	Values
Imports, crude..................................	390,047,665	41,415,854
Imports, prepared..:............................	790,650	258,849
Total imported.................................	390,838,315	41,674,203
Exported.......................................	12,873,082	1,729,354
Consumed......................................	377,965,233	39,944,849

The use of cocoa or cacao in this country approximates one-fourth that of coffee. Much of it is used in confections. The cost is approximately one-fourth that of the coffee used. There was a remarkable increase in the consumption of cocoa during 1917.

PART VIII

WINE

Antiquity of Wine.—Wine is the fermented juice of the grape. It is the oldest and most important of fermented beverages. It could not escape early discovery, because fruits of all kinds, when crushed and left to natural causes, undergo the alcoholic fermentation. Primitive man must, therefore, have been acquainted with the properties of the fermented juice of fruits, especially of grapes. Wine was known in the remotest historial times, as evidenced by references to it in the earliest preserved literatures.

Similarity of the Name.—Another point which is of interest in this connection, is that the term "wine" is not only of ancient origin, but has been preserved practically unchanged in many languages. The Greek name for wine, *oinos;* Latin, *vinum;* German, *wein;* French, *vin;* Italian, *vino;* and the English, *wine* are all the same word. It is not often that the name of a common object is preserved in this way throughout so many years and through so many tongues.

Official Standard of Wines.—By authority of Congress the Secretary of Agriculture has fixed the following definitions and standards for wines.

1. *Wine* is the product made by the normal alcoholic fermentation of the juice of sound, ripe grapes, and the usual cellar treatment,[a] and contains not less than seven (7) nor more than sixteen (16) percent of alcohol, by volume, and, in one hundred(100) cubic centimeters (20°C.), not more than one-tenth (0.1) gram of sodium chlorid nor more than two-tenths (0.2) gram of potassium sulphate; and for red wine not more than fourteen hundredths (0.14) gram, and for white wine not more than twelve hundredths (0.12) gram of volatile acids produced by fermentation and calcu-

lated as acetic acid.   *Red wine* is wine containing the red coloring matter of the skins of grapes.   *White wine* is wine made from white grapes or the expressed fresh juice of other grapes.

2. *Dry wine* is wine in which the fermentation of the sugars is practically complete and which contains, in one hundred (100) cubic centimeters (20°C.), less than one (1) gram of sugars and for dry red wine not less than sixteen hundredths (0.16) gram of grape ash and not less than one and six-tenths (1.6) grams of sugar-free grape solids, and for dry white wine not less than thirteen hundredths (0.13) gram of grape ash and not less than one and four-tenths (1.4) grams of sugar-free grape solids.

3. *Fortified dry wine* is dry wine to which brandy has been added but which conforms in all other particulars to the standard of dry wine.

4. *Sweet wine* is wine in which the alcoholic fermentation has been arrested, and which contains, in one hundred (100) cubic centimeters (20°C.), not less than one (1) gram of sugars, and for sweet red wine not less than sixteen hundredths (0.16) gram of grape ash, and for sweet white wine not less than thirteen hundredths (0.13) gram of grape ash.

5. *Fortified sweet wine* is sweet wine to which wine spirits have been added.   By act of Congress, "sweet wine" used for making fortified sweet wine and "wine spirits" used for such fortification are defined as follows (sec. 43, Act of October 1, 1890, 26 Stat., 567, as amended by section 68, Act of August 27, 1894, 28 Stat., 509, and further amended by Act of Congress approved June 7, 1906): "That the wine spirits mentioned in section 42 of this act is the product resulting from the distillation of fermented grape juice to which water may have been added prior to, during, or after fermentation, for the sole purpose of facilitating the fermentation and economical distillation thereof, and shall be held to include the products from grapes or their residues, commonly known as grape brandy; and the pure sweet wine, which may be fortified free of tax, as provided in said section, is fermented grape juice only, and shall contain no other substance whatever introduced before, at the time of, or after fermentation, except as herein expressly provided; and such sweet wine shall contain not less than four per centum of saccharin matter, which saccharin strength may be determined by testing with Balling's

saccharometer or must scale, such sweet wine, after the evaporation of the spirits contained therein, and restoring the sample tested to original volume by addition of water: *Provided*, That the addition of pure boiled or condensed grape must or pure crystallized cane or beet sugar or pure anhydrous sugar to the pure grape juice aforesaid, or the fermented product of such grape juice prior to the fortification provided by this Act for the sole purpose of perfecting sweet wine according to commercial standard, or the addition of water in such quantities only as may be necessary in the mechanical operation of grape conveyers, crushers, and pipes leading to fermenting tanks, shall not be excluded by the definition of pure sweet wine aforesaid: *Provided, however*, That the cane or beet sugar, or pure anhydrous sugar, or water, so used shall not in either case be in excess of ten (10) per centum of the weight of the wine to be fortified under this Act: *And provided further*, That the addition of water herein authorized shall be under such regulations and limitations as the Commissioner of Internal Revenue, with the approval of the Secretary of the Treasury, may from time to time prescribe; but in no case shall such wines to which water has been added be eligible for fortification under the provisions of this Act where the same, after fermentation and before fortification, have an alcoholic strength of less than five per centum of their volume."

6. *Sparkling wine* is wine in which the after part of the fermentation is completed in the bottle, the sediment being disgorged and its place supplied by wine or sugar liquor, and which contains, in one hundred (100) cubic centimeters (20°C.), not less than twelve hundredths (0.12) gram of grape ash.

7. *Modified wine, ameliorated wine, corrected wine*, is the product made by the alcoholic fermentation, with the usual cellar treatment, of a mixture of the juice of sound, ripe grapes with sugar (sucrose), or a sirup containing not less than sixty-five (65) percent of sugar (sucrose), and in quantity not more than enough to raise the alcoholic strength after fermentation, to eleven (11) percent by volume.

8. *Raisin wine* is the product made by the alcoholic fermentation of an infusion of dried or evaporated grapes, or of a mixture of such infusion or of raisins with grape juice.

Native Wines.—The wines produced in the United States

are, by no means, so well appreciated as they should be. There are many reasons for this, some of which follow.

In the first place, the virgin soils of the country do not produce, by any means, a wine of such bouquet and excellent quality as are produced by lands which have been long in cultivation.

In the second place, the wines of a country are the joint product of the skill of the viticulturist and the natural environment of the country. In the environment are included the rainfall—amount and distribution—the amount and distribution of the sunshine, the temperatures of all seasons of the year, the natural ferments which convert the sugar of the grape into wine, the character of the cellars employed in the ripening of the wine, and the skill and care exercised in the vintage and in the wine cellars. It would naturally follow from this that in a country where wine making is 2,000 years old, you would find a skill and excellence in manipulation and a practical knowledge of wine making which could not possibly exist in a new country. Further than this, if the wine makers of an old country should be transplanted to a new country, they would find an entirely new environment, and while their care and skill might be the same, the wine which they would make might be vastly different from that which they made in their old homes.

Another cause to which may be ascribed the lack of appreciation of American wines is the general attitude of the wine drinker to the wine. The people of the United States are not wine drinkers, and hence wine is consumed chiefly by a certain class of the community having exceptional advantages of money and locality, and representing only a small part of the population. The people who would naturally drink American wines are those who are accustomed to drinking foreign wines, and hence a general idea has existed among them that American wines are inferior in quality.

Another reason which may be assigned for the lack of appreciation of American wines is the extent of the adulteration and debasement to which they have been subjected. In many parts of our country it is a settled conviction among the grape growers that wine cannot be made from the juice of the grape without certain additions, chiefly of sugar. When once the wine grower begins to add sugar or water to his grape juice, the temptation to increase the volume of the product is so great that he does not know exactly where to stop. The result is that in many localities American wines are only partly made from the juice of the grape, while sometimes as much as a third, or even half, of their alcoholic content comes from a totally foreign substance. It must be admitted that every dilution of the natural flavors and aromas of the grape juice by an artificial substance, such as the fermented product of dilute sugar solutions, can only result in a lowering of the quality of the wine. There is no valid excuse for adding sugar to grape juice, since in other countries wines are often made which contain only 6 or 7 percent of alcohol, representing a quantity of sugar equally as low as that ever reached by American grapes suitable for wine making even in the poorest seasons.

A final reason why American wines are not appreciated is that a large quantity of imitation wines has been made in this country and sold under the names of the genuine. Investigations have shown that it is the custom in some localities to take the pomace, after the expression of the grape must, or of the finished wine, and treat it with additional quantities of sugar and water, allow it to undergo a secondary fermentation, and sell it separately as wine, or mix it with the original product. When, in addition to this, it is considered that not only sugar, but other substances, namely, dextrose, saccharin, artificial coloring matters, tannin, etc., are added, it is easily seen that the wine is hardly any longer even an imitation, but mostly an artificial product.

In spite of all these unfavorable conditions, with which the American wine maker producing a pure article has to contend, and the American consumer who would like to drink American wines has to bear, it cannot be denied that there are very large quantities of wine made in the United States of the pure juice of the grape and possessing high qualities which would naturally commend it both to the ordinary consumer and to the connoisseur.

Mr. Percy T. Morgan, former President of the American Wine Association, stated in an address at the St. Louis Exposition, in 1904, that according to tradition, as is well known, almost five centuries before the discovery of the American Continent Scandinavian navigators visited this country and by reason of finding great numbers of native vines growing therein called it

"Vinland." In the early history of our country accounts are given of the manufacture of wine in Florida and North Carolina, where the vines were extremely luxuriant. In North Carolina, as is well known, is found the original Scuppernong grape, which is now grown to a wide spreading tree of great proportions. Wines were made, also, by the early French settlers in Louisiana, and in 1630 skilled vineyardists from France settled in Virginia. One of the earliest accounts of manufacture of wine io this country is as follows:

Early Method of Making Wine.—An early method of making wine in America is as follows:[1]

The method I have found best for making wine from grapes, is to let them hang on the vines until fully ripe, then to gather them, when dry, throw away rotten ones if any, open the cider mill so as not to mash the stems or seeds, put the pumice (or mashed grapes) on some clean, long straw, laid on the cider press floor, lap it in the straw, press it well, then take off the pumice, add some water, and after it has soaked a while, press as before; the latter will make as good wine by adding sugar as is commonly done in the country, but I prefer making it of the juice without water.

The last autumn I tried several ways of making wine: one cask of 34 gallons that first run from the press, I set to ferment in its then state, expecting to make that without sugar; another of the same size had 17 pounds white Havanna sugar added, the remainder was mixed with the second pressing, and had the same proportion of sugar; the first ceased fermenting in half the time of the others; when the fermentation subsided, I drew them off (one cask at a time) into a tub and rinsed the cask with water and fine gravel, then put in about one-eighth of the quantity, of French brandy (good apple brandy, will make the wine as good, but not so like foreign wine), and having burnt a sulphur match (about half as much as would kill a hive of bees), after the match was burnt out I stopt the bung again, shook it to incorporate the liquor with the smoke, and finally filled the cask.

The first cask when racked, I found too tart, I believe owing to the wet summer, on which account I added sugar as above, and the like proportion of brandy; in about a month I racked all again, and found this last-mentioned cask far better and clearer than the others, from which I conclude it is better to let grape wine first ferment, and when racked, to add sugar to the palate, by which means wine may be made palatable from sweet or sour grapes.

Taking into consideration with what ease and expedition grape vines may be propagated; the great expense and uncertainty of being supplied from foreign countries, and the base and dangerous practice of adulteration by many of the venders of wine, I am induced to urge the propagation of grape

[1] Philadelphia Society for Promoting Agriculture, February 22, 1808, read by Mr. Joseph Cooper, of Coopers' Point, on March 8, 1808.

vines in preference to other fruit, especially in such places as shades are wanted, as they may be trained in such manner as fancy or convenience may direct, and more speedily than any durable fruit-bearing tree; and if properly trimmed and trained, will exceed the same kind of vines which grow on trees, in production of fruit in quantity, size and flavor, beyond most people's imagination.

Experiments of Longworth.—Nicholas Longworth, of Cincinnati, became greatly interested in the cultivation of vines on the banks of the Ohio River. To such an extent did he patronize this industry that he may be regarded as the father of the native wine industry from native grapes in the United States. Vines were cultivated on the Pacific Coast as early as 1769. Great progress was made in the United States by the introduction of the Catawba grape by Major Adlum, about the year 1824. The Horticultural Society of Cincinnati took a great interest in vine culture, as is shown by their reports issued as early as 1830. The principal American varieties which were developed at this early time were the Fox, the Catawba, the Ives Seedling and the Norton Seedling. From these original stocks have sprung nearly all the Eastern varieties of grapes.

The Original Scuppernong.—In the Roanoke Island in North Carolina the old mother Scuppernong vine still exists, as was pointed out in the first volume ("Foods and Their Adulteration"). Around the Scuppernong vine is entwined a romance of early times in America, which has been celebrated by Miss Sallie Southall Cotten in a novel entitled the "Legend of the White Doe." This is a story of the fate of Virginia Dare, the first white child born of English parents on the soil of North America. The disappearance of this child, together with that of an entire party of colonists, is a mystery which history has not solved. The legend runs that this beautiful young girl was turned by Indian sorcery into a white doe, which, being shot by a silver arrow, returned to human form, and dying, her heart's blood fertilized a seedling vine, the fruit from which yielded a deep red wine instead of the white juice usual to this grape.

Wines of Distinctive Type.—It is rather strange that American producers should establish foreign names for American wines, when it is well known that the American wines have a distinct character of their own. They are not like the wines of the Rhine or Mosel, nor yet those of the Côte-d-Or, nor do they resemble in any particular character the grand classed wines of the Medoc and the Sauternais. It is a well-established fact that plants of the same kind produce fruits in different localities which are entirely distinct in character. The mild peppers which grow in Hungary and produce a spice known as "Paprika," when grown in the United States produce a spice which is far more pungent and more distinctively of a high pepper character than that of the Hungarian plant of the same species. It is well known, again,

Fig. 18.—Scuppernong grape vine, Roanoke Island. (*Courtesy B. W. Kilgore.*)

that the Albermarle pippin does not reach its perfect production of character and flavor outside of a certain very limited area. While all grapes are essentially the same, they differ in acidity, amount of sugar, flavor, aroma, and character, due to the environment in which they are grown. In this country this environment is so powerful in character as to produce very distinct types of fruit, and these in turn make perfectly distinct types of wine. There never was another country so favored, therefore, to develop a wine of a distinctive type, under a distinctive name, and thus secure a trade wholly its own. I think it may be said with perfect truthfulness that one of the principal reasons which have retarded the growth of the American wine industry was the mistaken notion that wines could only be sold if labeled with the names of foreign wines whose character had been fully established and whose merits were fully recognized.

Native White and Red Wines.—In many parts of the United States both red and white wines are made, in which the type of the environment is distinctly marked. The white wines of New York, California, and Ohio are examples of this distinct character. The red wines of Virginia are other samples equally as well marked. In California also the type of wine varies with the part of the country in which it is produced. The vines which grow in the northern part of the State produce a wine which is entirely distinct from that grown in the southern part of the State. It is the part of wisdom to adopt for these distinctive type of wines in each locality, names by which they may be known and recognized in the trade. Thus their merits will be known under names which will enable the consumer to ask for them and be supplied with the exact character of article demanded.

That American wines are capable of reaching a high state of excellence is shown by the fact that they received high rewards in the Paris Exposition of 1900. The percentage of wines from this country that received rewards was fully as great, if not greater, than that from any other country. In this connection too, it must be taken into consideration that the judges giving these rewards not only were not partial to the American wines, but on the contrary many of them had distinct prejudices against them.

Sparkling Wine.—Attention has been called above to the distinctive character of the still wines produced in the United States. This distinction of character is even more marked in the sparkling wines. There has been a very great growth in the production of sparkling wines in the United States, and there is no reason to believe that this growth cannot be indefinitely extended if the real merits of the American sparkling wine can be made known to the consumer. The prejudice existing against the use of American sparkling wine would rapidly disappear if it could be sold to the consumer under its own distinctive name, and without attempting to imitate in any way the character or the name of the foreign product.

There are many parts of the country where sparkling wine is made in large abundance, and this is especially true of New York. California also produces a considerable quantity of sparkling wine, while it is made in moderate quantities in other parts of the country.

The great care which must be exercised in making a wine of this kind, and the long delay from the time of vintage until the time of the sale, tend to restrict to a certain extent a more rapid progress of its production. The peculiar fruity flavor and aroma of the American sparkling wine, while at first perhaps distasteful to those who are accustomed to drinking the imported article, becomes after a judicious use, a pleasant and attractive feature of the product. The American sparkling wine is distinct, and a more prosperous future and more extensive use awaits it just as soon as its real merits become appreciated.

Possibilities of Wine Production.—It is the general opinion in this country that California presents remarkable natural facilities for the growth of the grape and the manufacture of wine. This is true in many respects, but I do not believe it is true that California is by any means the only part of the country where excellent wines may be produced. In one respect the California grapes are superior to those grown in other parts of the country, namely, the content of sugar. The natural result of this is that the California wines contain a larger percentage of alcohol than those produced in other localities. While this is advantageous in making them more easy to keep in a sound condition, it is in another respect a prejudicial feature. Wines are not valued on account of the alcohol which they contain, but because of their aroma, their fineness, their bouquet, and their palatability, all of which depend on other ingredients than alcohol. If a wine of fine character in all respects could be made with a content of only 5 or 6 percent of alcohol, it would be better both for the wine grower and the wine drinker. The American viticulturist must learn the method of making a wine light in alcohol, and yet which is perfectly sound and of sufficient keeping properties to enable it to be put on the market at proper ages and kept until consumed. That this highly desirable result may be secured with grapes containing much less sugar than those produced in California there is no reason to doubt. Nevertheless, California presents advantages of climate, soil and environment which have placed it easily at the head of the wine-producing regions of the United States, and will probably keep it there for an indefinite number of years. There is very much more wine made in the State of California alone than in all the other parts of the United States combined.

Exemption from Taxation.—It is a curious incident that in the taxation of alcoholic beverages, for the purpose of raising revenues to pay the expenses of the Government, wines in general have been to within a short time exempt from tax. The tariff law of 1915 places a tax of a few cents per gallon on wine. The law of October 3, 1917, doubles the tax on all still wines as well as champagnes and sparkling wines. Although the wines contain from two to four times as much alcohol as beer, they are not subject to the Internal Revenue tax except as stated above unless fortified.

In the early history of the country a tax was laid on wine. When Thomas Jefferson was President of the United States he was an advocate of the revocation or abolition of the duty on wine, and his views are expressed in the following language.

I rejoice, as a moralist, at the prospect of a reduction of the duties on wine by our National Legislature. It is an error to view a tax on that liquor as merely a tax on the rich. It is prohibition of its use to the middling classes of our citizens, and a condemnation of them to the poison of spirits, which is desolating their homes. No nation is drunken where wine is cheap; and none sober where the dearness of wine substitutes ardent spirits as its common beverage.

Artificial Restriction of the Use of Wine.—Another reason why the use of wine in this country is a very restricted one, aside from the sentiment which is so prevalent in our country in opposition to the use of alcoholic beverages, is the numerous profits which are laid upon the wines by the merchant handling them. Even in California, where good red wine has been sold as cheap as 10 cents a gallon, it is quite difficult for a traveler, in a restaurant or a hotel, to get a quart bottle of wine for much less than 75 cents. In the eastern part of our country, where imported wines are principally sold, a pint of white or red wine can hardly be procured in any hotel for less than 50 cents. When it is recalled that the very best types of wine, aside from the grand classed growths, can be laid down in New York from a foreign country for not more than 75 cents a gallon, the profit upon the material, when it is finally consumed, is seen to be very great.

Claims of Excessive Acidity.—It has frequently been urged by the makers of wines in New York and Ohio that products of good quality could not be manufactured from the juice of the grapes alone because of the too great acidity. The investigations of Alwood under my direction have shown that such claims are wholly groundless. Not only can wine of fine quality be made in these localities from the unadulterated juice of the grape, but of much finer quality than the sugared wines which have been commonly manufactured. For many years in succession absolutely pure wines were made under conditions which obtain in the localities mentioned, and in all instances they have matured and ripened into wines of very superior character. The composition of these wines has also been determined in a great many instances.

Professor Alwood calls attention to the fact, as illustrated by his experience, that the composition of the fruit itself does not give a sufficient basis to form a final decision as to whether the fruit will produce a potable wine. The fruit which he used for the experimental purposes was purchased in the open market from the stocks which were brought to the wineries, as a rule. In a few instances the grapes were purchased directly from the growers. The quality of this fruit was exactly the same as that at the disposal of the wine makers of the District. The processes of manufacture are entirely analogous to those practised in the neighborhood, with the exception that neither sugar, water, nor other substance was mixed with the juice of the grapes.

The comparative analysis of the fruits and of the wines made therefrom show that the fermentation was quite complete, the wines, as a rule, being decidedly of the dry order. All of the grapes produced sufficient alcohol for preservation, and this shows that the claim that additional alcohol must be derived from added sugar is without scientific support. The samples of Delaware grapes show very high alcoholic content, far beyond all needs for conservation of the wine. The data obtained conclusively show that grapes grown in Ohio and New York will produce sufficient

TABLE II.—COMPOSITION OF PURE WINES MADE FROM EIGHT VARIETIES OF
AMERICAN NATIVE GRAPES, IN SANDUSKY, OHIO, AND CHARLOTTESVILLE, VIRGINIA

Variety	Year	Alcohol by volume	Alcohol, gms. per 100 c.c.	Total solids, gms. per 100 c.c.	Sugar-free solids, gms. per 100 c.c.	Sugar as invert, gms. per 100 c.c.	Total acid as tartaric, gms. per 100 c.c.
Catawba	1908–						
Fruit used for wine........	1911	....	....	21.63	2.43	19.20	0.943
Pure wine from above fruit..	......	11.39	9.03	2.25	2.04	0.222	0.849
Clinton	1908–						
Fruit used for wine........	1911	....	....	24.94	3.24	21.68	1.27
Pure wine from above fruit..	......	11.85	9.41	3.26	3.06	0.202	1.009
Concord	1909–						
Fruit used for wine........	1910	....	....	20.19	2.75	17.44	0.756
Pure wine from above fruit..	......	10.12	8.04	2.58	2.39	0.185	0.752
Cynthiana	1911						
Fruit used for wine........	......	....	....	24.93	3.27	21.66	0.754
Pure wine from above fruit..	......	12.14	9.63	3.59	3.21	0.328	0.606
Delaware	1909–						
Fruit used for wine........	1910	....	....	24.65	2.39	22.26	0.780
Pure wine from above fruit..	......	13.49	10.70	2.10	1.99	0.112	0.630
Iona	1909						
Fruit used for wine........	......	....	....	21.80	3.09	18.71	0.934
Pure wine from above fruit..	......	11.76	9.33	2.04	1.95	0.091	0.832
Ives	1909–						
Fruit used for wine........	1911	....	....	18.55	2.59	15.97	0.649
Pure wine from above fruit..	......	8.51	6.75	2.81	2.56	0.250	0.768
Norton	1907–						
Fruit used for wine........	1908	....	....	24.94	3.53	21.41	1.105
Pure wine from above fruit..	1909–1911	11.68	9.27	3.71	3.37	0.335	0.810

alcohol for all purposes in wine preservation. There has never
been any excuse for adding sugar to grapes grown in California,
because the grapes are sufficiently rich in sugar to produce all
the alcohol that is necessary, and even more. These investigations of Alwood show that the makers of wine in New York
and Ohio can no longer scientifically claim the privilege of stretching their wines with sugar. Sugared wines are essentially adulterated, even if labeled as such. The rigid investigations of Alwood
have also shown that there is no basis for the belief that the acidity
of properly made American wines is excessive. In regard to the
keeping qualities of pure Virginia wine, I have lately (1917)
tested a bottle of pure Virginia red wine made by Alwood in 1904

and found it has kept perfectly and has qualities which entitle it to rank among the good red wines of the Médoc.

FRENCH WINES

Classification.—The celebrated wines of the Médoc, the Graves and Sauternais regions are classified, according to established rules and regulations authorized by the Boards of Trade and officials of the Government, into a number of different grades. These grades are called by the French name "crus," which literally means growth or rank. There are five recognized classes altogether. All wines that are classed in these families by the authorities which have been mentioned are known as "grand wines." The official classification of these wines is as follows:[1]

TABLE OF THE PRINCIPAL CHATEAU WINES IN THE MÉDOC, THE GRAVES AND THE SAUTERNAIS REGIONS

FIRST CRUS
Chateau Margaux.
" Lafite.
" Latour.
" Haut-Brion.
SECOND CRUS
Chateau Mouton-Rothschild.
" Montrose.
" Cos-d'Estournel.
" Gruaud-Larose.
" Branne-Cantenac.
" Gruaud-Larose-Sarget.
" Léoville-Lascases.
" Ducru-Beaucaillou.
" Léoville-Barton.
" Léoville-Poyférré.
" Lascombes.
" Pichon-Lalande.
" Pichon-Longueville.
" Rauzan-Segla.
" Rauzan-Gassies.
" Durfort-Vivens.

THIRD CRUS
Chateau D'Issan.
" Cantenac-Brown.
" Desmirail.
" Lagrange.
" Kirwan.
" Malescot-St. Exupery.
" Giscours.
" Palmer.
" Ferrière.
" La Lagune.
" Calon-Ségur.
" Marquis d'Alesme-Becker.
FOURTH CRUS
Chateau Talbot.
" Beychevelle.
" Duhart-Milon.
" Branaire-Ducru.
" La-Tour-Carnet.
" Marquis-De-Terme.
" Le Prieure.
" St. Pierre-Bontemps.

[1] For this classification and much of the descriptive matter, I am indebted to "Bordeaux and Its Wines" by Cocks and Feret, published by Nathaniel Johnston and Sons, Bordeaux. The author has personally inspected nearly all the vineyards producing the most famous wines and many of those of other crus.

13

Chateau Rochet.
"   St. Pierre Sevaistre.
"   Poujet.

FIFTH CRUS
Chateau Cantemerle.
"   Mouton-d'Armailhacq.
"   Batailley.
"   Pontet-Canet.
"   Dauzac.
"   Belgrave.
"   Grand-Puy-Lacoste.
"   Le Tertre.
"   Lynch-Bages.
"   Lynch-Moussas.
"   Calvè-Croizet-Bages.
"   Pedesclaux.
"   Clerc-Milon.
"   Cos-Labory.
"   Ducasse-Grand-Puy.
"   Camensac.

EXCEPTIONAL CRUS OF THE GRAVES
Chateau Haut-Bailly.
"   La Mission-Haut-Brion.
"   Pape-Clement.
"   Smith-Haut-Lafitte.

Chateau Olivier.
"   Camponac.
"   Pomarede.
"   Du Vallon.
"   Haut-Brion-Larrivet.
GRAND WINES OF THE SAUTERNAIS

FIRST CRUS
Chateau Yquem.
"   Rieussec.
"   La-Tour-Blanche.
"   Rabaud.
"   Climens.
"   Guiraud.
"   Suduiraut.
"   Du Vigneau.
"   Coutet.
Clos   Haut-Peyraguey.
Chateau Lafaurie-Peyraguey.

SECOND CRUS
Chateau Filhot.
"   Mirat.
"   Suau.
"   Des Rochers.
"   Grillon.
Cru de Malle.

FIG. 19.—Chateau Figeac in the heart of the St. Emillion country.
(Photo by Author.)

Chateau Cantegril.	Chateau Lamothe.
" Lafont.	" Doisy.
" Raymond-Lafont.	" d'Arche.
" d'Arche-Lafaurie.	" La Montagne.
" Liot.	" Broustet.
" Caillou.	" Vedrines.

FIG. 20.—Chateau Yquem Vineyard looking north toward the Garonne. The hills in the distance are on the left bank of the river.

Description of the Principal Wines of the Médoc and Sauternais Countries.—The term Médoc is applied to that part of the wine region of Bordeaux lying on the left bank of the Garonne and Gironde extending from near Bordeaux almost to the sea. It is a narrow strip of varying width and somewhat irregular outline, on which is grown the most renowned red wines of that country.

The region of the Sauternais, on the contrary, is situated above Bordeaux on the river Garonne, and mostly upon the left bank of that river. It is continuous with the region in which the celebrated Graves wines are made.

Chateau Margaux.—One of the most famous, and one of the vineyards closest to Bordeaux is the Chateau Margaux. It is the property of the Count Pillet-Will, and belongs to the first-classed growth. The beginning of this famous chateau wine is noted in the 15th century, at which time at the point where the chateau now stands there was a vineyard of about 80 hectares,

composed of the finest strains of grapes. The situation of the
vineyard is exceptionally favorable to the growth of grapes, and
the care with which it is treated is one of the reasons why it has
assumed such a commanding position in the first rank of the

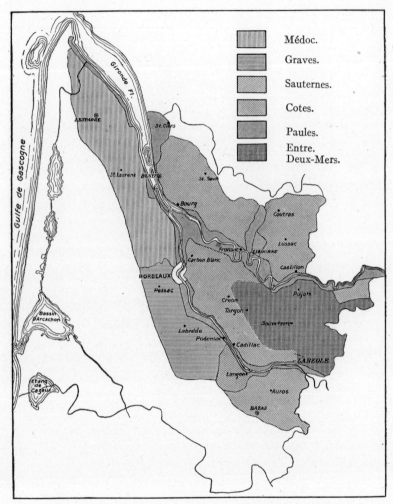

FIG. 21.—Map of the areas producing different kinds of wine.

classed growths. It produces a wine uniting the highest qualities
and marked especially by its bouquet, perfume, and fineness.
This wine is appreciated throughout the entire world, and has
always obtained the highest rewards at national and international

expositions.   The vineyard of the Chateau Margaux at the present
time produces about 225 tuns[1] of first-class wine, and 40 tuns of a
lower grade of wine which does not bear the name of the chateau.
The building of the chateau is almost colonial in its simplicity
and presents a piece of architecture simple in its beauty.

FIG. 22.—A VINE-PLANT IN THE SAUTERNES DISTRICT.
(Courtesy of Barton and Guestier.)

Chateau Lafite.—Equally, if not more, famous than the Chateau
Margaux and also belonging to the first-classed growth is the wine
of the Chateau Lafite. This vineyard is the property of
the Brothers de Rothschild, the celebrated bankers of Paris.
The buildings are situated in a beautiful grove of trees,
and the cellars are seen on the left of the chateau buildings.

[1] The tun = 192 gallons.

The wine also has a characteristic quality of dryness, fineness, and superb excellence, and unites in itself all the qualities which make a red wine famous and preferred by every connoisseur. The bouquet of the Lafite wines is especially well known throughout the professional world. The vineyard of the Chateau Lafite contains about 70 hectares, and produces an average of about 170 tuns of first-class wine, about 20 tuns of the second class, and 20

FIG. 23.—Chateau Margaux. (*Photo by Author.*)

tuns of wine harvested on neighboring vineyards, and, therefore, not capable of bearing the name of the chateau.

Chateau La Tour.—The red wines produced at this chateau must not be confounded with the celebrated white wine of the Sauternais region produced by the Chateau La Tour Blanche. As in the case of the former wine, this wine is named by reason of a tower built separate from the chateau and surmounted by a spherical dome. The wines of the Chateau La Tour are also the first growths, ranking in the same category with the two wines already mentioned. They enjoy a universal reputation of excellence and they are especially sought after by the English and the Germans. As in the case of the other wines, they have universally received the highest rewards at the expositions at which they have been exhibited. The vineyards of the Chateau La Tour are not so large as those of the Chateau Margaux and Chateau Lafite, and produce about 100 tuns of wine of the first class, and 25 tuns of wine of the second class. The proprietors of this vineyard are Messrs. de Flers, de Beaumont and de Courtivron.

Chateau Haut-Brion.—The wines of this celebrated vineyard are almost, if not equally, as famous as the three wines just mentioned. They are mostly red in color and the red wines belong to the first-classed growths. The Chateau Haut-Brion is one of the

FIG. 24.—A CELLAR FOR NEW WINE (*Chateau La Tour*).
(*Courtesy of Barton and Guestier.*)

celebrated vineyards which is closest to Bordeaux, being only five kilometers to the southwest; in fact, almost within the confines of the city. It is rather a large vineyard, consisting of 165 hectares, and this vineyard produces an amount of 110 tuns of wine of the first class, and about 20 tuns of wine of the second class. The owners of this celebrated chateau are the heirs of Eug. Larrieu. The wine of the Chateau Haut-Brion was already highly esteemed at the beginning of the 16th century, and has always maintained its high reputation on account of its excellence and fineness.

Grand Wines of the Médoc of the Second-class.—The four wines which have just been mentioned are the only wines in the Médoc which are entitled to the first rank, according to the system of classification which has been adopted there for many years. There are a number of wines, however, which are called grand wines, which are almost, if not quite, equal to the four mentioned, in excellence of quality, and these are ranked as grand wines of the second growth or class.

FIG. 25.—CHATEAU HAUT-BRION (FIRST "CRU").
The heirs of Eugène Larrieu, owners. (*Courtesy of Barton and Guestier.*)

Among the most important of this second growth are the following:

Chateau Mouton-Rothschild.—The proprietor of this vineyard is Henri de Rothschild. The vineyard of Mouton-Rothschild adjoins the Chateau Lafite, and is situated upon a gentle undulating surface where the soil is well suited to the best cultivation of the vine. This vineyard is a model in its culture and all its appointments. It produces annually about 150 tuns of wine of the second class. The wines of this vineyard have an exquisite bouquet and a most agreeable taste, and their fame has already augmented to a considerable degree their price, and they are regarded by experts as almost, if not quite, equal to the grand wines of the first class.

Chateau Montrose.—The proprietor of this vineyard is Mr. Charmolue. It contains about 67 hectares. It is a very prolific vineyard, its mean annual output being about 220 tuns. This vineyard is considered as a model one by competent authorities and has received a gold medal from the Minister of Agriculture for its peculiar excellence of appointment and cultivation. This

FIG. 26.—In Médoc, bearers emptying their "hottes" (baskets strapped on their backs), full of grapes, into the vans. (*Courtesy of Barton and Guestier.*)

medal was given to the vineyard in 1871, and since then it has constantly kept the vineyard in the public eye and it has maintained its excellence throughout the entire time. Its wines are much sought after by the highest grade consumers in Europe and other countries.

Chateau Cos-d'Estournel.—This is also the property of Mr. Charmolue. This is one of the oldest and most beautiful vineyards in the Médoc. Its vines are extremely old, and many of them have escaped the ravages of the phylloxera. It is situated upon a gravely soil, with a most favorable exposure to the sun, and contains only the finest selected varieties of grapes. Its extent is 63 hectares, and it produces annually a mean of 200 tuns of delicious wine, soft and fine and of a bouquet which entitles it justly to its universal reputation.

Chateau Gruaud-Larose.—The vineyard of Gruaud-Larose has had a very checkered history, and the old vineyard was divided at the time of the French Revolution into certain parts which now maintain separate names. One of these parts of the old vineyard is the property of an association called the Civil Society of Gruaud-Larose-Faure-Bethmann. The wines of the vineyard are extremely fine, very elegant, and have a particular bouquet and exquisite perfume which distinguish them from the other growths of the Médoc. It also enjoys a great reputation in France and abroad. The situation of the vineyard is admirable in its exposure to the sun and the character and surface of the soil. Its annual product is about 100 tuns.

Chateau Branne-Cantenac.—This vineyard is the property of the families Berger and G. Roy. The wine of the vineyard has an extreme fineness, is distinguished by its peculiar bouquet, and by its taste and perfume. It is justly considered as being a close rival with wines of the first growth. It has its place among the most agreeable and delicate wines of the Médoc region. The beautiful vineyard is admirably situated and has an extent of 50 hectares, producing a mean annual output of 125 tuns. For many years the wine of this vineyard has enjoyed a particular favor in commerce. In 1901 this wine obtained the gold medal and the diploma of honor from The Wine Growing Society of Gironde.

Chateau Gruaud-Larose-Sarget.—This is another one of the divisions of the old vineyard of Gruaud-Larose, and is owned by the widow of Baron Sarget. It is beautifully situated on a gravely soil, in the St. Julien district, and produces annually about 120 tuns of wine which is very agreeable in its character, and which has obtained numerous prizes, among others a grand gold medal at the Universal Expositions in Paris in 1867, 1878, 1889, and 1900.

Chateau Léoville-Lascases.—This vineyard is the property of Count Lascases. It is a part of the ancient domain of Leoville and is planted with the finest varieties of grapes, and is beautifully situated upon a gravely, undulating soil which is extremely favorable for the growth of the grape. It produces annually about 140 tuns of wine of a superior quality and bouquet.

Chateau Ducru-Beaucaillou.—This vineyard is the property of the celebrated wine merchant of Bordeaux, Mr. Nathaniel Johnston. It has a most favorable situation on a gravely soil and the varieties of the vines are of the finest. It produces a wine which is very much sought after on account of its bouquet and its generous body. It has obtained diplomas of honor and gold medals in various expositions, and it produces annually about 100 tuns of wine.

Chateau Léoville-Barton.—This is one of the celebrated vineyards of the St. Julien region, as well as its neighbor Chateau Langoa. The Chateau Langoa, however, belongs to the third-class growth, while the Chateau Leoville-Barton belongs to the second class. Its owners are the celebrated firm of Barton and Guestier. This vineyard produces annually about 90 tuns of wine, while that of Langoa produces about 130 tons.

Chateau Léoville-Poyférré.—This is one of the most celebrated of the wines of the second growth of the St. Julien region. The vineyard is owned by Mr. Edward Lawton. The vineyard produces a wine which is remarkable for its fineness, its taste, and its bouquet. The annual production of this wine is 125 tuns.

Chateau Lascombes.—This celebrated vineyard is near the Chateau Margaux, and is composed of the choicest varieties of grape vines, and produces a wine exceptionally fine and very highly esteemed. It approaches in its quality its neighbor, the grand wine of Margaux. It annual production is about 35 tuns.

Chateau Pichon-Longueville Lelande.—This vineyard belongs to the Countess Lalande. It was a part of the old possessions of the family of Pichon-Longueville, which owned it for more than 200 years. The original vineyard is now divided into two parts, one of which is the property of Baron Raoul de Pichon-Longueville, and has a production of about 80 tuns a year; and the other belonging to the Countess Longueville, which produces about 75 tuns a year. The wines of the Pichon-Longueville vineyard enjoy a great reputation and are one of the most important of the region of the Pauillac. The other part of the Chateau Pichon-Longueville Lalande is property of the Baron

Pichon-Longueville, and as before stated, has been for more than 200 years in the family. The wine produced here is of the same character as on the other half of the estate which has just been described.

Chateau Rauzan-Segla.—The owner of this vineyard is Mr. Durand Dassier. It is most favorably situated, planted with choice varieties, and cultivated with the utmost care. It is also a neighboring vineyard to the Chateau Margaux. It produces a wine which unites in a marked degree all the qualities of the grand wines of the Médoc. It has a beautiful color and marvelous boquet, and enjoys a great reputation, especially in England. The annual output of this vineyard is about 60 tuns.

Chateau Rauzan-Gassies.—This vineyard belongs to Madam E. Rigaud. It is also a neighbor of the Chateau Margaux, and occupies a beautiful situation. Its wines are fine in color and agreeable in bouquet, and are extremely well known.

Chateau Durfort-Vivens.—This vineyard is the property of Mr. G. Delor. It is a part of the fine domain situated upon the gravely soils of the canton of Margaux, planted with choice grapes, and produces a wine of marked fineness, bouquet, and exceptionally fine color. The total extent of the estate is 200 hectares. It produces from 80 to 100 tuns of wine of the second class, and from 500 to 600 tuns of a lower class.

Wines of the Third-classed Growth of the Médoc.—These wines approach very nearly in character those of the second growths. In fact, it is sometimes considered that if there should be a re-classification of the French wines, some which are now in the second growths would be transferred to the first, and some which are in the third class would be transferred to the second. Among the most famous of the third-growth wines are the following:

Chateau d'Issan.—This is the property of Mr. Gustave Roy. This estate is divided into two parts. The first, by its exceptionally fine situation and exposure, produces about 100 tuns of wine which is very fine and is classed at the height of the third growth of the Médoc; it is known as the Grand Wine of the Chateau d'Issan. The other part of the estate gives a mean production of 250 tuns of wine which is very fine and which is very much appreciated.

Chateau Cantenac-Brown.—This vineyard is owned by Mr. Armand J. Lalande. The estate embraces a superb vineyard of about 60 hectares, planted exclusively with the finest varieties of grapes, and which produces a wine which is very remarkable on account of its softness, its color, its fineness, and its bouquet. Its qualities are such as to make it easily classed among the best of the third growths of the Médoc. The wines of Cantenac-Brown are highly appreciated in the Bordeaux, as well as abroad, especially in England. In England and Holland these wines enjoy a remarkable reputation. The mean production of the Chateau Cantenac-Brown is about 140 tuns.

Chateau Desmirail.—This chateau is the property of Madam Sipiere. The extent of this vineyard is about 35 hectares, with a mean production of 70 tuns of a wine having a great distinction, although it is rather strong. It is one of the neighboring vineyards of the Margaux.

Chateau Lagrange.—This vineyard, which is in the St. Julien region, is owned by Mr. and Mrs. Muicy Louys. The wines of this vineyard are quite marked in character, having a fine bouquet, and combining all the qualities of the best wines of the Médoc, and can be compared with the growths which are most esteemed, both of the Margaux and Cantenac regions. The grand vineyard of the Chateau Lagrange, by its exceptional situation and by the care with which it is constantly attended, has conquered most justly a universal reputation. The extent is 120 hectares, and the annual production is 250 tuns of red wine and 80 tuns of white wine.

Chateau Kirwan.—This is a vineyard which belongs to The City of Bordeaux, and is among the well-known wines of the third growth. This vineyard was willed to The City of Bordeaux by Mr. Camille Godard in 1881, but it was left in such a way as not to come fully into the possession of the City until 1895. It is 38 hectares in extent, and its mean annual production is 150 tuns.

Chateau Malescot St. Exupéry.—This vineyard is the property of Mr. J. D. Lerbs. The extent of the vineyard is 125 hectares. It is beautifully situated and the wines have obtained numerous medals at various expositions. The wines of the Chateau Malescot unite with an extreme fineness a delicious bouquet.

Chateau Giscours.—This vineyard is the property of Madam E. Cruse. It is beautifully situated and planted with the finest varieties of grapes. It has an extent of 60 hectares, and produces about 100 tuns of wine annually.

Chateau Palmer.—The owners of this vineyard are Messrs. Pereire. It is situated partly in both of the communes of Cantenac and Margaux and has an extent of 86 hectares, planted with the best varieties of vines, and has lately been enlarged by the addition of a vineyard of 30 hectares. The production of the Chateau Palmer is about 200 tuns annually, of a wine fine in character and well appreciated by all who use it. It enjoys a fine reputation in the commerce of the country.

Chateau Ferriere.—This vineyard is owned by Mr. Henry Ferriere. It has a mean production of about 25 tuns of fine wine, highly appreciated both in France and in foreign countries.

Chateau La Lagune.—This vineyard is owned by Mr. Louis Seze. It has an extent of 80 hectares, of which 46 are planted with grapes of the best varieties; its mean annual production is 100 tuns. The wines of this vineyard find almost their sole market in England, but also to some extent in Holland and Germany where they are particularly well thought of. Their keeping qualities are remarkable, and it has been established that samples taken after 40 years are found to be perfectly preserved. The wines of the Chateau La Lagune have obtained numerous medals at various expositions.

Chateau Calon Segur.—The proprietors of this vineyard are Messrs. Hanappier and Gasqueton. The vineyard is 55 hectares in extent, is well situated in the upper Médoc, and has a mean harvest of 150 tuns of wine. Although classed among the third growths, its wines have a fine bouquet and are ranked among the best of the wines of the region of St. Estephe.

Chateau Marquis d'Alesme-Becker.—The proprietor of this vineyard is Mr. Arthur de Gassowski. This vineyard was well known in the 17th century. It is also a neighbor of the Chateau Margaux, and produces a wine remarkable for its fineness and bouquet and well appreciated in France and abroad. Its annual production amounts, however, to only 30 tuns.

Other Grand Wines of the Médoc.—There are other wines of the Médoc which are entitled to the name of Grand Wines, belonging to the growths of the lower classification. Those classified among the fourth growth are as follows:

Chateau Talbot.—Producing 140 tuns a year.

Chateau Beychevelle.—Containing 250 hectares and producing 150 tuns a year.

Chateau Duhart-Milon.—Embracing 50 hectares and producing 225 tuns per year.

Chateau Branaire-Ducru.—Producing a wine which by some professionals is classed as those wines of the second growth.

Chateau La Tour Carnet.—Comprising 200 hectares, of which 40 are devoted to the production of the best wines of the vineyard. Its annual output is about 100 tuns.

Chateau Marquis de Terme.—It is also of the Margaux region and has an annual yield of 90 tuns. The wines of this vineyard are highly esteemed in many parts of the United States.

Chateau Le Prieure.—Embracing 20 hectares, with an annual production of 45 tuns.

Chateau St. Pierre Bontemps.—Embracing an extent of 46 hectares, yielding an annual production of 70 tuns.

Chateau Rochet.—Producing a mean harvest of 75 tuns per year.

Chateau St. Pierre-Sevaistre.—Having a mean production of 80 tuns per year.

Chateau Pouget.—With a mean production of 30 tuns a year.

The last division of Grand Wines of the Médoc are those that belong to the fifth class of growth. The principal members of this group are as folows:

Chateau Cantemerle.—This is a large vineyard, of 380 hectares, producing an annual harvest of 200 tuns.

Chateau Mouton d'Armailhacq.—With an annual production of 250 tuns.

Chateau Batailley.—Embracing 50 hectares, with a mean annual production of 120 tuns.

Chateau Pontet-Canet.—This vineyard, the property of Madam Cruse, produces one of the best known wines, and one, perhaps, which is more imitated than any other red wine used in the country. This vineyard has an extent of 70 hectares, and the wine produced is among the very best of the fifth growth; in fact, its reputation is quite equal to some of the wines of the second or third growth. It is a wine which has a fine character, very dry, with excellent styptic properties, and of a character which fosters its constant use as a table wine whenever it can be had. The production of the wine is extremely limited as can be deduced from the small size of the vineyard, and by reason of the fact that the name is so often imitated in the United States purchasers should be sure that the wines which bear this name are the genuine article. One of the most common forms of imitating the name is to call it simple "Canet," which, of course, is quite as objectionable as to use the full name.

Chateau Dauzac.—A vineyard having a harvest of 150 tuns a year.

Chateau Belgrave.—A beautiful vineyard with an output of 60 tuns a year.

Chateau Grand-Puy-Lacoste.—With a production of 140 tuns a year.

Chateau du Tertre.—Embracing 125 hectares, with a production of from 80 to 100 tuns per year.

Chateau Lynch-Bages.—Producing a wine with a fine reputation especially in England.

Chateau Lynch-Moussas.—With an annual production of about 90 tuns.

Chateau Calvé-Croizet-Bages.—Embracing 21 hectares, with a mean annual output of 80 tuns.

Chateau Pedesclaux.—Producing annually about 30 tuns of wine particularly valued in England.

Chateau Clerc-Milon.—A very small vineyard near Chateau Lafite, and producing annually 15 tuns.

Chateau Cos-Labory.—A vineyard giving an excellent wine, and about 60 tuns a year.

Chateau Grand-Puy-Ducasse.—Producing about 120 tuns a year.

Chateau Camensac.—Producing about 100 tuns a year.

Principal Wines of the Graves Region.—The territory in which the Graves wines are grown is situated chiefly upon the left bank of the Gironde and Garonne a short distance above Bordeaux. These wines are famous for their bouquet, their fineness, and their general palatability. These wines are mostly white, although some very good red wines are produced also in the Graves region, and range in quality from an ordinary white wine to one of exceptional qualities. They are principally dry wines, but some of them contain sugar though not to such an extent as the Grand Wines of the Sauternais. The Chateau Haut-Brion producing the most costly wine of Bordeaux is situated in the Graves territory. Some of the other principal vineyards of the Graves region are the following:

Chateau Haut-Bailly.—The owner of this celebrated vineyard, perhaps one of the most noted in the Graves region, is Mr. Bellot des Minieres. The vineyard surrounds a magnificent chateau, situated on rather high land, which is exposed in an admirable manner to the rays of the sun. The culture which the grapes

Fig. 27.—Chateau Langoa (third "cru"). The property of M. Barton.
(*Courtesy of Barton and Guestier.*)

receive is of the best character, and the vineyards are well fertilized. The vines grown are the choicest varieties for producing white wine. The mean production of the vineyard of Haut-Bailly is about 80 tuns.

Chateau La Mission Haut -Brion.—This vineyard is owned by Mr. V. Coustau, and is one of the most renowned of the Graves region. It is, however, a small vineyard, comprising only about 15 hectares, and produces an average quantity of 30 tuns of wine per year remarkable for its fineness and bouquet. The prices received for this wine are equal to those of the second

14

growths of the Médoc. The wine of the Mission Haut-Brion enjoys a great reputation, both in America and England. It must not be confounded, however, as it sometimes is, with the Haut-Brion wine which is a classed growth of the Médoc.

Chateau Pape Clement.—The proprietors of this vineyard are the heirs of Mr. J. Cinto. The reputation of the Chateau Pape Clement wines extends back to an early period and is fully justified by the excellent quality of the product. It easily ranks with the Grand Wines of the Graves. This vineyard covers 43 hectares and produces about 90 tuns of wine annually.

Chateau Smith-Haut-Lafitte.—This vineyard belongs to Mr. S. Duffour-Dubergier. It is a model of wine culture and has obtained many rewards in the expositions where it has been shown. The vineyard contains 68 hectares and produces a wine which is remarkable for its individuality and fineness, and is easily classed among the first of the Grand Wines of the Graves region. The production of this vineyard amounts to about 200 tuns of wine per year. This wine must not be confounded with that of the Chateau Lafite.

Chateau Olivier.—The proprietor of this vineyard is Mr. A. Wachter. This is one of the oldest vineyards in the region and its history is known back to the 11th century. It is beautifully kept and contains only the best varieties of vines, producing annually about 90 tuns of wine. Its wines are of remarkable bouquet and are highly prized in England, Germany, Belgium and Holland.

Chateau Camponac.—This vineyard is the property of Mr. Eschenauer of Bordeaux. It is contiguous to the Chateau Pape Clement. It is a very small vineyard, containing only 8 hectares. It is, however, highly productive, having a mean output of from 20 to 25 tuns per year.

Chateau de Haut-Pomarede.—The owner of this vineyard is Dr. A. Delguel. It is one of the oldest domains of the Gironde and was noted on the maps of the 13th century. It has an extent of 75 hectares, and produces both red and white wine. Some of these wines have at times obtained a very high price; for instance, in 1887, when the old wines sold for 1,600 francs a tun.

Chateau Le Vallon.—This vineyard is owned by Mr. Charles Hanappier. It is a vineyard which is planted almost exclusively with the Cabernet variety of grape. It is a small vineyard producing only about 20 tuns a year. The chateau itself is one of the most renowned of the chateaus built by the celebrated architect Louis and is built in the gracious style of Louis XVI. It is surrounded by a fine park of 41 hectares, in addition to the vineyard.

Chateau Haut-Brion-Larrivet.—The owners of this vineyard are the sons of Theo. Conseil. This beautiful domain has an area of 125 hectares, of which 50 are devoted to the cultivation of the wine. The reputation of this vineyard has long been established. The production of the vineyard is about 80 tuns per annum.

Grand Wines of the Sauternais Country.—The regions producing the Sauterne wines are also found chiefly on the left bank of the Gargonne above the Graves region. Some of the most celebrated wines of the world are made in this neighborhood and rank in reputation and in price with the First Grand Wines of the Médoc. The region of the Sauternais is a gently rolling one, with a gravely soil which has been in cultivation since historic times. The extreme care which is used in the cultivation of the vines would be a revelation to the American farmer and gardener. Each vine apparently receives as much attention as a favorite child would get in the family circle. The principal vineyards of the Sauternais are described below.

Chateau Yquem.—This is the most celebrated of all the vineyards of the world, if we exclude Chateau Lafite and Chateau Margaux, and one or two others of the same rank. It is owned by the heirs of the Marquis B. de Lur-Saluces. The wine of the Chateau Yquem is not only the leading wine of the first grand classed growths of the Sauternais region, but is certainly, without exception, the most celebrated white wine produced in the world. The chateau itself has fallen to disuse and is not now inhabited, and is only opened on rare occasions. The Chateau Yquem wine naturally brings a price upon the markets of the world which is in harmony with its reputation. The vintages, however, as is the case in all of these grand wines, are not equally good. Some years produce wines of exceptional merit, and these wines naturally command the highest prices. Other seasons the wines are of a lower quality, and sometimes are sold simply as wine without having the name of the chateau attached to them, since they are not considered of a quality suitable to bear so illustrious a name. Some of the vintages of the Chateau Yquem have sold for as high as 10,000 francs a tun. The mean annual production of the Chateau Yquem is only 90 tuns. When one considers the total amount of Chateau Yquem wine, so-called, which is drunk in the world, it seems that it must be the result of a miracle that 90 tuns of wine could supply such an extraordinary demand. Two additional vineyards belong to the same ownership and

Fig. 28.—Interior court Chateau Yquem.   (*Photo by Author.*)

mangement as the Chateau Yquem, namely, the Chateau Fiihot and Chateau Coutet.

Chateau Rieussec.—This celebrated wine is the property of Mr. Paul Defolie, and although the wine is of the first grand growth of the Sauternais is practically unknown in the United States. The annual production of this vineyard is almost as great as that of the Chateau Yquem, namely, 80 tuns.

Chateau La Tour Blanche.—Next to the Chateau Yquem the most famous vineyard of the Sauternais is the Chateau La Tour Blanche. This is named by reason of a large tower which stands near the entrance to the grounds. The extent of this beautiful vineyard is only 40 hectares, and the production is somewhat limited, being only half that of the Chateau Yquem, namely, about 40 tuns a year.

Chateau Rabaud.—This vineyard is owned by Messrs. Drouilhet de Sigalas and Promis. It has an extent of 42 hectares and is well situated to receive the full benefit of the rays of the sun. It produces a mean annual output of 80 tuns of delicious wine, remarkable for its aroma and its exquisite bouquet. It is also classed in the first rank of the first growth.

Chateau Climens.—This vineyard is owned by Mr. Henry Gounouilhou. It is beautifully situated and produces a wine having a very fine quality and enjoying a fine reputation in the commerce of the country. It is only a small vineyard, containing 27 hectares, and produces about 50 tuns of wine a year.

Chateau Guiraud.—The owners of this vineyard are the heirs of Mr. P. Bernard. This vineyard adjoins that of the Chateau Yquem and has an extent of about 60 hectares. It produces a wine of remarkable quality, distinguished by its softness and its bouquet. It has also a fineness well established and has excellent keeping qualities. The mean production of the vineyard is 80 tuns annually.

Chateau Suduiraut.—This chateau is the property of Madam Petit de Forest. It has a content of 110 hectares and produces annually 100 tuns of fine wine.

Chateau de Rayne-Vigneau.—The owner of this vineyard is Count de Pontac. It has an area of 100 hectares, of which 60 produce white wines. The annual production is about 65 tuns.

Chateau Coutet.—This has already been mentioned as one of the vineyards belonging to the heirs of the Marquis B. de Lur-Saluces.

Clos Haut-Peyraguey.—The owner of this vineyard is Mr. E. Grillon. It is a very small vineyard, producing only 20 tuns of wine a year.

The last of the first-classed growths of the Sauternais region is the Chateau Lafaurie-Peyraguey, owned by Mr. F. Gredy. It is a small vineyard, producing only 20 tuns a year, but the wine is of a very high quality.

All of the foregoing wines belong to the first-classed growths of the Sauternais region. In addition to these, the following wines belong to the second class:

Chateau Filhot.—Producing 90 tuns a year.

Chateau de Mirat.—This vineyard, undoubtedly, should be classed among the first growths, as from the earliest historical times the product of the vineyard was of the greatest importance, and at one time it was classed among the first growths. In 1855, however, in the new classification made, by reason of the fact that the vineyard had been reduced in area, the committee assigned it to the second class. The mean production of the vineyard is 50 tuns per year. It is one of the finest of the fine wines of the region, and the vineyard contains 150,000 grape vines. The product of this vineyard is sent chiefly to Russia.

Chateau Suau.—With a mean annual production of 20 tuns.

Chateau des Rochers.—With a mean annual production of 25 tuns.

Chateau Grillon.—With a mean annual production of 25 tuns.

Cru de Malle.—With a mean annual production of 35 tuns.

Chateau Cantegril.—With a mean annual production of 30 tuns.

Chateau Lafon.—With a mean annual production of 15 tuns.

Chateau Raymond-Lafon.—With a mean annual production of 10 tuns.

Chateau d'Arche.—With a mean annual production of 10 tuns.

Chateau Liot.—With a mean annual production of 25 tuns.

Chateau Caillou.—With a mean annual production of 15 tuns.

Chateau Lamothe.—With a mean annual production of 10 tuns.

Chateau Doisy.—With a mean annual production of 10 tuns.

Chateau d'Arche-Lafaurie.—With a mean annual production of 20 tuns.

Chateau La Montagne.—This is a large vineyard, containing 160 hectares, with a mean annual production of 100 tuns.

Chateau Broustet.—With a mean production of 12 tuns.

Chateau Vedrines.—With a mean annual production of 60 tuns.

Saint Emillion Wines.—The region producing these wines covers a stretch of land northeast of Libourne in the angle formed by the Dordogne and its tributary the Isle. Its most celebrated vineyard is Chateau Ausone. This is thought to be the country seat of Ausonius, the Latin poet who praised the wines of the

vicinity in his verses. The town of Saint Emillion is of great
historic and architectural renown. Chateau Figeac is one of the
most beautiful in the country. The territory of the Saint Emillion wines embraces five communes. There are more than 60
vineyards that produce first-classed wines.

Wines of Burgundy.—There are three geographic divisions
which are recognized as the home of the Burgundy wines. These

FIG. 29.—Grape gathering at Saint-Emillion.   (*Courtesy of Barton and Guestier.*)

are Lower Burgundy, Higher Burgundy and Beaujolais. Each
of these divisions is subdivided into smaller areas. One of the
most famous in these sub-divisions is the Côte-d.Or,' belonging to
Upper Burgundy. In these regions are produced the celebrated
wines known as Chambertin, Musigny, Clos de Vougeot, Rom-
anées and Richebourg. In the Côte de Beaune are found the

wines known as Savigny, Beaune, Pommard and Volnay. The varieties of grapes which are grown in the Burgundy region are principally pinots and gamays, from which the red wines are made, and gamay blanc, aligoté and pinot blanc for the white wines. From the pinot grapes the finest of the red and white wines are made. The gamay varieties yield the ordinary wine. The different types of red and white wines also vary with each region. Thus we have the Montrachet, the Meursault, the Chablis and

FIG. 30.—Vintage in the tubs—tramping the grapes. (*Courtesy of Barton and Guestier.*)

the Pouilly, all of which are made from the pinot blanc, or white pinot. The Burgundy white wines should be drunk at a temperature a few degrees below that of the dining room, while the red wines should be kept in the dining room long enough to acquire the temperature of the room.

Differences in Quality.—The wines differ, notably among themselves, although they have the common traits of the family. The character of the soil and the exposure of the fields also play an important rôle in the quality of the wine, and I should add, also, the methods of culture, the pruning of the grapes, the fertilizers used, and especially the climatic conditions of the year. The makers of these wines recognize that the sun is an indispensable collaborator in order to secure good wine. The peasants of Burgundy call the sun the Burgundian, a respectful homage to this great brother of theirs, who alone has the ability to give value to their labors and their pains.

One of the reasons given for the superior excellence of the Burgundy wines is that they are grown near the northeast limit of grape culture in France, and on hills with their general

Fig. 31.—Making casks.   (*Courtesy of Barton and Guestier.*)

exposures to the southeast. Experience has established that fruits are usually much more fragrant when they are grown near the limits of culture. It is the same for the grape. Thus we are able to improve the bouquet of wines by extending them as far as possible to the north or in growing them upon high plateaus. The exposure of the vineyards of Burgundy to the southeast permits them to receive the greatest degree of radiation from the sun, and thus allows them to reach maturity with the largest possible saccharine richness.

FIG. 32.—Pouring the wine into casks. (*Courtesy of Barton and Guestier.*)

FIG. 33.—Head cellar man tasting a vintage wine. (*Courtesy of Barton and Guestier.*)

These natural causes must be added to the perfection of the cultural processes and the methods of vinification, which have been brought to the highest point of perfection by the continued experience of more than 2,000 years, in which the technique of the father has been handed down to the son.

ITALIAN WINES

Antiquity of the Vine.—It is difficult to determine historically whether Greece or Italy first developed the cultivation of the vine. Greek civilization is older than the Italian and historically wine was known in Greece before it was in Italy. Although Italy stretches over a large extent both in longitude and lattitude, it has the unique distinction of being a wine-producing country in all of its sub-divisions including the islands of Sicily and Sardinia. Next to France it produces the greatest amount of wine of any other country, but unlike France, it has very few vineyards which are world renowned.

Wine During the Era of Horace.—Nearly all Latin poets sing of wine, but Horace most of all. It is in his writings that we find mention of those varieties of wine which were most in vogue at the beginning of the Cæsarean epoch.

In fact, the greater part of the information which we have received respecting the wines of Italy, 2,000 years ago, are found in the writings of Horace. This poet has a good word to say for most of the wines which he was permitted to drink at the houses of the great, while the very common wine which he grew himself, "cheap Sabine," does not escape notice.

Massic was one of the finest of the Italian vintages at that time. It was used especially in celebrations or birthdays, marriages, etc., and was regarded by the poet as the mellowest wine of the time. There was on the other hand a cheap and sour wine which would compare very well with the French "Vin ordinaire." The poor were accustomed to drink this wine on holidays out of a cup of Campanian earthenware. In one case the wine had such a strong effect that the drinker fell into a deep lethargy and was only awakened from his sleep by a doctor on the following day. Falernian was one of the favorites in high-classed wines of ancient times. It is supposed now to have been produced in the neighborhood of Naples though there is some doubt respecting this point. Another wine which had a high reputation was Alban. At a certain dinner party which Maecenas attended he was greeted by the host as follows: "If you, Maecenas, prefer Alban or Falernian to those on the table, we have both." Drinking was carried on to great excess at some of the Roman feasts. Vibidius said to Balatro, a fellow guest, "Unless we drink ruinously, we shall die unavenged:" so they emptied their large cups which were whole bowls of wine; all of the guests followed except two on the lowest couch; they spared the flagons. Apparently in those days, as at the present many of the guests would turn their goblets down at dinner. Chian and Lesbian were also varieties of wine much in repute.

Horace pays quite a tribute to incipient drunkenness. He says: "What is there tipsiness does not effect? It unlocks hidden secrets, it bids hopes be realized, it impels the coward to the field, it lifts the load from off the anxious mind, it teaches new accomplishments! Whom have not flowing cups made eloquent? Whom have they not made free in pinching penury?

Horace shared the illusions which many people have shared since his days in believing that alcohol produced eloquence and elegance of speech, but in point of fact, it takes off the brakes and allows the vehicle to run down hill.

Horace called a wine which was grown on his famous Sabine farm (only two acres) "cheap Sabine," but he offers it to Maecenas when he was his guest. Cacuban, Calenian, Falernian and Formian were all superior vintages of Italian wine. These Maecenas undoubtedly had at home. Horace assured Maecenas that his very simple common wine was not even flavored with any of the higher growths. Horace indeed seemed opposed to all adulterations and blending to make a common wine pass for a higher grade. In his ode praising the charms of Glycera on her altar he offers a bowl of last year's unmixed wine.

A wine of high alcoholic content grown in Egypt was used in Rome on certain state occasions. It was called Mareotic. It evidently tended to produce delerium tremens. The wild delusions produced were transformed into real terrors in the case of Cleopatra pursued by Cæsars' destroyers.

Horace was opposed evidently to price fixing, at least in so far as wine is concerned. His guests invited to the feasts might be as learned as they liked but they should here neglect to say how much one should pay for his Chian wine, nor set up a food administrator. He tells Maecenas that water drinkers cannot write poems that live according to the theory of Cratinus (an old and dissolute Greek dramatist). As soon as Bacchus entered crazy poets, the dear muses generally had a bad breath in the morning. Homer constantly praised wine and was doubtless a wine bibber. To the sober be given the dry business of the Forum. They shall not write poetry—but drinking wine will not create a poet. Here at least, Horace strikes a logical form. We are loathe to believe that any of his delightful poems save perhaps those in praise of Bacchus were inspired in any way by reason of the wine he drank.

Character of Modern Italian Wines.—Generally Italian wines resemble more nearly the common wines of Spain or California than they do those of France. They are characterized by rather a high alcoholic content, due to the climate in which they are grown, which resembles that of California and tends to produce a grape rich in sugar. Both red and white common wines are produced, and are largely consumed in the localities in which they are grown. From the edition of L' Italia Vinicola of the 10th of March, 1918, I have taken a number of the names of these common wines grown in several of the localities of Italy, illustrating the general production of the whole country and also giving the names under which these typical wines find their way in the market.

GENOVA
Calabria
Avellino
Pachino
Castellamare

PIEMONTE
Alba Barbera
Barolo
Barbaresco nuovo
Dogliono

TORINO
Chieri da bottiglia
Pecetto di Torino
Caluso rossi

ALESSANDRIA
Tortona rosso
Sezzadio rosso

ASTIGIANO
Asti barbera
Asti grignoli
Asti moscato
Asti communi

MONFERRATO
Alta villa
Conzano
Frasinello
Olivola
San Salvatore

NOVARA
Gattinara
Sizzano
Fara
Borgomanero

LOMBARDIA
Cellatica
Salo

MONTORA
Montonara
Viadana
Gazzuola rosso

VERONA
Valpolicella
Verona rossi
Soave

VICENZA
Longara
Costazza

PADOVA
Bagnoli
Ponte di Brenta
Este

VENEZIA

Venezia Corbini

Venezia Rabosi

MODENA

Modena rossi

Carpi

Mirandola

BOLOGNA

Bologna communi

Bologna fini

RAVENNA

Luogo

Faenza

Bagnacavallo

UMBRIA

Montifales bianchi

Montifales rossi

TOSCANA

Pisa piano

Agnano

Crespina

Cecina

Lucca vini corrente

Pescia

Firenze fini

Castelfiorentino

Rufina de pasto

Arezzo rossi

Capezzine

Bibbiena

Radda Chianti

Castelnuova

Bolsena

ADRIATICA

Castellamare

Chieti

Mirabello Mosti

Cerignola

Barletta

Canosa

Ruvo

Brindisi

Gallipoli

Taronto

Martinafranca

MEDITERANEA

Rionovo

Barile

Genzana

Casino

Formia

NAPOLI

Napoli rossi

Isola di Capri

Serrara Fontona

Calore

Salerno

Vallo

SICILIA

Nicolosi Pedara

Messina

Marsala bianchi

Alcomo

Palermo

Catania

SARDEGNA

Cagliari Rossi

Sorso

Pirri

Bosa rossi

Oristano bianchi

WINE DISTRICTS OF ITALY

Piedmont.—The culture of the vine is practised in all parts of Italy and its adjacent insular dependencies. In the northwest Piedmont is noted for its wines. During the past half century it has claimed attention in the Italian markets. Before that time its vogue was wholly local. The chief regions of Piedmont from the viticultural point of view are Asti and Alexandria. Both still and sparkling wines are made of muscat grapes in many parts of Piedmont. The red wines of Asti have a vogue in foreign countries. They are naturally less alcoholic than wines produced farther south. In the province of Alexandria much of the harvest of grapes is sold for direct consumption. Brandy is also made to some extent here. In earlier days much wine was exported from Piedmont chiefly into France. It was widely believed at that period that much of this wine assumed the names of the celebrated classed wines of that country. This condition no longer obtains. Italian Vermouth, in the days when no restrictions were placed upon it, was largely fabricated in Piedmont

Lombardy.—To the east of Piedmont is found the vineyards of Lombardy. Before the days of irrigation, viticulture was largely practised in this region. Vines do not take kindly to irrigation, and for this reason Lombardy is not so prominent in wine productions as formerly. Nearly all the product of Lombardy consists of light red table wines and have marked tannic and acid qualities. These wines have long been used in Switzerland where they are much liked. The principal market of Lombardy wines is at Milan.

Venice.—The province of Venetia is of some importance in Italian viticulture. The most extensive vineyards are near Verona. The principal wine merchants are found here. Both red and white table wines are produced. The alcoholic content varies from 7 to 14 percent. The most alcoholic are those of Valpolicella and Prosecco.

Liguria.—The province of Liguria has for its principal city Genoa. Little of the surface in this region is suitable for vine culture. It is too hilly and precipitous. Nevertheless Genoa has quite a commerce in wines, but chiefly that of Sicily and Sardinia. The alcoholic content of the Liguria wines varies from 10 to 12 percent. The acidity is relatively low.

Emillia.—To the east of Liguria and south of Lombardy lies Emillia one of the largest in extent of the divisions of Italy. In it is found the valley of the Po and it extends eastward to Bologna. The wines have a low alcoholic content, medium acidity and are both red and white. The principal wine markets are found at Bologna. It is among the largest producing provinces of the country.

Umbria and the Marches.—This region borders the Adriatic and lies southeast of Emillia. It was well known as a wine product region by Ancient Rome. The musts in ancient times were concentrated and mixed with the wines to conserve them and make them sweet. The straight wines of this region are very low in alcohol varying from 7 to 10 percent. The quantity produced is above that of Emillia.

Tuscany.—West of Umbria and bordering the Mediterranean lies Tuscany. This province is famous for its light colored table wines. Many of these wines are made of mixtures of different kinds of grapes. Two of these varieties are the same as used in the Bordeaux vineyards viz., the Malbec and Cabernet. The wines thus produced have some of the properties of their more famous cousins of the Gironde. The vines are planted principally on the hill sides rather than on the plains. The phylloxera pest which laid waste the vineyards of France did not spare those of Tuscany. From 1879 to 1888 many vineyards were attacked. Vigorous treatment with bisulphide of carbon and grafting on the roots of American vines which were immune to this pest prevented disastrous losses. The most abundant production of Tuscan wines is found near Florence.

Latium.—The region of Italy embracing Rome and extending southeastward almost to Naples has from earliest historical periods produced wines of some importance. The alcoholic content of these wines for the most part is high ranging from 10 to 13 percent. In order to preserve the wines during the hot season numerous subterannean cellars are provided where a cooler and more favorable environment for the wines of finer quality is found.

Southern Adriatic.—This extensive region is divided into three sub-sections forming the Abruzzia and the Campobosso. The Abruzzia is distinctly a hilly territory forming a part of the Apennines which reach an altitude in some places of 8,000 feet. Between these two provinces is found the Pouilles which is partly hilly and partly level. The vines are cultivated from the shores of the Adriatic to an altitude of about 3,000 feet.

In these regions where the rainfall is meager irrigation is practised to some extent. Ancient aqueducts show that the practice of irrigation has been followed for more than 2,000 years.

Southern Mediterranean.—On the west coast the area extending from Naples to the extreme tip of the peninsula is called the Southern Mediterranean. The surface of this region is hilly and mountainous, the southern Appennines rising to an altitude of about 3,000 feet. Vines are grown almost to the summit of these mountains.

Some of the most famous wines of the Augustan age of Rome were produced in this region. Especially that famous product Falernian is said to have come from this part of Italy. The wines made in the vicinity of Vesuvius are still deemed some of the best found in the country. Lacrimae Christi, one of the most famous wines of modern Italy is thought to be the very wine known as Falernian at the time of Horace. This wine has a high alcoholic content, at times above 14 percent. This province produces more wine than any other region of Italy.

Sicily.—Many excellent wines are grown in Sicily. Both near the main land at Messina and Milazzo and at the other end of the Island at Marsala and Massera are found wines of excellent character. Marsala wine is perhaps better known to the commercial world than any other Italian wine except Lacrimae Christi and Asti. In the vicinity of Etna red wines are produced in great quantities. Cream of tartar is made in large quantities in Sicily and exported for the manufacture of baking powders and other purposes. Considerable quantities of wine are distilled in Sicily producing a brandy of fair quality. All the Sicilian wines have a very high alcoholic content. Following are the data for the most important kinds.

Messina......	15.00 percent
Marsala................................:.....	15.25 percent
Syracuse...................................	16.00 percent
Muscat....................................	15.30 percent
Albanello.................................	17.30 percent

The island of Sicily produces more wine than any of the regions of the main land.

Sardinia.—Sardinia and Sicily have about the same area and are both well suited to viticulture. The wines like those of Sicily have generally a high alcoholic content indicating a percentage of about 30 of sugar in the grape. The manufacture of brandy is also carried on to a considerable extent in Sardinia. The special wines of the Island are not well known in foreign countries. The most important of the special wines are: Vernaccia, Muscat, Malvoisie, Nasco, Canonao, Monica and Giro.

GERMAN WINES

Mosel Wines.—In the United States the wines of Germany are all included (but not correctly) under the general name of Rhine wines. With almost as much propriety they might be designated as Mosel wines, of which Berncastler Doctor is chief.

Rhine Wines.—The vineyards on the Rhine below Coblenz belong to the lower Rhineland region. As Cologne is approached

15

the vineyards become fewer in number and finally disappear altogether. In the neighborhood of Linz, on the lower Rhine, a very beautifully colored dark red wine is produced, and in the neighborhood of Remagen a lighter red wine is found Above Remagen, debouching in the Rhine valley, the well-known valley of the Ahr is found, and in this valley a large number of wines are produced, some of which are red. The wine region of the Ahr Valley ends at the village of Hönningen. Cologne was early known as a wine market of considerable magnitude.

The Valley of the Rhine and the Nahe.—The vineyards of the Rhine region are not confined to the banks of the stream alone. In the valleys which adjoin the Rhine, the vineyards spread out to a great extent, and in these valleys of the Rhine and of the Nahe some excellent wines are produced.[1] On the Rhine itself, near Bingen, are found some of the best vineyards. Kreuznach, in the Nahe Valley, is one of the best wine-producing areas of Germany. It is not so much known for any great single vineyards as it is for its products as a whole.

In the region of Caub and Bacharach the cultivation of the vine has expanded to a very large extent. At Bingen the Nahe empties into the Rhine, and following up the Nahe you come to Kreuznach, which is the center of a great wine industry. Bingen is better known to the people of the United States by reason of the famous poem entitled "Bingen on the Rhine," but to the wine grower Bingen is a most interesting place because of its proximity to some of the great vineyards which lie just across the river from the town and because of its being at the point of the confluence of the Nahe. As you come up the Rhine from Bingen you come into the Rheingau, on the right bank of the river, the most famous of the regions producing Rhine wines. Around Hochheim also are well-known vineyards, and the term "Hock" as applied to Rhine wines probably arose from the name of the village of Hochheim. Eltville is another locality known for its vineyards; so is Eberbach. In the Rheingau is found the Riesling variety of vine in all of its excellence and vigor.

The real Rheingau begins at Walluf. Nearby is Hattenheim, near which the Steinberg wines are produced. Near Assmann-

[1] The wine region known as the Nahe lies on the left bank of the Rhine between Bingen and Köln (Cologne).

shausen the best known of the German red wines are produced, of a Burgundy type. Wiesbaden is the largest city in the Rheingau, and Bieberich is not far away. Mainz, on the opposite side of the river, is the great commercial center of the wine industry of this region.

Rheinhesse.—In the Rheinhesse are Nackenheim, Nierstein, Oppenheim, Worms, Laubenheim, Bodenheim, and Ingelheim.

The Rheinpfalz.—If one crosses the Rhine at Mannheim he reaches Ludwigshafen, in the Bavarian Rheinpfalz. This is the region of Southern Germany, in the broad valley of the upper Rhine, extending over the area embraced by the towns of Odenwald and Schwarzwald on one side and Haardtgebirge and Vogesen on the other side. The Haardtz mountains bound this region upon the west. It is a wonderful extent of landscape of great beauty, where the land is level and fertile. The towns are numerous and prosperous and the vineyards some of the best known. Among the principal towns noted for their wine in the Rheinpfalz are Forst, Deidesheim, Neustadt, Wachenheim, Dürkheim and Landau. The wines of the Rheinpfalz are marketed chiefly from Ludwigshafen, Mannheim and Frankfurt am Main.

Wines of the Neckar and Main Regions.—On the right bank of the Rhine opposite the Rheinpfalz are the wine regions of the Neckar and Tauberland. A famous town of this region is Heidelberg, noted equally for its enormous wine tun and for its university. The culture of the vine here is extremely ancient. As early as the year 628 A.D. are historical accounts of vine culture. At Heidelberg the Neckar escapes from the hills and comes into a plain of considerable magnitude. Visitors to Heidelberg never fail to see the wonderful ruins of the castle, where the great tun is still preserved.

The part of the Neckar valley which lies in Baden has attained the greatest name for wine production. The varieties of grapes known as Gutedel and Sylvaner, as well as Riesling, are cultivated. There is also made here a beautiful red wine, which is manufactured out of a mixture of white and red grapes. In the nearby Tauber valley of Baden, especially in the neighborhood of Werthheim, the culture of the grape has attained wide proportions. So-called white wine is the especial product of the Tauber valley. On the banks of the Main, between Werthheim and Freudenberg, white wine is chiefly produced, although Freudenberg red wine is highly valued. The chief towns where commerce in wine is carried on in these regions are Tauberbischofsheim and Werthheim.

Where the river Main runs its zigzag course through the hill country, especially between Bamberg and Aschaffenburg, a wine production of excellent quality has long endured. The center of the culture of the grape is the city of Würzburg. Other important centers of the industry are Buchbrunn and Saaleck. The Stein wine and the Flaschen wine are produced in the neighborhood of Würzburg. In late years the producers of wine in this region have been turning more and more from the white to the red wine type. Especially around the villages of Bürgstadt, Miltenberg and Klingenberg.

Wines of Württemberg.—The chief commercial cities of this region handling wine are Stuttgart, Lichtenberg, Heilbronn and Weinsberg. Considerable wine is also produced in Baden, and especially about the towns of Müllheim, Badenweiler, Bühlerthal, Offenburg and Freiburg.

Large quantities of wine are also made in Alsace and Lorraine, though we usually think of these localities more as causes of war than as producers of wine. In Alsace the Riesling grape gives a wine of fine bouquet, and the Tokay grape makes a sweet, strong wine, richer in alcohol than the Riesling products. Metz is the chief city in this locality. In Lorraine red wines are chiefly made. They have a fine aroma and belong to the Burgundy type. In fact, they are not far from the Burgundy region of France. Some sparkling wines are also made in Lorraine.[1] Fine varieties of red wine of the claret type are also produced there. Sparkling wines are not made in very large quantities in Germany, though the production in recent years has increased. They are made of the same grapes that produce the still wines. The production of sparkling wine has spread, especially in Alsace and Lorraine, in the last half century. The principal centers of production and commerce in these wines are Coblenz, Frankfurt am Main. Hockheim am Main, Kreuznach, Mainz, Neustadt in the Haardt, Rüdesheim on the Rhine, Schierstein on the Rhine, and Würzburg.

[1] The wines of Alsace and Lorraine are now to be credited to France.

Character and Vogue of German Wines.—In regard to vogue and excellence many of the German wines hold rank almost equal to those of France. The German wines are associated intimately with the two noted rivers, the Rhine and the Mosel. The valleys which lie adjacent to these rivers are also devoted largely to wine culture. The most famous vineyards of Germany are those on the hills of the Rhine and the Mosel. These rivers give their names to this class of wines. Unlike the majority of French wines, the German wines are mostly of the so-called white character. They are either made from grapes of green or yellow tint or from the expressed juice of red grapes. Red wines are not common in the German vineyards, nor are many of the grape wines of Germany of a red color. The banks of the Rhine from Frankfort to Cologne present the German vineyards at their best. Every available foot of the soil has been brought under cultivation. The hillsides are protected by walls so that the cultivation of the vine can be continued as far toward the top as the climate permits. The very highest of the hills reach beyond the zone of cultivation, by reason of the inability of the grapes to properly mature at the high altitudes. The level lands near Heidelberg and around also have vineyards which produce most excellent wines, some of them of very high character.

Hock.—The Rhine and Mosel wines, which usually are called Hock, are of a light amber tint varying in depth, of a moderate alcoholic content, from 8 to 11 percent, and carry a delightful aroma which is quite characteristic of this class of wines. Some of them are of rather high acidity, but others are less marked in this characteristic. The methods of harvest and manufacture of the wine are much the same as the general methods already described. Each vineyard, however, claims some peculiar manipulation which is considered to be particularly related to the character of the wine produced. There is one peculiarity of the great vineyards, however, which cannot be denied, namely, the particular character of the yeast which conditions the fermentation. There is no doubt of the fact that the yeasts which are native to these vineyards, and which have been perpetuated for hundreds of years, have acquired a distinct biological character which is of high value in determining the character of the wine. It is not so much the chemical constituents of the grapes as it is the environment or nature of the fermentation to which the peculiar character of the wine is due. In other words, if the grapes which grow in one of these grape vineyards were carried to an ordinary vineyard and subjected to the conditions of fermentation which there obtain, the character of the wine produced would be far inferior to that which is the result of the conditions which obtain in the vineyard itself. This fact is applicable to all the great vineyards that make wines which are famous in history and commerce.

PALATINATE OR RHENISH BAVARIA

The viticultural districts of the Palatinate are situated at the foot of the Haardt Mountains, which form the western limit of the Rhine Valley in Rhenish Bavaria (left bank of the Rhine).

The Wine of the Palatinate is reputed for its good quality, high degree of alcoholic strength, maturity and body.

RHENISH HESSE

On the left bank of the Rhine, between Worms (Liebfraumilch) and Bingen (opposite Rudesheim). The Wines of Rhenish Hesse are soft in character and usually free from acidity.

WINES OF THE RHEINGAU

On the right bank of the Rhine. The Rheingau is enclosed between the Tannus Mountains on the north and the River Rhine on the south; it forms a bay in the mountain, twice as long as broad and filled with undulating hillocks.

WINES OF FRANCONIA

Vineyards surrounding the Bavarian town Würzburg on the upper Main. Full bodied wines but only matured in good years. Filled in boxbeutel. *Steinwein.* Full bodied, very pronounced character.

NEW GERMAN WINE LAW

Under the old law it was a practice even with some of the larger houses to ship under well-known brands large quantities of wine which not only did not come from the district, the name of which was chosen, but was quite different in character, and needless to say, inferior in quality.

The enforcement of the present law (passed 1909) is guaranteed by holding the proprietor responsible for all violations, and all intentional violation is punishable by imprisonment.

HUNGARIAN WINES

Tokay Wine.—The term "Tokay Wine" is associated by the connoisseur with a wine of extreme fineness of character and of superior excellence. It is, however, not really a wine in the ordinary sense of that term, since the real Tokay is more of a liqueur, in this sense resembling the famous Chateau Yquem.

There are two principal kinds of Tokay wine, one an extremely concentrated variety, called Tokay Essence, and the other a much less concentrated wine, called Tokay Ausbruch. The latter is the wine in general use, the Tokay Essence being generally used to mix with Tokay Ausbruch.

The analyses of the two forms of Tokay wine, taken from the work of Koenig, follow:

Tokay Essence

Sp. Gr.	1.1244
Alcohol by volume	8.33 percent
Extract	31.24 percent
Total acids	0.60 percent
Vol. acids	0.10 percent
Sugars	25.61 percent
Ash	0.36 percent

Tokay Ausbruch

Sp. Gr.	1.0354
Alcohol by volume	14.09 percent
Extract	12.72 percent
Total acids	0.60 percent
Vol. acids	0.10 percent
Sugar	9.01 percent
Ash	0.27 percent

It is seen that the Essence is only a partly fermented wine, the greater part of the sugar remaining still in its natural state.

If the whole of the sugar in the Tokay Essence were fermented the Essence would have almost, if not quite, 20 percent of alcohol. In the Tokay Ausbruch it is noticed that the fermentation has gone much further than in the Tokay Essence—the alcohol has been increased and the sugar diminished.

The grapes which are used in the manufacture of Tokay wine are chiefly produced by the variety known as Hungarian Blue.

In another sense the manufacture of Tokay wine resembles that of the Château Yquem, since the grapes are allowed to remain very late on the vines—until they are dry and somewhat shriveled. The vintage usually takes place about the end of October.

The ripe, somewhat shriveled grapes, in the process of manufacture, are placed upon a grooved table. These grapes, before being placed on the table are mashed, and the juice that runs naturally from them is used for making the Tokay Essence wine. Naturally, the quantity thus produced is very small, and it is considered of the highest value. After all the juice has run out that is possible, the grapes are again trodden in the vat with the naked feet and the small portion of the essence which has been extracted as above described is added to the must and the whole allowed to ferment. When the fermentation is completed the wine is pressed from the grapes and forms the famous Tokay Wine, or Tokay Ausbruch. During the fermentation, which takes a number of days, according to the temperature, the must is kept constantly stirred and the skums which rise to the top are removed. In separating the wine from the pomace too much pressure is not to be used; in fact, the most of the wine runs away under the force of gravity alone.

Tokay wine is never as brilliant as the ordinary red wines of commerce. On account of the large quantity of extract which it contains it remains more or less cloudy, even when perfectly finished. The presence of this extraneous matter doubtless adds much to the wonderful bouquet and aroma which this wine possesses. It has a soft and oily taste, with a very astringent effect, due to the presence of large quantities of tannin extracted from the skins and seeds.

So-called Tokay wines are sometimes made in other countries, and especially in the United States; but they have none of the properties of the genuine article and are not even a type of that article. There is not much uniformity in the composition of the so-called domestic Tokay wines. They usually have from 16 to 20 percent of alcohol, from 8 to 12 percent of solids, from 6 to 10 percent of invert sugar—that is, sugar of the grape, and from 0.2 to 0.3 percent of ash. The quantity of acid ranges from four to six parts per thousand. It is undoubtedly a violation of the proprieties to call a domestic wine "Tokay," either directly or by implication.

SPANISH WINES

Sherry—Early History.—Since the 16th century sherry has been a well-known wine in Spain and also in England. Shakespeare frequently refers to the wine, as for instance in Falstaff where he speaks of "good sherris sack."

The center of the sherry district is the town Jerez de la Frontera. In 1873 the vines in this region which were producing sherry were valued at $2,000 an acre, and the new sherry wine was worth at the time of its making as much as four shillings (one dollar) per gallon. In the year 1873 there was imported into England more than 8,000,000 gallons of sherry wine.

The town of Jerez has about 60,000 inhabitants and is ranked as one of the richest in Spain, and its wealth was due largely to the selling of its wine to England. Since the above date the wine industry of Jerez has gradually declined, due without doubt particularly to the adulteration of the article, which spoiled its taste and diminished the demand for it as well as supplanted the real with the imitation article. The average price of sherry has also declined with its diminishing demand and with the increasing adulteration. To such an extent has this decline taken place that the sherries are worth but little more than half at the present time the price they brought 40 years ago.

The Bodegas of Jerez.—The wine cellars of Jerez are known as bodegas, but they are not always cellars. In Jerez the bodegas are usually a building of one story covered with tiles and supported on brick walls. The bodegas are kept scrupulously clean, and everything is whitewashed periodically except the casks. The floor is kept covered with clean sand. There is always a very large storage space between the casks and the roof, affording an abundant supply of air. One thing, however, in a sherry cellar is never cleaned and that is the casks themselves, for these retain the organisms which are continued from cask to cask so that the character of the wine remains constantly the same. The casks are usually never completely emptied, but refilled with new vintages and thus a continuity of character is maintained throughout many years. Especially is the dust which forms on the outside of the cask respected as a matter to be venerated, just as old wine bottles are valued largely by the amount of grime and dirt that they carry.

Although great care is taken, as already described, to preserve the dust on the outside and the germs in the interior of the cask, yet it is perfectly certain that great changes have taken place in the character of sherry wines since the Shakespearean times. In fact, changes modifying considerably the character of the wine have been made within the last 100 years.

Dry Sherry.—In point of fact the only real sherry wine is that which is dry. The sweetened sherries are liqueurs rather than wines, and the real connoisseur who understands the character of a wine never asks for anything except the dry.

At the present time the drinkers of so-called sherry throughout the world get a very different article from that which is given them directly from the casks in Jerez.' I recall a visit I once made to the bodegas of Jerez many years ago and the indescribable delicacy and aroma of the dry wines served directly from the casks.

Mixing and Blending.—Unfortunately for connoisseurs in different parts of the world, the true sherry dry wines that are found in the casks at Jerez are not sent into commerce. The commercial sherries are usually mixtures of pure natural wines which are well matured, and often of quite distinct types. It is seldom even an unblended wine is shipped, although there are sometimes connoisseurs in different parts of the world who have acquired the taste for a particular kind of wine and give orders for it.

Object of Blending.—The object of blending the wines is, of course, the same as that for the blending of brandies and cognac—

to produce large quantities from year to year of a wine having practically the same character, so that the customers who drink that brand of wine may not be calling from time to time on the dealer to explain why one vintage has one taste and another, another. This, of course, is an unfortunate condition of the consumer's mind, since he ought to know, and perhaps does know, that every year produces a wine of a different character, even if it bear the same name, and he should expect this. It requires a great deal of skill on the part of the blender to make out of different kinds of wines produced from year to year, this standard blend and, of course, he does not always succeed.

Types of Sherry.—It is hard to say what the types of unmixed or unblended sherry are because so few of them ever reach the consumer.

In general, sherries in their natural or unblended state may be divided into two great groups, namely, wines which develop very markedly the characteristic aroma and taste of the sherry district, and second, wines which for some reason fail to develop this characteristic aroma and taste, but remain only good wholesome beverages affording a fine body for blending purposes. Naturally, the wines of the fragrant and aromatic taste are those which are developed particularly in the old casks. Just what the organism is which produces the characteristic flavor and aroma is not known, but it is supposed to be a species of fungus and is called by the Spaniards "flor." It looks very much like the ordinary mycoderma vini which develops in wineries. The growth of this flor is encouraged and regulated just as our mothers used to transmit the mother of vinegar from barrel to barrel. It may be said that the peculiar sherry taste which distinguishes this wine from all others is due solely to the development and growth of this fungus. The fungus does not grow in wines which are too strong in alcohol and so it is not expected to appear in those of a very high alcoholic content.

The two groups of wine are classed, first, as the Fino Type, a wine which in its early years has been matured with a film of the flor on its surface and possesses, therefore, a pronounced ethereal taste and bouquet. The second type is the Oloroso Type, a wine which from the completion of its first fermentation has produced so much alcohol that the flor cannot grow upon it and it is, therefore, devoid of that peculiar ethereal taste and aroma which is produced by the flor.

Divisions of Fino Sherries.—Several distinct varieties of wines are found included in the fino type. Manzanillo is a sherry belonging to the fino class. This name is given to the wine grown in the neighborhood of San Lucar de Barrameda, a town which is situated about 18 miles northwest from Jerez. This wine is somewhat light in alcoholic content, but it possesses very characteristic flavors which designate it as a sherry wine of high degree. Manzanillo is not quite clear. Its literal meaning is chamomile. Another meaning is a small apple, but neither of these meanings seems to have any bearing on its application to the wine. There is a small town about 40 miles from San Lucar, called Manzanillo, and this probably may have had something to do with the selection of the term.

Amontillado.—Everyone who is familiar at all with sherry knows the best quality of sherry wine is sold under the name of Amontillado. It is also a sherry wine of the fino group, and is a very distinct type of that group. A wine, however, is not entitled to be called Amontillado until it has acquired a considerable degree of age and has become transformed in its character by the secondary changes due to ageing. It is, therefore, characteristic of a wine which has entered well upon a distinct change due to ageing. The name Amontillado means "from Montilla" the town near which the wine is made.

Oloroso Sherries.—This type of wine is produced when at the beginning of the development the alcoholic content of the wine is so high as to prevent the growth of the fungus or flor which exists in the fino type of sherries. The oloroso wines, therefore, present those characteristics of aroma and flavor by which we easily determine whether or not a wine is of the sherry type.

Fortification.—It is seen from the above that the true sherry wines, those that are produced in the vintage growths and which belong to one of the two types which have been described, are really natural wines and have never had their alcoholic strength increased by fortification of any kind. The characteristic flavors of these wines are due to the development under the cellar management to which they are subjected and especially to the solera system peculiar to the district.

Fortified Types of Sherry.—Other wines made in the district and which are also called sherry wines, are fortified either by the addition of sugar or alcohol, or both. These wines, however, are not supposed to be suitable for drinking purposes. They are exported or used at home for mixing or compounding purposes only. The sherry wines, so called, are often found imported into this country in which there is often present from 20 to 35 percent of saccharine matter. In other words, they are not wines at all, but very concentrated liqueurs.

Of the types which are made in Jerez or vicinity there are three which are quite well known by the names Vino de Color, Pedro Ximenes, and Paraxete. These are the chief wines used in the compounding of commercial sherries. Muscat wines are also used in the sherry district, among the most familiar of which is the Rota Tent, which formerly was shipped largely to England to be used for communion purposes. This wine, or any of the other kinds made in the district, does not enter at all into the composition of sherry.

Vino de Color.—This wine is one of the oldest in Spain and is, undoubtedly, the wine mentioned by Shakespeare as "sherris sack." The wine is very closely related to a wine made at Malaga and shipped under the name of Malaga de Color, or Brown Malaga. These wines are not really pure wines because they are made by mixing together a portion of the fermented grape juice with another portion of the concentrated grape juice reduced to the condition of a thick syrup. Very old Vino de Color commands a high price because but little of it goes a great way in flavoring the sherries of commerce. It is a product which enters into the composition of almost all the sherries of commerce and to which they owe chiefly their color. Although with increasing age fine vintage sherries would become dark in color, in all but exceedingly high-priced wines the Vino de Color is the color agent used. Of recent years this brand of blended sherry is shipped in large quantities to whisky-producing countries to be used in the compounding of whiskies. The wine itself is practically undrinkable in its natural state and strictly cannot be regarded as a product to which the term wine unmodified could be properly applied.

Pedro Ximenes.—This is a wine made in the sherry district which is derived from a grape which bears this name. The grapes are partially dried on mats in the sun before they are crushed. A very concentrated must is thus obtained, which ferments slowly and produces a product which is so sweet that it is practically a syrup. It is, however, very fragrant by reason of the concentrated essences of the true grape which it contains. It is a wine which is not much used for drinking directly, except in the most minute quantities, but it has a high value as a compounding agent because of the fruity taste it gives to the mixture.

Paxarete.—This is a variety of the Pedro Ximenes and is made from the same grape. This name is derived from a small town where it is manufactured. The Pedro Ximenes vineyards were almost destroyed by phylloxera a third of a century ago and the old wines of this class are, therefore, now almost unknown. Like Pedro Ximenes it was not extensively used as a beverage itself, but was valued for its compounding qualities.

Two sweet wines made in the Jerez neighborhood, by stopping the fermentation by the addition of alcohol, at a certain stage, are also well-known brands. They are Dolce Apagoda and Mistela.

The Solera System.—The method by which the young wines described above are brought through a course of training, it may be said, to convert them into the real sherries is a most interesting chapter in the history of sherry. Almost all wines with which we are commonly acquainted, such as the red and white wines of commerce in all countries, are matured in casks by the refining and racking processes which depend upon natural sedimentation.

The process in ordinary wines lasts from two to five years, at which period the wines are said to be ripe and ready for bottling. During this time the shrinkage in the casks is made good by the addition of other wines of the same kind, so that the casks are kept practically filled. From 4 to 12 rackings are sometimes practised in the complete refining and racking of ordinary wines.

In the process of developing a sherry wine an entirely different method is pursued. This process is known in Spain as the Solera System, and is a method peculiar to Jerez and its neighboring districts of San Lucar and Montilla.

Significance of Name.—The name Solera is given in Spain to a series of casks of wines in process of maturation, so arranged as to provide for progressive fractional blending, thus insuring the continued production of a wine of even type in spite of differences which may occur between the produce of successive vintages. In the case of other wines the process of mixing is practised at the times the wines are completed and just before they are sent to market. In the making of sherry wine, however, the mixing begins at the first and is continued progressively until the wine is finished. It is evident, therefore, that a wine subjected to the Solera process cannot be of any one vintage, though the final product may be all from one vineyard. In this way there is eliminated the distinctions which occur in wines of different vintages, so that the product when finally completed is practically of a uniform character, in that the peculiarities of many seasons of vintages are blended into one expression. Inasmuch as there is a regular recurrence of peculiarities in different seasons, the final product must be practically uniform.

A Typical Solera.—A typical solera may be represented, for sake of illustration, by a solera containing 50 casks, divided into five groups. Each group, therefore, will contain 10 casks, and each group of casks may represent one year. In this case the wine would be matured in five years, which is, of course, too short a time for making an Amontillado. It may be further assumed that the solera has been in working order for a number of years. At the time of the first filling with new wine, therefore, each of the first series of casks, having had 25 gallons taken out of them, will receive 25 gallons of new wine. After one year 25 gallons of the old wine will be taken out of the fifth group, which will be filled from the fourth, the fourth from the third, the third from the second, and the second from the first. The first group will then be ready to receive 25 gallons more of new wine, and this process goes on indefinitely. In point of fact, the usual custom is to withdraw from the oldest group of casks twice during the year, so that half of the contents of the old wine will be drawn off each year. Enough of new wine is kept in casks to make the second filling in the first group. Thus, a solera of 50 casks will give 50 gallons from each one each year of finished wine, or 250 gallons a year.

WINES OF PORTUGAL

Character and Name.—A wine almost as celebrated as sherry is produced in Portugal near the city of Oporto, from which

the name "Port" is derived. Under the agreement of different
nations to respect each other's trade marks and copyrights and
under the regulations of the food law which forbid any false
information respecting origin the term "Port," when used without
qualification is applied only to the wines coming from Portugal
and from that part of it specified above. The region of Portugal
in which the wine is produced is comprised in the area of the
Douro River shown in map. The illustration shows the center of
the district where the finest wines are produced. I am indebted

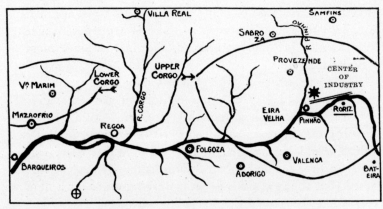

Fig. 34.—Map showing areas producing Port wine. (*From Challoner.*)

to Challoner[1] for many interesting facts connected with the production of Port wine. The name "O Porto" in Portuguese
means "The Gate." The name is attributed to the landing
of the Gauls in Portugal many centuries ago. The port of
Oporto is the mouth of the beautiful river Douro, along the banks
of which the wonderful vineyards which characterize this part
of the country are found. Not only grapes, but olives, almonds,
figs, peaches, nectarines, maize and tobacco are grown in this
region.

Climate.—The climate of Portugal is thought by many to be
the finest in the world, and the number of very old people is
quite as remarkable as in Bulgaria. The soil is rich, the sun is
warm, and the perfect climate greets the efforts of the farmer and
vintager in the manifold duties to which he devotes himself.

[1] "Port, Oporto and Portugal," by Frederick Challoner.

There is only one thing lacking, that is a plentiful supply of
moisture.   The average Portuguese farmer is a small proprietor.
He is an easy-going person, jogging along as his father and grand-
father have done before him.   Nothing matters to him.   He
never wishes to do what he can put off until tomorrow.

A Typical Vineyard.—The vineyard shown in Fig. 36 is the
most famous of the vineyards of Portugal.   The illustration shows

FIG. 35.—Utilizing the hillside in Portugal.   (*Courtesy of F. Challoner, London.*)

the approach to the residence and outbuildings and also the nature
of the walls which have been built to retain the soil of the vineyard.   Much of the excellence of this vintage is due to the care
exercised by the manager and his skilled assistants who serve all
their lives in one place.   It is known to history as long ago as
1715.   In 1867 the International Jury of Paris awarded the gold
medal to the proprietors of this vineyard as being the most handsome in the Douro.   As long ago as the year 1446 special privileges
were granted by the King of Portugal to English merchants.   In
1790 a British Factory House, or Club House, was constructed.
It is now known as the British Association.   The chief part of the
Port wine produced in the Douro valley is handled by the English
merchants.   The demand in England alone, prior to the war, for
the best Ports practically absorbed all of the production of the
most famous of the vineyards of Portugal.   Very few brut Port

16

FIG. 36.—A typical Port vineyard.   (Courtesy of F. Challoner, London.)

wines reach the outside world. The Port of commerce is usually blended from many or all of the principal wines of the Douro district to make a wine of uniform character for general export purposes. A little vintage wine may be secured, however, by the right parties who know where to buy.

Gathering the Grapes.—The mode of gathering the grapes is almost the same throughout the whole of the Douro valley. Of course in some instances greater care is exercised by some growers than by others. In the typical Port vineyard every detail of the harvest is carefully directed and superintended by

FIG. 37.—Carrying grapes to the Lagar for pressing. (*From Challoner.*)

an experienced master vintager. The grapes are gathered in large baskets. When these are full they are carried off on the shoulders of the laborers, some times 20 or 30 altogether. The illustration shows the laborers carrying the full baskets. Usually the leaders of this procession improvise a kind of a band. The instruments employed are usually a drum, an accordion and a triangle. The mountain sides are so steep that even if it were advisable the grapes could not be transported in wagons.

Handling the Grapes.—When the grapes arrive at the winery they are placed in large square depressions, about three feet deep, and in them they are pressed. The grapes are pressed entirely by the feet of the workmen. This is a period of great festivity, the workmen singing and shouting continually for 6 solid hours, when they are allowed to rest until their turn comes again after 4 or 6 hours of sleep. After the juice is expressed it is subjected to fermentation, the progress of which is determined by experience and by a spindle which determines the density of the juice. As the sugar ferments the juice grows lighter and lighter until at the end it is probably no heavier than water. The expressed juice, after the first fermentation is over, is run into cisterns, some of which are very large, holding as high as 14,000 gallons. The harvest having taken place in October, the fermented juice is allowed to remain undisturbed in these cisterns until the following January. A careful examination is then made of the wine, in order to determine its degree of excellence and maturity. After the first fermentation is over it is usually transferred to other cellars and placed in smaller containers for the continuation of the ripening process.

The varieties of grapes which are most generally employed in the manufacture of genuine Port wines are the following:

RED WINE GRAPES

Malvazia Vermelha	Souzão (deepest color)
Alvarilhao	Tinta amarella
Bastardo	Tinta bastardeira
Bocca de Mina (most delicious)	Tinta de Castello
Cornifesto	Tinta Cão (very good)
Douzellinho do Castello	Touriga (the finest)
Mourisco preto	Tinta Francisca (very good)

Rabo de Ovelha

WHITE WINE GRAPES

Agudelho	Malvazia
Alvaraça	Mourisco Branco
Arinto	Muscatel
Douzellinho	Promissao
Gouvéio	Rabigato

TABLE GRAPES

Dedo de Dama	Ferral (rose color)
Muscatel de Jesus	All the Chasselas

The Tinta Coa, Touriga and Tinta Francisca are very fine grapes, but as they yield less juice than others they are not much used by the smaller growers.

Character of the Wine.—It is difficult to describe the character of Port wine. By many it is regarded as heavy and of a character

to promote gout. In olden days when an English gentleman
was rated it was as a "One-bottle," "Two-bottle" or "Three-
bottle" gentleman, according to the number of bottles of Port
he would consume during and after his dinner. This heavy
drinking of Port is no longer possible, both because of the increased number of drinkers and the better knowledge of the
principles of temperance which has been spread abroad. Port
wine is of a deep purple-red color, with abundance of mineral product and a certain quantity of unfermented sugar and other un-
fermentable substances, making a rather heavy extract. Among
connoisseurs genuine Port, properly aged and handled, is reckoned
second to no other wine, unless it be a few of the grand vintages
of Bordeaux. If the Chateau Margaux, Chateau Lafite and
Chateau Yquem and two or three of the Burgundy vineyards
are excluded there is no other wine of a rank superior to the real
vintage Port such as is produced in the Alto Douro.

   Good and Bad Vintages.—As with every other vineyard the
Port vintage wines have their good and bad years. Some of the
good and bad vintages which have been particularly recognized
during the last 120 years are as follows:

PORT VINTAGES

1800 bad
1802 good
1803 good
1804 good
1805 ordinary
1806 very good
1812 very fine
1815 very fine
1820 very fine
1821 fine
1822 fine
1827 very fine
1830 fine
1834 excellent
1840 very fine
1842 fine
1844 very fine
187 excellent
1850 good dry—oïdium appeared
1851 very fine, some went wrong

1853 very fine, fruity, some went
   wrong
1854 good at first, but went sour,
   oïdium being at its worst
1858 very good, big, dry, the first
   healthy wine since 1850
1861 good
1863 excellent, rich
1868 very fine
1869 some good
1870 very fine
1872 some good
1873 some good
1875 good, matured quickly
1878 very fine
1881 good, light, flavory
1884 good, like 1875
1887 very good, not big, "Jubilee
   Port"
1888 only fit for the still

PORT VINTAGES (*Continued*)

1889 good, light, not vintage year	1900 fair
1890 good, best Quintas, very good	1902 some fine Quinta wines
1893 bad, mildew began	1904 very good
1894 some rather good	1908 very fine
1896 very fine, well bred (Noval, a great	1911 some fair, "drought year"
success)	1912 fine (Noval, a great success)
1897 useful	1913 failure

1834, 1847, 1863, 1870, 1878, 1887, 1896 and 1908 are the outstanding vintages of the century.

1912 promises to be one of the best of recent years.

Poets love Port. Mr. Comyns Carr, the dramatist, delivered an address in Liverpool entitled "Authors I have known"[1] in which he relates the following story:

Irving told me that while preparing for rehearsal of one of Tennyson's plays he went to visit the poet. At dinner Tennyson said "Irving, you like a glass of Port, don't you?" Irving said, "Yes, I do," whereupon the poet poured him out a glass of Port and finished the bottle himself. A long time after this when Irving was giving "Becket" he again visited the poet, who was then under a doctor's strict orders. After dinner the Port was again put on the table and Tennyson said: "Irving you like a glass of Port?" "Yes," said Irving, "I do." Irving in telling the story, said: "Tennyson took the glass, and I finished the bottle." The next morning Irving was awakened by Tennyson himself, who said: "How are you Irving?" to which Irving replied: "Oh, very well, very well," whereupon Tennyson added: "Are you? You drank a lot of Port last night.

Browning also was fond of Port. He took it at soup and finished a bottle during dinner. He told Mr. Carr a story about his father, who also loved Port, so the liking seems to have been inherited. Browning's father, seeing the poet drinking a glass of water, turned to him and said, "Water, Robert? For washing purposes I believe it is useful; for navigable canals I am informed it is indispensable, but for drinking—Robert, God never intended it."

Adulteration of Port.—The principal adulterations of Port wines are found in the mixing, or blending, or compounding, of wines which have some similarity to Port and to which the name "Port" is given. In the United States a number of wines have been called "Port," usually with a geographical restriction, as, for instance, "California Port," "Old Virginia Port," etc. It is hardly proper even to allow this designation in the limitation

[1] The Wine and Spirit Gazette, January 4, 1908, page 32.

showing that it is of American origin. The character of these so-called Port wines is so different from that of the genuine that they cannot be recognized even as imitations or as types of a wine of this distinguished character. As in all other cases of the imitation of foreign names it would be far better if American wine growers would create a nomenclature of their own and drop all imitations of foreign character or quality.

MAKING SPARKLING WINES

First Fermentation.—In the process of fermentation the sugar of the grape is transformed into alcohol with other various congeneric substances, and into carbon dioxid, commonly known as carbonic acid gas. When the process of fermentation is in full vigor the liquor in the fermenting tank boils as if heat had been applied thereto. This boiling is due to the escape of the carbon dioxid. In the making of still and sparkling wines the first or principal fermentation is the same in both cases. In the making of still wines there is no attempt made at any time of the fermentation to confine the gas which is allowed to escape in the manner just described. In the making of a sparkling wine, the subsequent fermentations following the first partial fermentation are conducted in the bottle in which the wine is to be sold. This bottle is stoppered in such a way as to confine the gas so that it becomes dissolved in the liquid contents of the bottle. When the cork is removed from a bottle of this kind the gas which has been kept under pressure immediately begins to escape, producing the boiling or foam, or frothing, which is characteristic of carbonated beverages.

The general term "sparkling wines" is applied to the products manufactured as just described.

Technique of Manufacture.—Sparkling wines are often made from the same grapes used in the manufacture of still wines. After the first fermentation a distinct difference in the treatment of the wines is necessitated where sparkling wine is to be made. The wine from the first fermenting cask is placed in bottles which are made of peculiar structure in order to bear the pressure to which they are to be subjected. The bottles are thick, of well-annealed glass, and the bottom is drawn into the interior of the bottle so as to form a kind of cone. This form adds great strength to the bottom of the bottle, and thus enables it to resist the pressure to which it is subjected. After the bottles are filled the cork is securely tied so that it will resist the pressure of one or more atmospheres without being blown out. The bottles thus prepared are placed in racks slightly inclined with the mouth down, and in this position they undergo the secondary fermentations, producing quantities of carbon dioxid which are dissolved in the liquid contents of the flask. During this time of fermentation there is a clarification of the wine similar to that which is produced when wine is drawn off into barrels. The solid matters which are separated from the wine consist of particles of organic matter and of tartar (acid potassium tartrate). They fall by gravitation and become aggregated in the neck of the flask, resting directly upon the inner end of the cork. In order that none of this separated material may adhere to the sides of the bottle, and thus spoil its appearance, the bottles while undergoing this secondary fermentation are shaken from time to time in order to detach any particles of solid matter and allow them to fall to the bottom. This shaking is done very skilfully by an attendant, who, at the same time, turns the bottle slightly so that the lower inclined side of it shall from time to time be a part of the whole circumference.

By the above patient and delicate manipulation the secondary fermentation is finally concluded, all of the mud, so called, is collected on the cork, and the supernatant wine is clear and bright. Usually about 18 months are necessary to produce this complete clarification.

Discharging the Mud.—When the wine is clarified, as has just been described, and the mud all accumulated in the neck of the flask, the next step is the removal of the mud, wasting as little as possible of the wine. Formerly this was done by a skilled attendant cutting the cords which held the stopper in place and allowing the pressure from the inside of the bottle to blow out the cork, followed by the whole of the mud. By a quick and skilful movement the bottle which had up to this time been held with its mouth down is raised to an upright position, thus avoiding any further loss of wine.

So skilful do the operators become in this delicate manipulation that the whole of the mud is removed with the loss of only a very small quantity of the wine.

Final Treatment.—A modification of this process consists in freezing the neck of the bottle which contains the mud which is then withdrawn as a frozen cork with but little loss. The wine at this point has naturally lost all of its gas, since the cork is opened, and it is practically good still wine. It is necessary therefore to produce a third fermentation in order that the gas may be present in the wine at the time it is consumed. To this end there is introduced into the bottle a solution of pure sugar and brandy. The quantity added is sufficient to make up the loss produced by blowing out the mud, being always a comparatively small quantity for each bottle. The object of dissolving the sugar in the brandy is to add to the wine a slightly greater content of alcohol than would be produced by the natural fermentation of the grape. As a rule, grapes which are used in making sparkling wine are not very high in sugar content, usually not exceeding 20 or 22 percent. The object of the additional sugar dissolved in the brandy is to furnish the material on which the third fermentation may act, thus converting the sugar with the evolution of the carbon dioxid into alcohol. When the brandy and sugar have been added the cork which is finally to be sold with the bottle is inserted and tied down, usually with wire. The bottle is then again placed on a fermenting rack, mouth down, and allowed to remain until the final fermentation is completed, which usually takes from 6 months to a year.

Total Time Employed.—It is thus seen that the total time employed for making properly a bottle of sparkling wine is about three years. Very often a longer time is required, and sometimes a shorter.

Breakage.—In spite of the great care with which the bottles that hold sparkling wine are made there is always a large percent of breakage during fermentation, which has been estimated as high as 10 percent. It is not unusual to hear the sharp reports of breaking bottles in the cellars where sparkling wine is maturing.

Names of Sparkling Wine.—Sparkling wines have various names according to the country in which they are made. In France the great region for sparkling wines is what is known as the champagne region, and all wines made in these departments

of the Marne are entitled under the French law to bear the word
"Champagne." Sparkling wines in other parts of France are
not permitted to bear the word "Champagne." The term "Champagne," has been used in many countries improperly, but broadly
to designate a sparkling wine. The term, should be strictly
reserved out of respect for the French law and custom for the wines
made in the Department of the Marne, already alluded to. Fortunately the most of the great brands of sparkling wines are sold
under firm or fanciful names, so that the name champagne is
not necessarily employed.

In ordering a sparkling wine one would not be simply ordering
champagne, but would designate by name any particular kind of
sparkling wine which he desired to use. There can be no real
justification of the continuance in this country of using the term
"Champagne" as indicative of American sparkling wines.

American Sparkling Wine.—Large quantities of very excellent
sparkling wine are made in the United States, principally in New
York, California and Ohio. The kinds of grape used in making
American sparkling wine except in California is chiefly the
Catawba. This is a grape which has a very rich agreeable fruity
flavor, and this flavor is imparted to the wine made therefrom.
The American sparkling wine is therefore totally unlike the French
in flavor and quality, and resembles it only in the manner in which
it is manufactured and in its content of carbon dioxid. This is
another reason why American sparkling wine should not be
called champagne, because, aside from methods of manufacture,
it is nothing like the actual champagne made in France. The
name applied to sparkling wine in France other than in the champagne is *Vin Mousseux.* The German name is *Schaumwein;* and
the Italian name is *Vino spumanti*, while in the United States the
names of the locality from which it comes, or the names of the
makers, are almost universally employed.

Quality of Sparkling Wines.—It is evident from the methods
of description in regard to the manufacture that a sparkling wine
is not likely to be as fragrant, as wholesome, nor as desirable as a
still wine. A still wine is matured with free access of the air,
hence the character of the congeneric substances developed during
the secondary fermentation and during the ripening is quite
different from those which are developed in a closed bottle, to

which no oxygen is allowed access. Sparkling wine, therefore, is a less mature product than good still wine, and has the flavoring characteristics of a raw or immature product.

The great use of a sparkling wine is for table purposes. It is in many countries considered the necessary accompaniment of the after-part of the dinner, especially that part devoted to the toasts and the speeches. It is not a wine, however, to be drunk as a table wine in the sense that a good red or white wine is consumed. A single glass of it is quite sufficient for any one person at any one meal.

Cost of Sparkling Wine.—It is not surprising in view of the description given of the care, time and expense attending the manufacture of a sparkling wine that its cost is usually out of all proportion to its merit as a drink. While by no means the most costly of wines, the sparkling wines, as a class, are much higher priced than any other kinds. With the duties attached thereto in this country the ordinary sparkling wines imported from abroad sell at from four to five dollars a bottle, while the very highest grades of vintage wine made from the grapes of a single vineyard and not blended with any other materials may be double that which has just been mentioned. These prices do not begin to compare with the prices which a very old still wine of any of the first growths of France would fetch in the market. American sparkling wines, free of the duty of the imported article, and only lately subject to any excise duty, are made and profitably disposed of at a price something like a dollar a bottle.

Capacity of the Bottles.—The ordinary bottle of sparkling wine is one-fifth of a gallon. They come in half bottles, namely: one-tenth of a gallon, and also in magnums, which may hold a half gallon or a gallon apiece. A magnum of a wine of this kind is rarely ever used except upon occasions of anniversaries, celebrations, such as weddings, family reunions, and on birthdays.

Color of Sparkling Wines.—Nearly all sparkling wines are white, or better, amber tinted. There are very frequently found red wines which have been manufactured according to the sparkling standard. Among these a red wine made in France, and known as sparkling Burgundy is very highly esteemed. The sparkling red wines cannot begin to compare in flavor and delicacy, or taste and aroma, with the still red wines made from the same grapes.

Adulteration of Domestic Wines.—Under the provisions of the
Food and Drugs Act there was issued, in August, 1909, an official
decision which I had prepared defining the proper labeling of
wines. The decision runs as follows:

"Wine" without modification is an appropriate name solely for the product made from the normal alcoholic fermentation of the juice of sound ripe
grapes, without addition or abstraction, either prior or subsequent to fermentation, except as such may occur in the usual cellar treatment for clarifying
and ageing. The addition of water or sugar, or both, to the must prior to
fermentation is considered improper, and a product so treated should not be
called "wine" without further characterizing it. A fermented beverage
prepared from grape must by addition of sugar would properly be called a
"sugar wine," or the product may be labeled in such a fashion as to clearly
indicate that it is not made from the untreated grape must, but with the
addition of sugar. . . . . . No beverage can be made from the marc of
grapes which is entitled to be called "wine" however further characterized,
unless it be by the word "imitation." The words "Pomace Wine" are not
satisfactory, since the product is not a wine in any sense, but only an
"imitation wine" and should be so labeled.

Vigorous opposition was made to this definition under the
Food and Drugs Act by the makers of wines, especially in New
York, Ohio and Missouri. So powerful was this representation
that a new decision was rendered against my protest under date of
May 13, 1910, revoking in part the former decision and amending
it by the addition of the following:

However, it has been found that it is impracticable, on account of
natural conditions of soil and climate, to produce a merchantable wine in
the States of Ohio and Missouri without the addition of a sugar solution to
the grape must before fermentation. . . . . . Having regard to the
fact that fermented beverages have been produced in the States of Ohio and
Missouri by the addition of a sugar solution to grape must before fermentation
and sold and labeled as "Ohio Wine" and "Missouri Wine," respectively, for
a period of over 60 years, it is held a compliance with the terms of Food
Inspection Decision 109 if the product made from Ohio and Missouri grapes
by complete fermentation of the must under proper cellar treatment, and
corrected by the addition of a sugar solution to the must before fermentation
so that the resultant product does not contain less than five parts per thousand acid and not more than 13 percent of alcohol after complete fermentation, are labeled as "Ohio Wine" or "Missouri Wine" as the case may
be qualified by the name of the particular kind or type to which it belongs.
. . . . . The product made in Ohio and Missouri by the addition of

. of water and sugar to the pomace of grapes may be labeled as "Ohio Pomace Wine" or "Missouri Pomace Wine" as the case may be.

Thus, in less than a year after the true definition of a wine was published it was amended so as to permit of extensive adulteration. This was done, too, in the face of a number of protests of the makers of genuine wine in those states against libeling all of their products by calling the adulterated wines "Ohio Wine" or "Missouri Wine" as the case might be. The experiments conducted by Professor Alwood during this period had also proved beyond question that most excellent wines, absolutely pure, made solely from the fermented must of grapes, could be made in all of these localities without any trouble whatever. This concession to the claims of the adulterators has fixed a stigma upon the wines of Ohio and Missouri which probably will not be effaced in half a century.

Wine Adulteration Legalized.—By act of Congress wine is defined as follows:

That natural wine within the meaning of this Act shall be deemed to be the product made from the normal alcoholic fermentation of the juice of sound, ripe grapes, without addition or abstraction, except such as may occur in the usual cellar treatment of clarifying and ageing.[1]

Had our lawmakers stopped here they would have deserved the plaudits of their countrymen. But immediately following this ethical definition comes the pernicious proviso:

Provided, however, That the product made from the juice of sound, ripe grapes by complete fermentation of the must under proper cellar treatment and corrected by the addition (under the supervision of a gauger of storekeeper-gauger in the capacity of gauger) of a solution of water and pure cane, beet, or dextrose sugar (containing, respectively, not less than ninety-five per centum of actual sugar, calculated on a dry basis) to the must or to the wine, to correct natural deficiencies, when such addition shall not increase the volume of the resultant product more than thirty-five per centum, and the resultant product does not contain less than five parts per thousand of acid before fermentation and not more than thirteen per centum of alcohol after complete fermentation, shall

[1] Revenue Bill Approved September 8, 1916.

be deemed to be wine within the meaning of this Act, and may be labeled, transported, and sold as "wine," qualified by the name of the locality where produced, and may be further qualified by the name of its own particular type or variety: *And provided further,* That wine as defined in this section may be sweetened with cane sugar or beet sugar or pure condensed grape must and fortified under the provisions of this Act and wines so sweetened or fortified shall be considered sweet wine within the meaning of this Act.

These provisos open the flood gates for legalized adulteration. There is little left for the professional adulterator to desire. The whole purpose and aim of the pure food law in so far as purity of wine is concerned is set aside and denied. Almost everything needed for adulteration is specifically provided for by this Act. To be sure only 35 percent adulteration is permitted! But who shall supervise and restrict the mad scramble to make adulterated wines?

Fortunately an act of Congress does not make wrong right. It protects the offender but does not change the offense. All the products made under this indulgence are and will remain types of adulteration.

FORTIFIED WINE

Definition of Fortification.—The term "fortified" is used in connection with a wine not fermented exclusively from grape juice but from grape juice to which sugar has been added. This addition of sugar produces a larger quantity of alcohol than naturally would be found in the fermented juice. Wines fortified in this way are very commonly made in all countries where the climatic conditions are such as to produce grapes rather poor in sugar. Even in countries where ordinarily the grape contains a sufficient quantity of sugar to make a good wine there are occasionally wet seasons and cold seasons, or lack of sunshine, in which the grapes have a low content of sugar, and usually a corresponding high content of acidity. When the sugar content of the grapes is less than 14 percent it is quite the custom in many countries, including the United States, to add sufficient sugar to produce a wine of an alcoholic content of from 9 to 12 percent. In these cases it is seen that part of the alcohol in the wine is derived from sugar and the beverage is not much more of a wine than it is an undistilled rum.

A more common method of fortifying wine is to add the alcohol directly to the wine after fermentation. Presumably the alcohol which is added is brandy, but this presumption is not always verified by the actual facts of the case. Very frequently neutral spirits made from other sources, as for instance, Indian corn, the potato, or beet root and molasses, are employed for fortification. In the United States the spirit which is used for the fortification of wine is required by law to be distilled from grapes or the residues of the winery. This admits to fermentation pomace from which the grape juice is expressed, and in this way produces a variety of brandy which to most persons is wholly unpalatable. A brandy of this kind has an excessively high content of higher alcohols and ethers, and has a taste and flavor which are quite unpalatable to those not accustomed to its consumption.

Brandy which is used for the fortification of wines in the United States is usually freshly distilled in a rectifying column, and does not have the properties of a real brandy, due to the distillation of sound wines and proper ageing in wood.

Adulteration of Wines by Fortification.—One of the principal methods of adulterating wines is under the present system devised by act of Congress and by the regulations made thereunder by the Bureau of Internal Revenue for fortifying wine. We have already seen that the law as it now stands recognizes an adulterated wine as wine and permits it to be called by that name. The law also provides for the fortification of so-called sweet wine in a manner which produces a highly adulterated and injurious article. This was done under previous revenue laws and is continued under the law, which was approved on September 8, 1916. The language of the act in regard to fortified wines is found in Sections 42 and 43 of the above act, as follows:

SEC. 42.—That any producer of pure sweet wines may use in the preparation of such sweet wines, under such regulations and after the filing of such notices and bonds, together with the keeping of such records and the rendition of such reports as to materials and products as the Commissioner of Internal Revenue, with the approval of the Secretary of the Treasury, may prescribe, wine spirits produced by any duly authorized distiller, and the

Commissioner of Internal Revenue, in determining the liability of any distiller of wine spirits to assessment under section thirty-three hundred and nine of the Revised Statutes, is authorized to allow such distiller credit in his computations for the wine spirits withdrawn to be used in fortifying sweet wines under this Act.

SEC. 43.—That the wine spirits mentioned in section forty-two herein mentioned is the product resulting from the distillation of fermented grape juice, to which water may have been added prior to, during, or after fermentation, for the sole purpose of facilitating the fermentation and economical distillation thereof, and shall be held to include the product from grapes or their residues commonly known as grape brandy, and shall include commercial grape brandy which may have been colored with burnt sugar or caramel; and the pure sweet wine which may be fortified with wine spirits under the provisions of this Act is fermented or partially fermented grape juice only, with the usual cellar treatment, and shall contain no other substance whatever introduced before, at the time of, or after fermentation, except as herein expressly provided: *Provided*, That the addition of pure boiled or condensed grape must or pure crystallized cane or beet sugar, or pure dextrose sugar containing, respectively, not less than ninety-five per centum of actual sugar, calculated on a dry basis, or water, or any or all of them, to the pure grape juice before fermentation, or to the fermented product of such grape juice, or to both, prior to the fortification herein provided for, either for the purpose of perfecting sweet wines according to commercial standards or for mechanical purposes, shall not be excluded by the definition of pure sweet wine aforesaid: *Provided, however*, That the cane or beet sugar, or pure dextrose sugar added for sweetening purposes shall not be in excess of eleven per centum of the weight of the wine to be fortified: *And provided further*, That the addition of water herein authorized shall be under such regulations as the Commissioner of Internal Revenue, with the approval of the Secretary of the Treasury, may from time to time prescribe: *Provided, however*, That records kept in accordance with such regulations as to the percentage of saccharine, acid, alcoholic, and added water content of the wine offered for fortification shall be open to inspection by any official of the Department of Agriculture thereto duly authorized by the Secretary of Agriculture; but in no case shall such wines to which water has been added be eligible for fortification under the provisions of this Act, where the same, after fermentation and before fortification, have an alcoholic strength of less than five per centum of their volume.

Amount of Brandy Used in Fortifying Wines.—During the fiscal year 1916 there were used in the fortification of pure sweet wines 1,257,399 proof gallons of brandy as against 4,505,218.7 proof gallons used for like purpose during the preceding fiscal year.

This noticeable decrease is doubtless due to the fact that by the act of October 22, 1914, the tax on brandy so used was increased from three cents per proof gallon, as fixed by the act of June 7, 1906, to 55 cents per proof gallon. The result of this increase in the rate of tax is also clearly indicated by the quantity of brandy used under these different rates during the fiscal year 1915, the quantity used at the rate of three cents per proof gallon during the first four months being 4,132,419.4 proof gallons, as against 373,199.3 proof gallons at the higher rate of tax used during the remaining 8 months.

By the act of September 8, 1916, the tax on brandy so used has been reduced to 10 cents per proof gallon, and it may, I think, be safely predicted that the quantity of brandy hereafter so used annually will closely approximate 5,000,000 proof gallons.

By the Act of October 3, 1917 the above tax has been increased so that now so-called grape brandy pays 20 cents per proof gallon when used for fortifying purposes.

Congress has thus placed a premium on the use of distilled spirits in fortified wine. The consumer of such distilled spirits, if used as such for beverage purposes, saves the difference between a tax of $3.20 and 20 cents per proof gallon.

Inasmuch as it is possible to raise the content of alcohol in fortified wine to 24 percent this half whisky ought, as the Commissioner of Internal Revenue hopes, to have a great vogue.

Extended Wine.—It has already been pointed out that wines may be extended or expanded or adulterated with various forms of sugar and water provided their volume be not increased more than 35 percent. If such wines are distilled it is evident that the spirit which is obtained is not a pure spirit of fermented grape juice but may be to the extent of 35 percent composed of a spirit distilled from sugar. It is therefore a mixture of brandy and of rum. Further than this the law permits that the spirit known as brandy or grape brandy may be produced from a highly adulterated article itself and shall include, as the law specifies, the product from grapes or their residues. The residues of grapes of course are the pomace. It is well known that a spirit distilled from pomace is highly deleterious and injurious to health. It is true that the spirit which is distilled under this law is happily manufactured in what is known as a rectifying still, so that a considerable portion of the injurious substances which it contains may be removed. Nevertheless, the spirit which is withdrawn on the

17

payment of a tax of 10 cents per gallon and is then allowed to be mixed with wines in the process of fortification is one wholly unfit for consumption. Thus the fortified wine made in this country, or the fortified sweet wine, is a beverage which must be looked upon with the gravest suspicion and the less of it that is used the better.

Taxation of Wines.—Under the law of September 8, 1916, a tax is laid upon all wines according to the following schedule: On wines containing not more than 14 percent of absolute alcohol four cents per wine gallon, the percentum of alcohol taxable to be reckoned by volume and not by weight; on wines containing more than 14 percent and not exceeding 21 percent of absolute alcohol 10 cents per wine gallon; on wines containing more than 21 percent and not exceeding 24 percent of absolute alcohol 25 cents per gallon. Under this law when so-called wines contain more than 24 percent of absolute alcohol by volume they shall be classed as distilled spirits. In other words, a wine which itself is adulterated may be fortified with a spirit of wine or brandy which itself is adulterated until the alcoholic content is increased to 24 percent and still be called a wine and pay a tax of only 25 cents per gallon. The moment an additional amount of alcohol is placed in the wine, say making it 25 percent of alcohol, it becomes a distilled spirit and pays a tax of $1.10 a gallon. It is evident, therefore, that the only difference between a wine and a distilled spirit in this case is that which is produced by legislative act. It is true that it is provided that the use of wine spirits for the fortification of sweet wines shall be under the immediate supervision of an officer of the Internal Revenue. The impossibility of complete supervision is easily recognized.

Under the Act of October 3, 1917, the above taxes are doubled.

Sulphurous Acid in Wines in France.—The use of sulphurous acid in the making of wines in France, as has been described, is quite universal, though there are many individual makers who would be glad to see the use of this re-agent entirely eliminated. That it has been abused is well recognized by the French themselves, and in consequence thereof, by virtue of a decree issued on the third of December, 1907, under the French Food Law, the limit of sulphurous acid in white wine was fixed at 350 milligrams per liter, either free or combined.

Discussion in regard to the amount of sulphurous acid allowed in wines has not existed much over 30 years. The first regulations due to the discussion, appear to have been made in Austria, in 1885. In 1895, Switzerland undertook the delimitation of sulphurous acid in its wines, and fixed the maximum quantity which should exist in the free state at 20 milligrams per liter, and the total, that is, the free and combined together, at 200 milligrams per liter. In France, soon afterward, the Committee of Hygiene of the Siene, in 1901, provisionally fixed the total limit of sulphurous acid at 200 milligrams per liter, without specifying the quantities which might be free and combined.[1] In spite of this arrangement, many protestations were raised against it in France, and in consequence the limits fixed by the Council on Hygiene were not enforced. According to a French account, the movement spread gradually from France to other states of Europe, almost all of which undertook the regulation of the quantity of sulphurous acid which should be used in vinification.

When the food inspection laws of the United States went into effect, in 1907, the French rule which was then in existence, of 200 milligrams to the liter, was enforced against imported wines. Owing to protestations on the part of French exporters the Government appointed a new commission, which, after deliberation, agreed to double the quantity tolerated by the Committee on Hygiene, fixing it at 400 milligrams per liter. This involved a discussion of the toleration of sulphurous acid in wine with the authorities of the United States, and in order to promote national comity the United States finally agreed to a maximum limit of 350 milligrams per liter.

The question was further discussed in France, both by hygienists and by the wine makers, with a view to securing larger tolerations, and it was urged that the sole excess of sulphurous acid which could possibly be considered as threatening health was that of the free state, and that the regulations should simply limit the quantity of this acid, and not that of the combined. It was suggested, in the interests of the manufacturers, that the limit of free acid be fixed at 100 milligrams per liter, but this suggestion has never been enacted into law.

Conviction for the Use of an Excess of Acid.—A charge of adultera-

[1] Feuille Vinicole de la Gironde, December 23, 1909.

tion has been made in the French courts against wine which contained more than 350 milligrams per liter.[1] This was an action which took place in the court of Paris on goods cited by the Central Laboratory for the Repression of Frauds. The defense urged, first, that the expert evidence was so contradictory in regard to the effects of an excess quantity as to be of no value; second, that the defendant was ignorant of the exact quantities of sulphurous acid contained in the wines under consideration; third, that the quantities exceeded by very little the amounts which were permitted by the regulation; fourth, that the wines were of the vintage of 1905, and having been made previous to the enactment of the law, were not subject thereto.

In spite of this ingenious defense, the court under the presidency of Judge Lemercier, condemned the defendant to a fine of 50 francs.

Apotheosis of Sulphurous Acid.—Such may be the appellation applied to a breakfast given by the Wine Syndicate of the Gironde to some of the eminent wine experts of Bordeaux. It was a true repast of delicacies, where the best wines of the Sauterne, almost all of them containing more than 350 milligrams of sulphurous acid per liter, were said to be by the experts, simply exquisite. The guest of honor, for 31 years chemist of the Syndicate, said:

"It is regretted that the detractors of these famous wines could not have been present, but let us hope that, little by little, they may be convinced. The question is too large, and the con-sequences too important, to be lost from view. It is important not to permit credit to be given to the errors of hygiene. Long live our golden Sauterne, vertible rays of the sun in bottles."

The two views given above show that the courts and the gour-mets are not in entire harmony. In my experiments on 12 young men I found that sulphurous acid and sulphites were highly deleterious. They are still permitted to be used freely in wines, molasses, syrups and dried fruits, in all parts of the United States where not inhibited by State laws. Our authorities are active in correcting the "errors of hygiene." Commerce is above health. Lucre is more precious than life.

[1] Annales des Falsifications, December, 1909, page 400.

PART IX

BEER, ALE, PORTER AND STOUT

Definitions.—The art of brewing an alcoholic beverage from cereals is one of great antiquity, though not so old perhaps as the art of making wine from grapes. The term "beer," "ale," "porter," or "stout," is applied to an alcoholic beverage produced chiefly from barley by converting the starch into sugar, mixing it with the extract of hops, fermenting the product with yeast, and ageing the fermented product in casks. The term "beer" is applied to a fermented product of barley or barley admixed with other fermentable substances, forming a mash of such density, that is, content of sugar, as will produce an alcoholic beverage containing from 3 to 5 percent of alcohol. "Ale" is applied to a similar product made from a more concentrated mash and containing in the final product from 4 to 7 percent of alcohol. The character of the fermentation is somewhat different in the two processes, the beer being made from a so-called bottom fermentation and ale from a so-called top fermentation. The yeast cells tend to sink in a mash of the consistency which produces beer, while they will tend to come to the top of a denser mass such as is used in the manufacture of ale. The terms "porter" and "stout" are given to products similar to beer and ale, made under slightly different conditions of fermentation.

Color.—Many shades of color are found in the products embraced under the above names. In some instances the finished product is only of a pale amber color. This increases through all shades of amber to red and even to black. The dark products are colored by using a certain quantity of malt which has been partly parched, thus giving the deeper color to the finished product.

Manufacture.—The process of manufacture of these fermented beverages is as follows: Barley or other cereal, is first converted into malt. This is done by spreading the cereal on a floor, usually of concrete, moistening it with water, and allowing it to remain in this condition until it has sprouted. The best malts are made where the temperatures are relatively low, as the sprouting is less rapid and the growth is retarded. During the sprouting process there is formed in the grain large quantities of a ferment which is known as diastase. Diastase has the property of converting starch into a sugar, maltose. During the sprouting, especially if it is a slow one, practically all the starch in the grain is converted into maltose. When the sprouts are approximately one-half inch in length, which requires from 5 to 15 days according to the temperature, they are removed to a drying kiln and dried at a low temperature until the moisture is reduced to 5 percent or below. The grains apparently have not changed much in size or even appearance by this process, except due to the sprouting, but on chewing the malt instead of the ordinary taste a sweet taste of the maltose is noticed.

Density of the Mash.—On treating the ground malt with water from 60 to 70 percent of it is dissolved, forming a sugar solution of varying strength according to the quantity of water used. The process of grinding and extracting the sugar with water is called "mashing." The density of the mash, aside from the undissolved solid matters (protein, bran, fat, germ, fiber), varies from 10 to 16 percent. The mash is now treated with sufficient quantity of hops, thoroughly stirred, and heated to a temperature of boiling. This temperature destroys the further action of the diastase and also kills all forms of ferment which may exist in the mash, so that it can be seeded with a pure culture of yeast to start the fermentation. Immediately after treating with the hops and cooling to room temperature the mass is treated with the proper quantity of yeast and allowed to ferment. As the fermentation takes place the temperature of the mass rises to a considerable extent, in fact, to such an extent that it is often advisable to keep the tubs cool or in a cold cellar. The escaping of the carbon dioxid produces an apparent boiling of the mass, forming a foam, and the production of large quantities of additional yeast cells. The fermentation is practically finished in from three to five days, according to the temperature.

Separation of the Yeast.—As soon as the liquid becomes still the yeast cells and other solid matters begin to separate. Those that are lighter come to the top and those which are heavier, which is the greater number, sink to the bottom of the vessel. When the separation is practically finished the beer may be drawn off into storage tanks, or, in order to insure greater limpidity, may be filtered into the storage tanks. These tanks are kept at a low temperature, usually in cellars, where the completion of the secondary fermentations and the ripening of the beer go on. When beer is stored for improving its quality in this way the finished product is known as "lager." The best beers are those which are ripened in this way for a period of from 9 to 12 months. Usually the period of storage is much less. Beers are often sold in from three to four weeks after manufacture, but these are more or less raw in character and not to be compared in quality with those stored for a longer time.

Unmalted Cereals.—When unmalted cereals are used as barley substitutes the quantity of malt made from barley which is used may be diminished in beer to as low as 10 percent. The quality of the beer made partly from unmalted cereals is not equal to that made from pure malt. For the conversion of the starch in the unmalted cereals a mash is made by grinding the cereals, and this is treated with the quantity of malt necessary to convert the starch of the unmalted cereals into sugar. This is accomplished best at a temperature of from 90 to 100°. After the sugars are formed in this way the rest of the process in making beer is that just described.

Substitutes for Malt.—The commonly employed substitutes for malt are unmalted barley; the starchy parts of Indian corn, known as hominy grits, rice; brewers' sugars, made by the treatment of cornstarch with an acid; and low-grade cane sugars. The beers that are made partly from brewers' sugars are not of as high a quality as beer made wholly from cereal. The best beer is that made only of barley malt and hops.

Varieties of Beer.—From the academic point of view there are only two kinds of beer, namely, beer made solely of barley malt and hops, and beer made from substitutes for barley, the hops being a common content of both classes. Another classification of beers is lager beer as distinguished from other beers. Lager beers are those beers which are kept for a considerable time in the cellars for ripening, as distinguished from beer, which in this case means a product which has no very great age. Those new beers are sometimes called "steam beers."

Geographical Names.—Beers are often classified from their country of origin, as, for instance, Munich beers, made in Bavaria; Pilsner beers, made at or near Pilsen, in Austria; English beers, commonly known as "ale," made in England; and domestic beers, made in the United States. Under the Food and Drug Law it is a misbranding to label a beer of domestic origin in such a manner as to induce the purchaser to believe that it is of foreign origin. Before the passage of the food law such misbrandings were very commonly practised throughout this country.

Composition of Domestic Beers.—The greater number of beers of domestic origin are made only partly from barley malt and the rest of the alcohol is derived from the substitutes for barley which are employed. Some of these beers are of high quality, while others are of low-grade quality, as the substitutes for barley are used only when they cost less than barley malt itself. The cities of Milwaukee and St. Louis are especially renowned as centers of beer making. There are other great centers where the brewing industry is practised, as, for instance, in Chicago, Cincinnati, New York and Philadelphia. In fact, until the states began to enact prohibition laws there was hardly a large city in the United States where beer was not made.

Sale of Beer.—The manufacture and sale of beer is strictly controlled by the Bureau of Internal Revenue of the Treasury Department. Beers are made under a license, and also each package of manufactured beer pays a special stamp tax. Dealers in beer also pay special tax privileges, so that the manufacture and sale of beer are under strict federal control. Beer is sold in two forms: first, draft beer, drawn directly from the wood, as it comes from the brewery; and second, bottled beer, usually pasteurized, in which condition the beer will keep for a long period without any injurious fermentation. In so far as flavor and taste are concerned the draft beers are, as a rule, much superior to the bottled beers. Both draft beers and bottled beers, following the taste of the people of our country for ice-cold drinks, are sold at or near the freezing temperature, much to the detriment of their quality and much to the injury of the stomachs into which these cold beverages are poured.

Composition of Beer.—The ordinary beer of commerce, made from pure barley malt and hops, has approximately the following composition:

		Composition of beer percent
Unfermented carbohydrates	Alcohol	4.0
	Extractives	6.0
	Protein	0.5
	Ash	0.5
	Water	89.0

Beers that are made of malt substitutes differ in chemical composition from the malt beers chiefly in the diminished content of protein or nitrogenous matter. The quantity of ash is also less, provided common salt is not added in the process of manufacture. In other words, the food value of beers is diminished when malt substitutes are employed.

Composition of Ale.—The chief difference in the chemical composition of ale and beer is in the increased quantity of alcohol, which for ordinary ales may be rated at about 6 percent.

Regulating the Content of Alcohol.—The Act of Congress relating to the control of foods and beverages during the war authorizes the President to limit the quantity of alcohol in beer. By proclamation the maximum content in beer has been limited to 2.75 percent by weight. This corresponds to 3.4 percent by volume. This is only a little under the usual quantity occurring in American beers. The limit of 2.75 percent by volume would be far more effective as a grain-saving measure. By a recent order beer making will cease Dec. 1, 1918.[1]

Adulterations and Misbrandings of Beer, Ale, Porter and Stout.—If beer is defined as a product made from barley, malt, and hops with the aid of water and heat, it follows that all beers that are not of this composition are to that extent adulterated. Personally I am strongly persuaded that the term "beer" without qualification should be applied to a product made solely from malt and hops. In practice in this country the term "beer" has a wide application, though in no case, do I believe, is it ever used where at least a small quantity of barley malt is not employed, unless indeed a qualifying term is used. Beer may be made of other

[1] By Act of Congress, approved Nov. 21, 1918, the sale of any alcoholic beverage is prohibited after June 30, 1919.

cereals altogether, and wheat beer is quite a common product in some parts of Europe, especially in Germany. Wheat beer may be made in this country, but it has not been brought to my notice.

MATERIALS USED BY BREWERS IN THE PRODUCTION OF FERMENTED
LIQUORS IN THE UNITED STATES, YEARS ENDING
JUNE 30, 1915 AND 1916

(Compiled by the Bureau of Crop Estimates, U. S. Department of Agriculture from records of the Office of Internal Revenue, Treasury Department)

Material	Unit of quantity	July 1, 1914 to June 30, 1915	July 1, 1915 to June 30, 1915
Malt..............................	Bushels	62,991,856	57,683,970
Hops..............................	Pounds	38,839,294	37,451,610
Rice..............................	Pounds	167,750,177	141,249,292
Corn or cerealine....................	Pounds	604,890,901	650,745,703
Grape sugar or maltose...............	Pounds	52,079,621	54,934,621
Glucose or syrup.....................	Gallons	7,185,563	2,742,854
Grits..............................	Pounds	116,619,510	109,371,482
Other materials(a)....................	Bushels	484,641	72,355
Other materials(a)....................	Gallons	6,630	19,112
Other materials(a)....................	Pounds	68,880,530	24,756,974
Total all items, estimated(b)...........	Pounds	3,274,261,921	3,004,754,590
Beer produced: Barrels of 31 gallons..............		59,808,210	58,633,624

(a) "Other materials" in 1915 include 159,855 bushels of barley, 333,030 bushels of maizone, and smaller quantities of meal, frumentum, malt extract, salt, moss, isinglass, etc. "Other materials" in 1916 include about the same list of commodities, but amount in all to a relatively small total.

(b) The various reported units were reduced to pounds by the following equivalent 1 bushel of malt = 34 pounds, 1 bushel of corn, cerealine, maizone, or grits = 56 pounds, 1 gallon of glucose or of "other materials" = 8 pounds, 1 bushel of barley = 48 pounds, 1 bushel of miscellaneous materials = 40 pounds.

It is difficult to get accurate data in regard to the materials used in the manufacture of beer in the United States because the law forbids the data to be published. The table published above gives the only official data obtainable.

The principal adulterations of beer from the standard above given are the admixture of malt substitutes. These have already been enumerated. In earlier times hops also had a substitute to a certain extent in beer manufacture in the use of other bitter principles, such as quassia, but this practice I believe has entirely disappeared. According to the taste of connoisseurs the quality of a beer begins to deteriorate the very moment any part of the barley malt, which should be sole source of the alcohol, is substituted by any other product. The substitution least objectionable is that of unmalted barley; after that, unmalted hominy grits and rice. The most objectionable substitutes are the brewers' sugars.

Preservatives are rarely used at the present time in beer. In some imported beers is found more or less bisulphite of lime as a preservative, but this is not used to any extent, as far as I know, by domestic manufacturers. The use of sulphurous acid or sulphites is permitted by the Department of Agriculture, although my own investigations showed without question that these products were markedly injurious to the health of strong young men. The matter of the harmfulness of the fumes of burning sulphur and its compounds was appealed from my decision to that of a Referee Board of Consulting Scientific Experts. This Board of eminent chemists had this matter under consideration for a long while and I have been told that a report on the subject has been submitted. This report has never been published, nor has its contents ever been disclosed. The bad effects of sulphur dioxid as experimentally determined on 12 healthy young men are described in my report, Bulletin 84, Part III, of the Bureau of Chemistry, Department of Agriculture, pages 761 to 1041. The decision I made was revoked in 1908 and at this date 1918, no intimation has been given by the persons who administer the laws that correction of this travesty of justice will ever be made. Meanwhile manufacturers who wish to use the fumes of burning sulphur in this food product, either directly or in combination in the form of sulphites, are permitted to do so without restriction. While imported beers, especially those from England, carry small quantities of sulphites, they were not usually present in imports from Germany prior to their discontinuance due to war. Our own domestic beers, as a rule, are free from sulphites. The beer which is packed in bottles, which forms quite a considerable part of the trade, is pasteurized and thus preserved, if kept in a cool place, for many weeks and even months without deterioration, though sooner or later a secondary fermentation takes place which spoils the flavor and character of the beer.

Antiquity of Adulteration.—Frederick Accum has an interesting article on the "Adulteration of Beer:" in his book published nearly 100 years ago. He says:

Malt liquors, and particularly porter, the favorite beverage of the inhabitants of London, and of other large towns, is amongst those articles, in the manufacture of which the greatest frauds are frequently committed.

The statute prohibits the brewer from using any ingredients in his brewings, except malt and hops; but it too often happens that those who suppose they are drinking a nutritious beverage, made of these ingredients only, are entirely deceived. The beverage may, in fact, be neither more nor less than a compound of the most deleterious substances; and it is also clear that all ranks of society are alike exposed to the nefarious fraud. The proofs of this statement will be shown hereafter.

The author of a Practical Treatise on Brewing, which has run through eleven editions, after having stated the various ingredients for brewing porter, observes "that however much they may surprise, however pernicious or disagreeable they may appear, he has always found them requisite in the brewing of porter, and he thinks they must invariably be used by those who wish to continue the taste, flavor, and appearance of the beer. And though several Acts of Parliament have been passed to prevent porter brewers from using many of them, yet the author can affirm, from experience, he could never produce the present flavored porter without them. The intoxicating qualities of porter are to be ascribed to the various drugs intermixed with it. It is evident some porter is more heady than other, and it arises from the greater or less quantity of stupefying ingredients. Malt, to produce intoxication, must be used in such large quantities as would very much diminish, if not totally exclude, the brewer's profit.

The practice of adulterating beer appears to be of early date. By an Act so long ago as Queen Anne, the brewers are prohibited from mixing cocculus indicus, or any unwholesome ingredients, in their beer, under severe penalties; but few instances of convictions under this act are to be met with in the public records for nearly a century. To show that they have augmented in our own days, we shall exhibit an abstract from documents laid lately before Parliament.

These will not only amply prove, that unwholesome ingredients are used by fraudulent brewers, and that very deleterious substances are also vended both to brewers and publicans for adulterating beer, but that the ingredients mixed up in the brewer's enchanting cauldron are placed above all competition, even with the potent charms of Macbeth's witches:

> "Root of hemlock, digg'd i' the dark,
> X   X   X   X   X   X
> X   X   X   X   X   X
> "For a charm of pow'rful trouble,
> "Like a hell-broth boil and bubble;
> "Double, double, toil and trouble,
> "Fire burn, and cauldron bubble."

[1] A Treatise on the Adulteration of Food, London, 1820.

The fraud of imparting to porter and ale an intoxicating quality by narcotic substances, appears to have flourished during the period of the late French war: for, if we examine the importation lists of drugs, it will be noticed that the quantities of cocculus indicus imported in a given time prior to that period, will bear no comparison with the quantity imported in the same space of time during the war, although an additional duty was laid upon this commodity. Such has been the amount brought into this country in five years, that it far exceeds the quantity imported during 12 years anterior to the above epoch. The price of this drug has risen within these 10 years from 2 shillings to 7 shillings the pound.

It was at the period to which we have alluded, that the preparation of an extract of cocculus indicus first appeared, as a new saleable commodity, in the price-currents of brewers'-druggists. It was at the same time, also, that a Mr. Jackson, of notorious memory, fell upon the idea of brewing beer from various drugs, without any malt and hops. This chemist did not turn brewer himself; but he struck out the more profitable trade of teaching his mystery to the brewers for a handsome fee. From that time forward, written directions, and receipt-books for using the chemical preparations to be substituted for malt and hops, were respectively sold; and many adepts soon afterward appeared everywhere, to instruct brewers in the nefarious practice, first pointed out by Mr. Jackson. From that time, also, the fraternity of brewers'-chemists took its rise. They made it their chief business to send travelers all over the country with lists and samples exhibiting the price and quality of the articles manufactured by them for the use of brewers only. Their trade spread far and wide, but it was amongst the country brewers chiefly that they found the most customers; and it is amongst them, up to the present day, as I am assured by some of these operators, on whose veracity I can rely, that the greatest quantities of unlawful ingredients are sold.

The Act of Parliament prohibits chemists, grocers, and druggists, from supplying illegal ingredients to brewers under a heavy penalty, as is obvious from the following abstract of the Act:

No druggist, vender of or dealer in drugs, or chemist, or other person, shall sell or deliver to any licensed brewer, dealer in or retailer of beer, knowing him to be such, or shall sell or deliver to any person on account of or in trust for any such brewer, dealer or retailer, any liquor called by the name of or sold as coloring, from whatever material the same may be made, or any material or preparation other than unground brown malt for darkening the colour of worts, or beer, or any liquor or preparation made use of for darkening the color of worts or beer, or any molasses, honey, vitriol, quassia, cocculus indicus, grains of paradise, Guinea pepper or opium, or any extract or preparation of molasses, or any article or preparation to be used in worts or beer for or as a substitute for malt or hops; and if any druggist shall offend in any of these particulars, such liquor preparations, molasses, etc., shall be forfeited and may be seized by any officer of Excise, and the person so offending shall for each offense forfeit 500 pounds.

Beer Without Hops.—In some instances the original method of making beer, namely: by the fermentation of the mash of cereals without the addition of any hops is still practised. In Germany a beer is made called Weiz Bier, which contains no hops, and is said to be made from wheat, hence its name. It is a very common practice in Germany to serve the Weiz Bier in a large vessel, which like a loving cup is passed around the table. It is used very largely at family, social and other intimate gatherings. Beer without hops is not known at all, or at least but little, in this country. I have never seen a sample of American beer brewed without hops.

Malt Extracts.—There are a great many preparations of malt with hops which are known as malt extract, or some similar name, which varieties of beer are usually made from a mash the fermentation of which is not carried to such an extent as is the case with real beer. In other words, there is a larger percentage of extract in proportion to the amount of alcohol. The malt extracts are more nutritious than beer in the same proportion that sugar is more nutritious than alcohol.

The quantity of alcohol contained in malt extracts is usually not very large, while the quantity of unfermented maltose, dextrose and of the other extractives is increased.

In so far as the name is concerned, the beverage which has just been described is no more malt extract than a regular beer. While perhaps there may have been no intention or deception, it can hardly be denied that to call a preparation of this kind a malt extract is to a certain extent a misnomer. Real malt extracts are highly prized as tonics for convalescents and invalids, and are largely prescribed by physicians. It is possible that many persons who use these tonics may not be aware of the fact that they are partially fermented, and are of the nature and character of a mild beer. A genuine malt extract, secured from the malt by the action of water, can be easily made and preserved by sterilization in bottles. Such an extract might also be made with hops, and a preparation of this kind would be a true extract of malt and hops, in which no alcohol could be found, but in which the nutrient properties of the malt and the bitter properties of the hops would be combined in one beverage. The name malt extract should be reserved for an extract of malt pure and simple, and the name for an extract of malt and hops be descriptive of the kind of beverage which is indicated. The official formula for making a true malt extract is given on page 156 of the Ninth Decennial Revision of the U. S. Pharmacopœia. It is a saturated extract of 1,000 grams of malt to make a liter of extract. It contains no alcohol.

Other Uses of the Term Beer.—The term beer is also applied to quite a number of other beverages, as, for instance, root beer, tonic beer, etc. The use of the name in these instances is not without objection. Even if those preparations contain alcohol they do not resemble beer in the method of manufacture, or if they do not contain alcohol, they are then misbranded.

Other Uses of the Word Ale.—The term ale is also used in the same significance as beer in the above paragraph. A great many so-called ales are made and sold which are not at all similar to the genuine article. One of the most common is perhaps ginger ale. Others are hop Scotch ale, hop ale, etc.

The same remark may be made respecting the use of "ale" as applied to the word beer above. While it may be claimed that no deception is intended to be practised the border line between verity and deception is very nearly reached in the use of names in this way. Further than this, the terms used with the objects described do not designate any definite compound. Root beer may be made in many different ways, and out of as many materials, and the same is true of ginger ale. These words used in this manner give no definite information, and hence to that extent are deceptive by at least concealing the truth. It would be far better for manufacturers in matters of this kind when they make a beverage which contains objectionable principles if they would give them the proper name, that is, call them by their true name as indicated by their composition, or coin a fancy name for the kind of beverage which gives no false indication of quality. Thus all suspicion of misbranding or attempted misbranding would be avoided, and the beverage in question would be placed upon much higher planes.

Near Beer.—This so-called beverage is usually a light beer which to a degree tastes like the genuine article but contains less than 2 percent of alcohol. If beer must be consumed it is advisable to keep its content of alcohol as low as possible. Important legal technicalities arise in those states where prohibition laws obtain, respecting the manufacture and sale of near beers.

Two analyses of near beers made under my supervision gave the following data:

	A	B
Alcohol by volume...	0.26 percent	0.34 percent
Total solids.........	5.44 percent	5.84 percent
Ash................	0.17 percent	0.10 percent

These beverages were hopped, carbonated and looked like beer.

SAKE

Description.—In Japan the national beverage corresponding to beer is manufactured from rice, and is known as sake. This beverage, although a beer in respect of its origin and method of manufacture, does not resemble beer either in its organoleptic properties nor in its content of alcohol. It is very much richer in alcohol than any beer made in the ordinary way and contains even more than light wines. The Japanese beer also contains very much less solid matter, so-called extract, than common beer. The greatest difference of all is that it is not hopped. The color and appearance of sake is more that of a light white wine, and its taste resembles to some extent that of some varieties of sherry. Sake is made from rice, water and koji. In making koji, according to an article in "Pure Products," the grains which have not been powdered but simply freed from their hulls are washed with water and allowed to soak for 12 or 18 hours and then are placed in a tank with a perforated bottom. In this tank they are mashed in the usual way by means of hot water or steam until the starch is reduced to a pasty condition although the grains themselves are not broken, but simply become soft and swell. Any germs which have been separated in the original pounding or crushing or rise to the surface are removed, and the grains of rice become very soft. After cooling to ordinary temperature this mass of soft grain is treated with fungus, consisting principally of the spores of *aspergillus oryzæ*. Different kinds of yeast are also found in this yellow powder. The soft rice is acted upon by these yeasts and molds producing a considerable rise of temperature. The heaps of small rice grains are stirred from time to time in order that the temperature may not rise too high. When this process is finished the resulting material is called koji.

Koji.—In the making of Japanese beer the koji has something of the same function as mold.   The actual brewing is accomplished by using boiled rice, koji and water.   The enzym of the koji acts on this mixture and converts the starch into sugar and thus prepares it for fermentation.   The fermented liquor is separated from the solid ingredients, filtered, sterilized and stored. Under the Japanese law also a small quantity of salicylic acid may be used for preserving the sake.   The beer thus produced is the national beverage of Japan.   Quite a great deal of it is sent into the United States for the use of the citizens of Japan who are residents of this country.   In Japan the price of the sake varies from 7 to 24 cents per liter according to its quality.   The sake is usually warmed and drunk from a small dish resembling a saucer.

Kefir.—Among the fermented beverages which are made from milk kefir has a high rank.   It is a beverage prepared from cow's milk by the addition of a special yeast, commonly called kefir grains.   Kefir, is nearly related to koumys, and has its origin in the same country.   It is made especially in the northern part of the Caucasian mountains.   Koumys, is or should be prepared exclusively from mare's milk, and with a different kind of yeast.

Yeast.—The yeast which is used in the preparation of kefir is found in grains of various sizes, and has an appearance something like cauliflower.   Their color, when dry, is yellowish or reddish, while, when moist, they are almost white.   This yeast is composed chiefly of the ordinary yeast used in the preparation of beer, namely: *saccharomyces cerevisiæ.*

Formerly this yeast was difficult to secure, it being carefully guarded as of sacred origin, and could rarely be bought.   During the last 20 years it has come gradually into commerce, until now it can be obtained in large quantities in the markets of the Caucasus.

In the country of its origin it was customary to give a quantity of this yeast as a dower to a daughter of the house upon her marriage.   The yeast itself is supposed to be derived originally from the same yeast which produces the koumys.   It is thought by growing a koumys yeast in cow's milk it has gradually been transformed into its present form of kefir grains.

Fermentation.—The action of the yeast upon the cow's milk produces fermentations of various kinds.   Among others the

18

alcoholic fermentation is worthy of note. There are also lactic ferments, and others, producing a highly desirable flavor in the beverage. In addition to this there is more or less peptonized or albuminous substances which render the beverage wholesome and palatable.

The fermentation starts rather rapidly, and in good conditions a very good fermented beverage is made in 24 hours. Much better beverages are produced, however, by fermentation which lasts three or four days, care being taken meanwhile to carefully shake the bottles every two or three hours in order that coagulation be avoided.

Quality of the Milk.—In the preparation of kefir the milk should be fresh, and a better product is made if it be skimmed and sterilized so that the fermentation when finished will be kefir grains alone. A richer product in alcohol is also secured if some additional milk and sugar are added before fermentation.

Further information regarding koumys is found in "Foods and Their Adulterations," 3d edition, pages 179–80. Yoghurt is a fermented milk produced by the action of the bacillus bulgaricus.

PART X

WHISKY

Foreword.—The whole subject of alcoholic beverages is one difficult of treatment, because of the growing conviction that even when continuously used in small quantities alcohol in the end will prove injurious. There is a strong tendency developing in this and other countries looking to the limitation of the traffic in alcoholic beverages both from a health and ethical standpoint. There is a rising tide of public sentiment favorable to entire prohibition. This is largely due to the fact that attempted limitations of the traffic have not been attended with the success expected. Even in dry territory the ease of access of alcoholic beverages has done much to neutralize the good effects of prohibitory laws.

There is a marked tendency in social entertainment to diminish the supply of alcoholic beverages. Not only are the more dangerous kinds, such as distilled liquors, cocktails, cordials and other mixtures, more or less under ban, but even the serving of wine and beer is growing more and more infrequent. When these beverages are served the quantities consumed are becoming gradually less. In fact, there are signs which are not without significance of the approach of a nation and perhaps a world-wide prohibition. The Czar of Russia before his deposition under the stress of military exigency forbade the manufacture and use of distilled alcoholic beverages (Vodka) in Russia. Lord Kitchener just before his tragic death appealed to the patriotic citizens of Great Britain to omit any alcoholic beverage of any kind in sending favors to the soldiers. France is considering a further inhibition of the manufacture and sale of certain distilled and compounded beverages. In the United States a number of states have voted for prohibition and five more were added to the dry number by the 1916 and 1917 elections. During the continuation of the war with Germany and Austria the manufacture of all distilled alcoholic beverages in the United States ceased at midnight, September 7, 1917.

What is Whisky?—Nevertheless, a work on beverages would be extremely incomplete without reference to those of an alcoholic nature which have been in use almost since the beginning of history, and which have been extensively consumed by mankind for centuries. The magnitude of the industry, the extent of the habit of using at least small quantities of alcoholic beverages and the general interest in the question still maintains the primacy of this subject among the beverages of the world.

There is none of them over which a fiercer controversy has taken place than that of whisky. "What is Whisky?" has become not only a local but a nation-wide problem.

This problem has entered into the administration of the food laws, it has been the subject of consideration by cabinet officers and Presidents of the United States, and has frequently been brought before the courts of the country for adjudication. A discussion of whisky as a beverage will therefore render unnecessary any very extensive discussion of other forms of distilled alcoholic beverages. For this reason this discussion is given here in some detail.

Origin of the Term Whisky.—There is a very common consensus of opinion, based upon historical error, however, in the belief that the term "Usquebaugh" as used in Ireland, and by Burns in some of his poems, is synonymous with, and identical in composition with, modern whisky. Nothing could be more erroneous than this, as can be shown by numerous historical works.

In all the works relating to the making of cordials and punches practically similar recipes are given for the material known as "Usquebaugh." It is not, therefore, historically correct to consider ancient or modern whisky as the same substance as usquebaugh. On the other hand, usquebaugh was a punch which could be made either with whisky or brandy, as the case might be, and it was a punch or a cordial, and not a straight distillate.

In 1731, we find a recipe for making usquebaugh, in a book entitled:

"A complete Body of Distilling, Explaining the Mysteries of that Science," by G. Smith, of Kendall in Westmoreland. London: Printed for Henry Lintot. In this book is a recipe for making fine usquebaugh, as follows:

	£.	s.	d.
5 gallons of proof Molossus spirits at 2s. 7d. ½........	00	13	01½
6 gallons of proof rectify'd malt spirits at 20d.........	00	10	00
Mace and Cloves each one ounce and a half...........	00	03	09
Nutts four ounces and a half ⎱ .... Cinnamon 3 ounces ⎰	00	05	01½
Coriander-seed, Ginger, each 3 ounces................	00	00	03
Cubebs 1 ounce and a half.........................	00	00	04½
Raisins 4 pound ½................................	00	01	06
Dates 3 pound....................................	00	04	00
Liquorice 2 pounds ¼.............................	00	01	06
Best English Saffron 4 ounces and a half.............	00	11	03
10 pound Lisbon sugar at 7d. ½....................	00	06	03
	02	17	01½

For Sale

	£.	s.	d.
10 gallons of fine Usquebaugh at 20s. per gallon.......	10	00	00

Scotch Whisky.—The name whisky is a corruption of the old Erse word *usquebagh* or "water of life"—*aqua vitæ*—only if *usquebagh* means the water of life, it should be written *Iskebeogh*. But in ancient Ireland *usquebagh* meant distilled spirits of any kind, whereas in Scotland whisky means— just whisky. While in England *aqua vitæ* was distilled, and was even subject to duty, in the 17th century whisky does not seem to have been a recognized product in Scotland until after the Union, whereas the Irish were distilling spirits from malt at least as far back as 1590. In 1776 the export of Scotch whisky to England began, and it has never ceased since. For the first half-century, however, Scotch whisky did not go direct into the hands— and mouths—of the English consumer. It went to rectifiers, who treated it with drugs and flavoring essences and sent it out as English gin or French brandy. It was not until 1825 that Scotch whisky became recognized on its own merits outside Scotland. And in those days "sma 'stills" of the smugglers were legion. They, indeed, formed the foundation of the malt whisky industry of the Highlands. With a restricted supply of material, and working in limited spaces in remote corners, these smugglers could only operate with local barley and malt, and by the process of distillation by pot still over a peat fire.

Thus an old authority describes the beginnings of the Scotch whisky industry. Morewood makes the following statement on page 617 of his celebrated work on distillation, in regard to whisky and usquebaugh:

The application of the letter X to whisky, ale, or porter, was, and continues to be, a distinguishing mark of its strength and purity; and lest the single character might not be sufficient to indicate the strength of some of our malt liquors, it has been doubled, as in the instance of double X porter, now so strongly recommended by the faculty, for its refreshing and strengthening qualities. To usquebaugh the letter X has never been applied, because this appellation was extended to aqua vitæ in its compound state after the admixture of raisins, fennel-seed, and other ingredients, to mitigate its heat, render it more pleasant, less inflammatory, and more refreshing. . . .

With respect to the nature or peculiarity of the spirits used in those times, it is now not easy to determine; but usquebaugh seems to have been a general name for all compounded spirits, and plain whisky as we have it at present, was not the common beverage, it being customary to infuse the liquor with some savory, or tasty ingredients.

The Earliest Recipe for Usquebaugh in Print.[1]—In a little book, Delights for Ladies, etc., 1602, is the following recipe for Usquebath, or Irish Aqua Vitæ:

To every gallon of good Aqua Composita, put two ounces of chosen liquerice, bruised and cut into small peeces, but first cleansed from all his filth, and two ounces of Annis seeds that are clean and bruised. Let them macerate five or six daies in a wodden Vessel, stopping the same close, and then draw off as much as will runne cleere, dissolving in that cleare Aqua Vitæ five or six spoonfuls of the best Malassoes you can get; Spanish cute, if you can get it, is thought better than Malassoes; then put this into another vessel; and after three or four daies (the more the better), when the liquor hath fined itself, you may use the same; some added Dates and Raisons of the Sun to their receipt: those groundes which remaine, you may redistill, and make more Aqua Composita of them, and of that Aqua Composita you may make more Usquebath.

It is therefore evident that while "whisky" is probably "usque," it is not now and never was an artificially compounded article. The unjust claim that has been made that only rectifiers can make genuine whisky is therefore without any historical justification.

Scientific Definition of Whisky.—Whisky is the distillate from the fermented mash of malt, or unmalted cereals, the starch of which is converted by malt, and which contains all the natural volatile principles which are present in the malt or formed at the time of fermentation congeneric with ethyl alcohol and which are volatile at the ordinary temperatures of distillation.

This general definition of whisky will be sufficient to distinguish it from pure ethyl alcohol or what is known as velvet, neutral or cologne spirit in the United States, and silent spirits in Great Britain.

[1] Drinks of the World, by James Mew and John Ashton, published in 1892 in London.

Neutral Spirit.—Neutral spirit, alcohol, cologne spirit, velvet spirit and silent spirit are all alike chemically, and consist of a very pure alcohol; regardless of the source from which it may be derived. In other words rum, brandy and whisky owe their properties as beverages not to ethyl alcohol, which is just the same in character and in quantity in all three, but to the associated principles which are natural to the parent substances or which are produced during fermentation and distilled with the ethyl alcohol, or developed on ageing. These principles are dfferent in all three cases. In the one case they give to the distillate the properties of brandy; in the second case the properties of rum; and in the third case the properties of whisky.

The Congeneric Principles in Whisky.—The question is pertinently raised here as to what these congeneric principles are in a whisky. Chemists have discovered in whisky from 20 to 40 different substances, and all of these are necessary to make a real whisky, even though they be present, as they are usually, in minute quantities.

In this work it will be necessary only to mention the principal congeneric products which are found in a whisky. The reader may consult technical and chemical works for details.

In addition to ethyl alcohol, which is the principle volatile constituent of whisky, there are found the following important constituents:

Fusel Oil.—In the fermentation of the mash of cereals, the starch of which has been converted into sugar by malt, there is formed not only ethyl alcohol in great abundance, but also small quantities of other associated alcohols. Among these the chief are butyl, propyl and amyl alcohol. The latter is the most abundant of these three. There are doubtless traces of other alcohols, and they are important even though present in minute quantity.

These alcohols together form the oily constituent which is characteristic of whisky, especially of old whisky, and they have collectively been called "fusel oil." This is rather an unfortunate term, because it is associated with the bad taste and bad odor of the commercial article known under that name. The bad taste and the bad odor of the article known to commerce as fusel oil are not due to anyone of the three alcohols mentioned, but to certain other volatile bodies mostly formed during fermentation.

The case is very much the same as that of wood alcohol, chemically known as methyl alcohol. Crude wood alcohol has such an excessively repugnant and nauseating taste and odor that it is employed in many countries for denaturing ethyl alcohol and making it unfit for consumption. The bad taste and odor of wood alcohol are not due to the methyl alcohol, which it contains, but to the impurities which are generated during the manufacture of the product.

When wood alcohol is purified and refined, so as to become a chemical unit, it is not unpleasant to the taste nor to the nostrils. Such a product is called Columbian Spirits.

Quantity of Fusel Oil.—The quantity of fusel oil in a whisky is not very large, but this is the principal component of the whisky aside from the ethyl alcohol and the water. The average quantity of fusel oil in a whisky may be stated as $25/100$ of 1 percent for alcohol of 100 proof; that is, 50 percent of alcohol by volume.

Ethers and Aldehydes.—The other important ingredients of whisky are in the order of magnitude—ethers, acids and aldehydes. The ethers especially are the bodies which give fragrance and bouquet to the whisky. It is well known that, when alcohols are oxidized, they form volatile and very fragrant products which are known as ethers. For instance, the common ether which is used as an anesthetic is produced from ethyl alcohol, usually by treating it with sulphuric acid. All of the alcohols form fragrant ethers, and this is particularly true of the propyl, butyl and amyl alcohols. These ethers exist in limited quantities at the moment of distillation, but larger quantities of them are formed during the process of ageing.

The principal acid which is found in whisky is acetic acid, although others may exist in small traces. This acid is found in the freshly distilled whisky, but increases in quantity during ageing. The acids of whisky have the faculty of uniting with the ethers to form ethereal compounds. For instance, acetic acid combined with ordinary ether forms ethyl acetate, which is a very fragrant compound, giving taste and bouquet to the mixture. There are traces of other organic principles in whisky, but these unite with other bodies and form compounds more or less stable, all of which tend to increase fragrance and bouquet.

The aldehydes are regarded as the most objectionable form of compounds in a whisky. The aldehydes are nearly related to alcohols, but only extremely small quantities are found in the freshly distilled liquor. These quantities of aldehydes tend to decrease during the ageing process, and they are converted into other fragrant bodies, which add to the bouquet and character of the liquor.

The chief constituents occurring in the products of alcoholic fermentation are as follows:

Ethyl alcohol;
Normal propyl alcohol;
Iso-propyl alcohol;
Alpha-normal butyl alcohol;
Beta-iso-primary butyl alcohol;
Tertiary butyl alcohol;
Alpha-normal primary amyl alcohol;
Beta-iso-primary amyl alcohol;
Gamma-iso-primary amyl alcohol;
Methyl-propyl carbinol;
Iso-primary hexyl alcohol;
Iso-primary heptyl alcohol;
Acetic acid;
Ethyl acetate;
Ethyl valerate;
Amyl acetate;
Amyl valerate;
Aldehyde;
Acetone;
Acetal (diethyl aldehyde);
Furfurol (furfuraldehyde);
Pyridin;
Ethyl caproic ester;
Ethyl caprylic ester;
Ethyl capric ester;
Ethyl pelargonic ester,
Ethyl butyric ester;
Ethyl acetic ester;
Amyl caproic ester,
Amyl caprylic ester;
Amyl capric ester;
Amyl pelargonic ester;
Amyl butyric ester;

Amyl acetic ester;
Ethyl formate;
Amyl formate;
Terpene;
Terpene hydrate;
Normal nonylalcohol;
Secondary nonylalcohol;
Stearic acid;
Palmitic acid;
Lauric acid;
Acetaldehyde;
Isobutyleneglycol;
Benzaldehyde-cyanhydrin;
Hydrocyanic acid;
Benzoic acid;
Benzoic esters;
Ammonia;
Organic bases;
Ethyl sulphid;
Amyl sulphid;
Mercaptan;
Ethyl sulphite;
Amyl sulphite;
Trimethylamine and other amines;
Collidin and homologues;
Beta-glykosin;
Pyrazin and derivatives;
Trimethyl pyrazin;
2–5 dimethyl pyrazin;
2–5 dimethyl piperazin;
Hydrogen sulphid;
Dimethylketol.

The common artificial essences which are used are: rye flavor, Bourbon flavor, brandy flavor, rum flavor, Scotch flavor and gin flavor. Beading oil is also employed.

The above is a formidable array and shows how very composite a substance whisky is. All of these bodies, however, taken together, numerous as they are, form, as a rule, less than 1 per-cent of the total weight of the whisky. Yet it is to this very small portion that the .whisky owes its character, property and value. It must not be supposed that all or any considerable number of these compounds are found in whisky. Some of them are peculiar to rum and brandy.

As an illustration of how small a quantity of constituents may give character the case of maple sugar may be cited. The peculiar flavor of maple is extremely minute, in fact, so small that it has not yet been isolated and weighed by any chemist, and yet it serves to give to the sugar its distinctive name, its character, and its quality. The real sugar in maple sugar is exactly the same that is found in sugar cane, or the sugar beet. In a sugar beet, however, the sugar is associated with bad tasting elements, so that raw beet sugar is not fit to eat. A beet sugar must be purified and refined before it may be consumed. If one were to purify and refine maple sugar, as is the case with beet sugar, it would be making a product identical in properties with refined beet sugar, and thus it would lose all of its valuable and distinctive characteristics. In the same way, if one were to refine whisky to produce pure alcohol from it, as is easily done, the pure alcohol would have exactly the same qualities as that derived from brandy, from rum, or from potatoes, and would lose all of its character, property and quality as a beverage.

From the above general description it is seen what the true difference is between a whisky and alcohol on the one hand and whisky and brandy and rum on the other hand.

Varieties of Whisky.—Many varieties of whisky are known, due either to the materials used in their composition, variations and methods of manufacture, or the country in which they are made. It is very remarkable that minute changes in the manu-facture of a distilled spirit, due to the kind of material employed, the method of fermentation, the shape of the still which is used, and the method of manipulation will change profoundly the character of the product.

American Whisky.—There are two principal varieties of American whisky. The first of these is made chiefly from Indian corn. There is just enough malt used with it to convert the starch or Indian corn into sugar. The other ingredient is rye, the starch of which is also converted into sugar by malt. When a corn and rye whisky is made in Kentucky it is called Bourbon. Roughly speaking the average composition of a mash for a bourbon whisky is Indian corn 65 percent, rye 20 percent, and malt 15 percent. There are many variations from this formula, but it may be regarded as a type of the raw materials from which the bourbon whiskies are made.

There is another variety of whisky made from Indian corn, which is known as corn whisky. Corn whisky is supposed to be made from corn alone. The Indian corn is sprouted, in other words, malted, by which process the starch therein is converted into sugar, then mashed, fermented and distilled. There is probably very little pure corn whisky made in the United States today. The corn whisky is commonly made by mixing Indian corn with barley malt, thus converting the starch into sugar. Corn whisky is often drunk fresh, that is, without being stored in wood, hence it is nearly colorless, like all freshly distilled whisky. It is the chief kind of whisky made by "Moonshiners."

The other important variety of whisky manufactured in the United States is rye. There is, I believe, very little pure rye whisky made in the United States, that is, whisky made from malted rye.

Official Definitions of Whisky.—Various official definitions of whisky have been given by different authorities. These definitions are not wholly concordant, some of them differing in many essential and basic points.

The object of the revenue laws being to raise money, the descriptions of the products are not important, and hence, as a rule, are not given. The revenue tax is laid on proof spirit; that is, on the total volume of spirit produced, calculated to a volume in which the quantity of ethyl alcohol is exactly one-half by volume that of the whole liquid.

Various kinds of spirits are made in distilleries, under the direction of the revenue law, but most of them have exactly the same rate of tax.

Before the enactment of the revenue law there was no supervision whatever of the manufacture of whisky, except that which was imposed by the conditions of trade. Often the whisky was used almost in the fresh state; at other times, through exigencies of the trade, it was kept for a greater or less length of time in wood. It was soon found that if freshly distilled whisky was placed in wood it began to acquire a better flavor, and became more palatable, more aromatic, and more desirable in every way. These changes were supposed to be due to the elimination of the more harsh and disagreeable components of whisky, by absorption, by oxidation or otherwise. Especially, was it supposed that the harsh taste and bad odor of the newly distilled whisky was due to the substance or substances known as "fusel oil." As this harsh taste and bad odor disappeared, it was reasonable to assume that the cause of it, namely, fusel oil, disappeared also. Modern chemical investigations show that substances which were supposed to be fusel oil, namely, the higher alcohols, do not disappear on ageing, except in the proportion as the contents of the cask may be evaporated. Even then, it has been discovered that the so-called fusel oils evaporate less rapidly than the ethyl alcohol, so that at the end of the ageing period the relative content of the "fusel oil" may be actually increased. The improvement in the character of the liquor on ageing must be due to some other cause. The chemist has discovered that there is a change taking place in the constituents of the distilled liquor, by which various forms of combination are secured, certain oxidations accomplished, certain ethers, esters, and acids developed, and as the result of the action of all these factors, the flavor, character and aroma of the distillate are vastly improved. There is, therefore, a sharp line to be drawn between new whisky, freshly distilled, and old whisky that has been properly matured and mellowed.

Old and New Whisky.—There is no exact legal definition of the difference between new and old whisky. A definition, however, is employed in the "Bottled In Bond" Act, where it is provided that distilled spirits may be bottled in bond, under the supervision of the United States officials, provided they are at least four years old. This provision of the law may be reasonably construed to mark the line between new and old whisky. It is, to be sure, an arbitrary limit fixed by the Legislature, but it also has a reasonable basis, since after four years of storage there is such a great improvement in the liquor itself as to warrant the transposal from the new to the matured class.

It is true that distilled spirits will improve in wood for a much longer period than four years. They are left sometimes in wood for 8, 12, 16, and even 20 years, and in many cases improvement is noticed throughout the whole time. The cost of the liquor which has been stored so long a time, and during which storage has lost a large portion of its volume, is so great that distilled liquors are kept only over these great lengths of time to supply the wants of connoisseurs who have sufficient income to warrant the purchase of such expensive liquors.

Definitions of Whisky.—This is a question which has puzzled many competent and conscientious investigators, and also a question which has been answered in many different ways.

In countries where food laws exist the distilled beverages naturally come within the scope of such laws, and hence, their proper naming and designations are matters of legal supervision.

In the United States such supervision has been exercised and the proper designation of "whisky" to a distilled spirit limited by various decisions of executive officers and courts.

The first decision of this kind was issued by the Secretary of Agriculture on the recommendation of the Bureau of Chemistry, on December 1, 1906. This was in compliance with the request of manufacturers, and was couched in the following language:

Question Propounded.—On account of the uncertainty prevailing in our trade at the present time as to how to proceed under the Pure Food Law and regulations regarding what will be considered a blend of whisky, I am taking the liberty of forwarding to you to-day two samples of whisky made up as follows:

Sample A contains 51 percent of Bourbon whisky and 49 percent of neutral spirits. In this sample a small amount of burnt sugar is used for coloring, and a small amount of prune juice is used for flavoring, neither of which increases the volume to any great extent.

Sample B contains 51 percent of neutral spirits and 49 percent of Bourbon whisky. Burnt sugar is used for coloring, and prune juice is used for flavoring, neither of which increase the volume to any great extent.

I have marked these packages "blended whiskies" and want your ruling as to whether it is proper to thus brand and label such goods.

My inquiry is for the purpose of guiding the large manufacturing interests in the trade that I represent.

The question was answered on my recommendation by the
Secretary of Agriculture in the following language:

The question presented is whether neutral spirits may be added to
Bourbon whisky in varying quantities, colored and flavored, and the resulting mixtures be labeled "blended whiskies." To permit the use of the word
"whiskies" in the described mixtures is to admit that flavor and color can
be added to neutral spirits and the resulting mixture be labeled "whisky."
The Department is of opinion that the mixtures presented cannot legally
be labeled either "blended whiskies" or "blended whisky." The use of the
plural of the word "whisky" in the first case is evidently improper, for the
reason that there is only one whisky in the mixture. If neutral spirit, also
known as cologne spirit, silent spirit, or alcohol, be diluted with water to a
proper proof for consumption and artificially colored and artificially flavored,
it does not become a whisky, but a "spurious imitation" thereof, not entirely
unlike that defined in section 3244, Revised Statutes. The mixture of such
an imitation with a genuine article cannot be regarded as a mixture of like
substances within the letter and intent of the law.

Appeal was made from this decision and the matter was finally
referred to the Attorney-General of the United States, by order
of President Roosevelt. The Attorney-General of the United
States issued a decision on this question on April 12, 1907, the
principal parts of which are as follows:

Attorney-General Bonaparte said, "I conclude that a combination of
whisky with ethyl alcohol, supposing, of course, that there is enough whisky
in it to make it a real compound and not the mere semblance of one, may be
fairly called, 'Whisky;' provided the name is accompanied by the word
'Compound' or 'Compounded,' and provided a statement of the presence
of another spirit is included in substance in the title. I am strengthened in
this conclusion by understanding from the papers you have referred to me
that it has been reached by the Department of Agriculture as well.

The following seem to me appropriate specimen brands or labels for
(1) 'straight' whisky, (2) a mixture of two or more 'straight' whiskies,
(3) a mixture of 'straight' whisky and ethyl alcohol, and (4) ethyl alcohol
flavored and colored so as to taste, smell, and look like whisky:

1. Semper Idem Whisky: A pure, straight whisky mellowed by age.

2. E Pluribus Unum Whisky: A blend of pure, straight whiskies with all
the merits of each.

3. Modern Improved Whisky: A compound of pure grain distillates,
mellow and free from harmful impurities.

4. Something Better than Whisky: An imitation under the Pure Food
Law, free from fusel oil and other impurities.

In the third specimen it is assumed that both the whisky and the alcohol are distilled from grain."

This decision of the Attorney-General of the United States was contested by the various interests, in the courts of the United States, and among others, two important decisions were made in the matter, one by the District Court for the Southern District of Ohio, Judge Thompson, presiding, and the other by the District Court for the Southern District of Illinois, Judge Humphrys, presiding, both sustaining the opinion of the Attorney-General.

These decisions of the court were not satisfactory to the rectifiers and President Taft referred the question to the Solicitor-General. Solicitor-General Bowers heard very extensive testimony on this case, took it under consideration and came to the conclusions stated further on. This decision was referred to the President of the United States, who, after reviewing the various opinions that had been held overruled his Solicitor-General and decided all questions in favor of the rectifiers, thus nullifying the decisions of the courts.

This decision of the President of the United States and the regulations made by the Internal Revenue Bureau in comformity therewith, are now official in this country, and the acts of the law officers having to do with the subject of whisky are guided accordingly, irrespective of the provisions of the Pure Food Law as upheld by the federal courts.

Opinion of Judge Thompson, District Court for Southern District of Ohio.

. . . The affidavits of the complainants do not show that the people—the consumers—were ever advised or knew that complainants' product branded "whisky" was in fact alcohol artificially colored and flavored, and affidavits submitted by the defendants, 7 of wholesalers and 10 of retailers, in Washington and Cincinnati, do show that no wholesaler or rectifier ever offered to sell to them "as whisky a product which such wholesaler or rectifier represented to be alcoholic colored or flavored," and that they "would not knowingly purchase as whisky such a product." . . .

The fact that this practice has been long continued does not establish a right which the court would protect. It misbrands the product in violation of the Food and Drugs Act and of section 3449 of the Revised Statutes of the United States, if that section is applicable to the case (76 Fed. Rep., page 367). It does not truthfully designate "the kind and quality of the contents of the casks or packages," as required by section 3449. The product is not genuine whisky in the making of which age modifies objectionable elements and develops the flavor which pleases the taste and adds so much to the value. It should be branded "imitation whisky."

Decision of Judge Humphrys, District Court for Southern District of Illinois.

That there is a product called *whisky* and also a product *imitation whisky* the law itself clearly contemplates, and section 3244, in defining what is meant by the business of rectifying, denominates the maker of imitation whisky and other imitation liquors as a rectifier. . . .

The convincing weight of testimony in this subject, given by such men as Professors Frear of Pennsylvania, Scovill of Kentucky, Tolman and Adams of Washington, D. C., Shepherd of South Dakota, Jenkins of Connecticut, Fischer of Wisconsin, and many other State Analysts and chemists of repute, is to the effect that neutral spirits reduced by water to potable strength from which most of the fusel oil has been removed, is not a like substance with whisky. . . .

The record also shows that diluted spirits treated with artificial coloring matter and essences are not sold to the trade as such, but are always presented under such labels, terms, and descriptions as import age and maturity, and which the consumer identifies with the genuine product whisky.

The fact that this practice had, to some extent, prevailed for many years does not show in the complainants any right which the court should protect. It shows rather that the Commissioner of Internal Revenue has been tardy in promulgating a regulation which he had legal power to enforce, even before Congress gave emphasis to the subject by the enactment of the Food and Drugs Act.

As a result of the testimony taken by Solicitor-General Bowers on the following questions, he came to the following conclusions:

I

What was the article called whisky as known (1) to the manufacturers, (2) to the trade, and (3) to the consumers at and prior to the date of the passage of the Pure Food Law?

My opinion upon, and answer to, this question is:

1. The article called whisky as known to the manufacturers at and prior to the date of the passage of the Pure Food Law was:

(*a*) What is often spoken of as "straight whisky," made from grain.

(*b*) Also, what is often spoken of as "rectified whisky," made from grain, when not a mere neutral spirit, as described in section (*d*), below, of the answers to this question 1.

(*c*) Also, a mixture of straight whiskies, or of rectified whiskies, or of straight whisky and rectified whisky, or of straight whisky and what is often known as neutral spirit (made from grain), or of rectified whisky and such neutral spirit (made from grain), or of straight whisky, rectified whisky and such neutral spirit (made from grain), if in the particular case the mixture satisfied the description of whisky given below in answer to question II.

(d) Also, neutral spirit—being a distillate from grain, which lacks a substantial amount of by-products (other than alcohol) derived by distillation from grain and giving distinctive flavor and properties—when, but only when, colored and flavored and sold by the manufacturer to a retailer; but the purchasing retailer in such case seldom knew that in fact he was getting neutral spirit, colored and flavored.

Such neutral spirit made from grain was not known to the manufacturer as whisky in the dealings of distillers with rectifiers; and I do not consider that it is proved to have been known as whisky in the dealings of distillers or rectifiers with wholesalers. A neutral spirit made from molasses, potatoes, or any other substance than grain, has not been known to manufacturers as whisky, except in very rare cases.

2. The article called whisky as known to the trade at and prior to the date of the passage of the Pure Food Law was:

(a) What is often spoken of as "straight whisky," made from grain.

(b) Also, what is often spoken of as "rectified whisky," if conforming to the description of whisky given below in answer to question II.

(c) Also, a mixture of straight whiskies, or of rectified whiskies, or of straight whisky and rectified whisky, or of straight whisky and what is often known as neutral spirit (made from grain), or of rectified whisky and such neutral spirit (made from grain), or of straight whisky, rectified whisky, and such neutral spirit (made from grain), if in the particular case the mixture satisfied the description of whisky given below in answer to question II.

(d) Also, neutral spirit—being a distillate from grain, which lacks a substantial amount of by-products (other than alcohol), derived by distillation from grain and giving distinctive flavor and properties—when colored and flavored; except that neutral spirit was not known to retail dealers as whisky, because such retailers seldom were aware that the article which they were buying or selling was in fact neutral spirit.

A neutral spirit made from molasses, potatoes, or other substance than grain has not been known to the trade as whisky.

3. The article called whisky as known to the consumers at and prior to the date of the passage of the Pure Food Law was:

(a) What is often spoken of as "straight whisky," made from grain.

(b) Also, what is often spoken of as "rectified whisky," if conforming to the description of whisky given below in answer to question II.

(c) Also, a mixture of straight whiskies, or of rectified whiskies, or of straight whisky and rectified whisky, or of straight whisky and what is often known as neutral spirit (made from grain), or of rectified whisky and such neutral spirit (made from grain), or of straight whisky, rectified whisky,

19

and such neutral spirit (made from grain), if in the particular case the mixture satisfied the description of whisky given below in answer to question II.

A neutral spirit derived by distillation from anything else than grain has not been known to the consumer as whisky, whether or not it was colored or flavored or both colored and flavored; and a neutral spirit derived by distillation from grain, but lacking a substantial amount of by-products (other than alcohol) which are derived by distillation from grain and give distinctive flavor and properties, has not been known to the consumer as whisky, whether or not it was colored or flavored or both colored and flavored.

The second question is:

II

What did the term whisky include?

My opinion upon and answer to this question is:

The term "whisky" included, both at and prior to the date of the passage of the Pure Food Law, and has since included, the spirituous liquor composed of (1) alcohol derived by distillation from grain; (2) a substantial amount of by-products (often spoken of as congenors) likewise derived by distillation from grain and giving distinctive flavor and properties; (3) water sufficient without unreasonable dilution, to make the article potable; and (4) in some cases—though such addition is not essential—harmless coloring or flavoring matter, or both, in amount not materially affecting other qualities of whisky than its color or flavor.

A mixture of two or more articles, being each a whisky within the foregoing description, was at and prior to the date of passage of the Pure Food Law, and has since been, whisky. A mixture of one or more whiskies, being each whisky within the foregoing description, with alcohol or a neutral spirit—being an article different from whisky through lack of a substantial amount of by-products derived by distillation from grain, and giving distinctive flavor and properties—is whisky, if the alcohol or neutral spirit is derived by distillation from grain, and if the mixture still conforms to the above general description of whisky; and so it was at and prior to the date of passage of the Pure Food Law.

A spirit derived from any other substance than grain was not at or prior to the date of the passage of the Pure Food Law, and has not since been, whisky; and a mixture of a whisky with such spirit was not at or prior to the passage of the Pure Food Law, and has not since been, whisky.

A neutral spirit derived by distillation from grain, but lacking a substantial amount of by-products derived by distillation from grain and giving distinctive flavor and properties, was not at or prior to the passage of the Pure Food Law, and has not since been, whisky.

*　　　　*　　　　*　　　　*　　　　*

From the above observations it will be seen that there has been a wide variance of opinion among established authorities respecting the correct use of the term "whisky." The President of the United States, W. H. Taft who, after giving the matter consideration, rendered a decision in which he summarizes an answer to the question, "What is Whisky?" as follows:

. . . The term "straight whisky" is well understood in the trade and well understood by consumers. There is no reason, therefore, why those who make straight whisky may not have the brand upon their barrels of straight whisky, with further descriptive terms as "Bourbon" or "Rye" whisky, as the composition of the grain used may justify, and they may properly add, if they choose, that it is aged in the wood.

Those who make whisky of "rectified," "redistilled," or "neutral" spirits cannot complain if, in order to prevent further frauds, they are required to use a brand which shall show exactly the kind of whisky they are selling. For that reason it seems to me fair to require them to brand their product as "whisky made from rectified spirits," or "whisky made from redistilled spirits," or "whisky made from neutral spirits," as the case may be; and if aged in the wood, as sometimes is the case with this class of whiskies, they may add this fact.

A great deal of the liquor sold is a mixture of straight whisky with whisky made from neutral spirits. Now, the question is whether this ought to be regarded as a compound or a blend. The Pure Food Law provides that "in the case of articles labeled, branded, or tagged so as to plainly indicate that they are compounds, imitations, or blends," the term "blend" shall be construed to mean a mixture of like substances, not excluding harmless coloring or flavoring ingredients used for the purpose of coloring and flavoring only. It seems to me that straight whisky and whisky made from neutral spirits, each with more than $99\frac{1}{2}$ percent ethyl alcohol and water, and with less than half of 1 percent of fusel oil, are clearly a mixture of like substances, and that while the latter may have and often does have burnt sugar or caramel to flavor and color it, such coloring and flavoring ingredients may be regarded as for flavoring and coloring only, because the use of burnt sugar to color and flavor spirits as whisky is much older than the coloring and flavoring by the tannin of the charred bark. Therefore, where straight whisky and whisky made from neutral spirits are mixed, it is proper to call them a blend of straight whisky and whisky made from neutral spirits. This is also in accord with the decision of the British Royal Commission in the case which I have cited upon a similar issue.

Canadian Club whisky is a blend of whisky made from neutral spirits and of straight whisky aged in the wood, and its owners and vendors are entitled to brand it as such.

Neutral spirits made from molasses and reduced to potable strength has sometimes been called whisky, but not for a sufficient length of time or under circumstances justifying the conclusion that it is a proper trade name. The distillate from molasses used for drinking has commonly been known as rum. The use of whisky for it is a misbranding.

There are other kinds of liquor in respect to which a decision is invoked, but it is thought that the principles above stated, and the directions above given in specific cases, will furnish a clear precedent for all other cases.

By such an order as this decision indicates the public will be made to know exactly the kind of whisky they buy and drink. If they desire straight whisky, then they can secure it by purchasing what is branded "straight whisky." If they are willing to drink whisky made of neutral spirits, then they can buy it under a brand showing it; and if they are content with a blend of flavors made by the mixture of straight whisky and whisky made of neutral spirits, the brand of the blend upon the package will enable them to buy and drink that which they desire. This was the intent of the act. It injures no man's lawful business, because it only insists upon the statement of the truth in the label. If those who manufacture whisky made of neutral spirits, and wish to call it "whisky" without explanatory phrase, complain because the addition of "neutral spirits" in the label takes away some of their trade, they are without a just ground because they lose their trade merely from a statement of the fact. The straight-whisky men are relieved from all future attempt to pass off neutral-spirits whisky as straight whisky. More than this, if straight whisky or any other kind of whisky is aged in the wood, the fact may be branded on the package, and this claim to public favor may truthfully be put forth. Thus the purpose of the Pure Food Law is fully accomplished in respect of misbranding and truthful branding.

*   *   *   *   *

The regulations of the Bureau of Internal Revenue which control the marking and branding of distilled spirits at the present time (1918) are based on the above opinion but go much farther in validating misbranding and adulterations.

Legislation in Regard to Whisky.—Some of the earliest legislation in America, relating to food products, was in connection with distilled spirits, or whisky.

In Hening's Statutes at Large, of the Laws of Virginia, beginning with the first session of Legislature in 1619, we find the following:

ATT A GRAND ASSEMBLY holden att James Citty the twentieth of November, 1645.
Present Sir William Berkeley, Kn't, Governour, etc.
Act 11, p. 300.
WHEREAS there has been great abuse by the unreasonable rates exacted by ordinary keepers, and retailers of wine and strong waters, *Be it enacted* that no person or persons whatsoever retailing wines or strong waters shall exact or take for any Spanish wines (vizt.) Canary, Mallego, Sherry, Muskadine, Alegant or Tent above the rate of 30 pound of tob'o per gall. And for Maderea and Fyall wine above 20 pound of tob'o per gall. And for all

ffrench wines above the rate of 15 lb. tob'o per gall. And for the best sorts of all English strong waters above the rate of 80 lb. of tobacco per gall. And for aqua vitæ or brandy above the rate of 10 lb. tob'o per gallon.

And if any person or persons retaileing wines or strong waters as aforesaid shall fraudulently mix or corrupt the same, vpon complaint and due proofe made thereof before two commissioners whereof one to be of the quorum, The said commissioners shall by warrant under their hands comand the constables to stave the same, And if any shall take more than such rates sett they to be ffined at double the value of such rates soe exacted.

Licensed Distilling.—As far as historical records go, the first license issued to a distiller in America was given in April, 1652, which is found in the same book quoted above.

At the Grand Assembly James Citty, the 30th Aprill, 1652.

It is ordered that Mr. George Fletcher shall have to himself, his heirs, ex'rs. and adm'rs. liberty to distill and brew in wooden vessels which none have experience in but himself for 14 years, and it is further ordered that no person or persons whatsoever shall make use thereof within this colony without agreeing with the said Mr. Fletcher under the penalty of 100-pound sterl.

Existing Legislation and Regulation.—It is hardly practicable to trace here the progress of legislation in regard to this matter, further than to make mention of what has led up to the present status of the laws relating to whisky and other distilled spirits in the United States.

The present Revenue Laws of the country were incident to the expenses produced by the conduct of the Civil War, and the first of these regulations was enacted in 1862. These laws laid a tax upon all distilled spirits, and defined the process of distillation and rectification. The first law relating to rectifiers was in the act of July 1, 1862, which provided,,

Rectifiers shall pay $25 for each license to rectify any quantity of spirituous liquors, not exceeding 500 barrels or casks, containing not more than 40 gallons to each barrel or cask of liquor so rectified; and $25 additional for each additional 500 such barrels or any fractional part thereof. Every person who rectifies, purifies, or refines spirituous liquors or wines by any process, or mixes distilled spirits, whisky, brandy, gin, or wine, with any other materials for sale under the name of whisky, rum, brandy, gin, wine or any other name or names, shall be regarded as a rectifier under this act.

Numerous additions and changes were made in the Revenue Laws from time to time, including the introduction of the bonding period. The first acts required the tax to be paid upon the spirits as soon as they were produced. Inasmuch as this did not permit the keeping of distilled spirits in wood for ageing purposes until after the tax had been paid, concessions were made in the collection of the revenues in this respect, allowing a certain time to elapse during which the spirits could remain in bond without payment of the tax. The first bonded period was for a year. This was subsequently changed in 1879 to three years, and finally to 8 years, which is as it remains at present.

The distinction between a distiller and rectifier was early established by law, in the following language:

Every person who produces distilled spirits, or who brews or makes mash, wort, or wash fit for distillation or for the production of spirits, or who by any process of vaporization separates alcoholic spirits from any fermented substance, or who making or keeping mash, wort, or wash, has also in his possession or use a still, shall be regraded as a distiller: Provided, That a like tax of four dollars on each barrel, counting 40 gallons of proof spirits to the barrel, shall be assessed and collected from the owner of any distilled spirits which may be in any bonded warehouse at the date of the taking effect of this act, to be paid whenever the same shall be withdrawn from such warehouse, under the provisions of the sixty-second (fifty-sixth) section of this act: Provided, That no tax shall be imposed for any still, stills, or other apparatus used by druggists and chemists for the recovery of alcohol for pharmaceutical and chemical or scientific purposes which has been used in those processes.

Rectifiers of distilled spirits, rectifying, purifying, or refining 200 barrels or less of distilled spirits, counting 40 gallons of proof spirits to the barrel, within the year, shall each pay $200, and shall pay 50 cents for each such barrel produced in excess of 200 barrels. And monthly returns of the quantity and proof of all the spirits purchased and of the number of barrels of spirits, as before described, rectified, purified, or refined by him, shall be made by each rectifier in the same manner as monthly returns of sales are made. Every person who rectifies, purifies, or refines distilled spirits or wines by any process, and every wholesale or retail liquor dealer or compounder of liquors who has in his possession any still or leach-tub, or who shall keep any other apparatus for the purpose of refining in any manner distilled spirits, shall be regarded as a rectifier.

Compounders of liquors shall each pay $25. Every person who, without rectifying, purifying, or refining distilled spirits, shall, by mixing such spirits, wine, or other liquor with any materials, manufacture any spurious, imitation, or compound liquors, for sale under the name of whisky, brandy, gin, rum, wine, spirits, cordials, or wine bitters, or any other name, shall be regarded as a compounder of liquors.

An Ancient Treatise on Whisky.—As long ago as 1740 the problem of whisky in its relations to health and politics was under discussion. A physician writing under the name of E. B. printed a treatise on the virtues of whisky considered in its physical and political uses, which was printed at Dublin in 1740. It was written especially in relation to the conditions existing in Ireland. A somewhat learned but rather antiquated discussion of the virtues of the grains from which whisky is made is given as an introduction to the treatise on whisky itself. Attention is drawn to the fact that most animals live on grain, although the carnivorous animals are dignified by being mentioned. Man has half his food flesh, and the immoderate use of flesh is regarded as the cause of most of our distempers and particularly as the basis of gout and scurvy. It follows, therefore, in the course of the argument that if corn is used in the sense of barley, wheat, oats and rye, the liquor made from it has the same virtues. In other words, and in the language of the author, "All nourishment drawn from corn, though variously prepared, must answer the same good end in some degree, and share the virtues of its source and original."

The definition of whiskey nearly 200 years ago is most interesting. It is as follows:

Whisky is a water highly impregnated with the most spirituous part of the corn, and highly exalted by fire; it is distilled from oats or barley malted, replete with all the good qualities and nutritious parts of the grain, and cannot but be exceeding nourishing.

Further on in the treatise on whisky the author says:

Whisky drank too new, may hurt; but this is no objection to the liquor itself, but against the abuse of it; because whisky may be kept to grow old, and to become mild as well as brandy and rum. If people will drink it hot from the pipe, that is no fault of the liquor.

In argument that whisky is more wholesome than brandy the author says:

But why should not a spirit distilled from corn be more salutiferous to a human constitution than a spirit extracted from the lees and sediments of the worst of wines, as brandy is?

The effect of whisky in those days seems somewhat peculiar, as the author says:

Whosoever has drank of it in its purity, either in punch, or otherwise, has found his head easy after it, and not distracted, hot and crazy as after brandy or rum, it is a lighter and more volatile spirit than either, or any other spirit, it exhilarates sooner, and evaporates the speedier, but it is not so hot and acrid, so active and penetrating as brandy and rum. It is a grand opiate, and a most excellent narcotic, a great opener and emollient; and that it is a mild and gentle sudorific, all who have drank it will concede: It exhilarates quickly, but like a merry companion, does not tease us by the prolixity of its visit. It leaves us cheerful, it actuates the spirits and puts them into a regular motion, not confusing and perplexing, heating and fatiguing them like the other spirits, whose very qualities must produce this hurry.

The compounding of spirits, that is, the art of the rectifier, seems to have been as objectionable in those days as at the present time, for the author says:

Whenever foreign spirits grow scarce or dear, as is often the case, we are then sure to have the pernicious trade go on of mixing spirituous liquors, and instead even of those foreign spirits (tho' bad enough) we get some compound much more unwholesome, the consequences whereof are, the loss of our health, and entailing distempers on our posterity; but as the neatest and best whisky is always of a moderate price, it is less liable to be adulterated than those foreign spirits; and it will be the interest of every person who deals in whisky to have it genuine, as well for the publick good, as their own private benefit.

Early Distillation in America.—In "Bishop's History of American Manufactures," 1608–1860, are quotations from many authorities concerning distillation as an industry in the American colonies. Bishop quotes from Stith's History of Virginia, London, 1753, the instructions to Sir Francis Wyatt in 1621 to "withdraw attention from tobacco and to direct it to corn, wine, silk and others already mentioned; to the making of *oil of walnuts*, and employing the apothecaries in distillation; and searching the country for minerals, dyes, gums, drugs, and the like."

There is a description of a "still and worm" recently set up in London, Conn., for distilling rum from the molasses procured there in exchange for the exports of the colony.

In 1739 there were in Newport 30 distilleries devoted to the manufacture of rum, "a staple article in the African slave trade, which tarnished the fair fame of its enterprising traders."

The distillation of grain was evidently started about this time. "In 1676, seven years before the settlement of Philadelphia, the Court ordered that no grain shall be distilled, unless it be unfit to grind and bolt: a measure proposed by the town of New Castle in 1671."

In 1633 the West India Company, through their Director, Van Twiller, caused the erection of mills and other buildings including a Brewery upon Farm No. 1, extending from the present Wall Street westward to Hudson Street.

From that time forth the place continued well supplied with the national drink. The distillation of brandy commenced there as early as 1640, which was probably the first instance of that manufacture in the Colonies.

It is probable that this brandy was what is known as British Brandy, or a distillate from grain, as it is not likely that there was any wine to be distilled as early as 1640. It appears, therefore, that a straight distillate from grain, now known as whisky, was first known in this country as brandy.

In the following year, drunkenness had become so alarmingly prevalent that, to abate the disorders arising from it, and to secure a better observance of the Sabbath, the municipal authorities of the town, in April of that year, prohibited the tapping of Beer during divine service, or after 10 o'clock at night, under a penalty of 25 guilders, or $10, for each offense, beside the forfeiture of the Beer for the use of the "Schout Fiscaal," or Attorney-General.

From the very beginning it appears that fraud was introduced into the trade, since the Burgomasters passed the following resolution:

Whereas, complaints are made that some of our inhabitants have commenced to tap Beer during divine service and use a small kind of measure, which is in contempt of our religion and must ruin the State, etc.

In 1676 the distillation of grain in the New Netherlands was again forbidden on account of the scarcity of breadstuffs. The use of grain for making beer was not, however, forbidden. In Albany as early as 1747 beer making was largely practised and barley largely grown near that city. An account is made of Harman Gansevoort who was one of the most famous brewers of that time. He believed in giving an additional flavor to his beer and it was related at that time that this wealthy brewer when he wished to give a special flavor to a good beer, would wash his old leathern breeches in it.

The purity of beer was protected by means of laws established by the Duke of York in 1664 for the Government of New Amsterdam, which provided that

No person whatsoever shall henceforth undertake the calling or work of Brewing Beers for sale, but only such as are known to have sufficient skill and knowledge in the Art or Mistery of a Brewer. That if any undertake for victualling of ships or other vessels, or master or owner of any such vessels, or any other person within this Government, do prove unfit, unwholesome and useless for their supply, either through the insufficiency of the Malt or Brewing or unwholesome cask, the person wronged thereby shall be and is enabled to recover equal and sufficient damage by action against that person that put the Beer to sale.

In Delaware it is stated that the Swedes made brandy from persimmons. It is evident at this time, therefore, that the word, "brandy" was applied very generally to distilled spirits other than rum.

In 1704 it was evident that the adulteration of rum and brandy of spirits was practised. The legislature of Pennsylvania in that year passed an act containing the provision that the seller of adulterated rum, brandy, or spirits, should forfeit the same and three times its value.

In 1787, Mr. French Coxe, in an address to the Friends of American Manufacturers, called attention to the fact that in Philadelphia the consumption of beer was much diminished by the general use of distilled spirits, which was made and imported in great quantities.

In 1761 the first distillery was built in Baltimore on the southeast corner of Water and Commerce Streets by Samuel Purviance, from Philadelphia.

In Virginia in 1649 it is stated that maize is not only commendable for bread but also for malting, and also that an "extraordinary and pleasing strong drink" was made from the West India (sweet) potato.

Peach brandy, of an excellent quality, was, during Colonial times, a household manufacture of considerable value, and more or less of it was regularly exported. It was, after simple fermentation, distilled into strong spirit.

Dr. Ramsey states in 1808,

The habit of the Carolinians is in favor of grog (a mixture of ardent spirits and water) when water is not deemed satisfactory. Hence, breweries are rare while distilleries are common.

In Georgia a large brewery was established by Oglethorpe in 1740.

His efforts to keep out the use of ardent spirits were found impracticable, and, it is said, his Scotch settlers and officers would withdraw from his presence to quaff their favorite whisky, at the smell of which he would denounce woe to the liquor, and which, if it came to his sight, he always destroyed.

The present arts of adulteration seem to have been quite early known in connection with the distilling and brewing trade.

Johnson, in 1645, "numbers among the trades of New England, divers shopkeepers, and some who have a mystery beyond others, as have the vintners."

In the second volume of this work we find allusions to the Whisky Rebellion of 1791. This rebellion arose on account of the act of Congress, passed March 3, 1791, imposing a higher duty varying from 20 to 40 cents a gallon, according to strength, and an excise duty of 11 to 30 cents, upon domestic spirits, distilled from molasses, sugar, or other foreign materials; and of 9 to 25 cents per gallon on that made from materials the growth or produce of the United States. Private stills were required to have written upon the doors of vaults and buildings the words "Distiller of Spirits." It is stated that there were 5,000 private distilleries in Pennsylvania which were affected by this action, and strong remonstrances against it were received, not only from Pennsylvania, but also from North Carolina, Virginia and Maryland, where numerous stills existed.

"A large body of Scotch and Irish distillers and farmers, in Westmoreland, Washington, Fayette, and Allegheny counties, questioned the power of the new government to impose so heavy a tax." That the distilled spirits made at that time were known as whisky in these private stills is indicated by the fact that this insurrection was known as the Whisky Rebellion. There is no record of the mixing of these whiskies with flavors, essences, cordials, extracts, and other substances at that time, while all the evidence at our disposal shows that such spirits went directly into consumption.

It is stated that as early as 1783 there were no less than 60 distilleries for molasses in Massachusetts. The quantity of molasses imported into the United States for the fiscal year

1791 amounted to 7,194,606 gallons, and the total exports of American spirits for the same time were 513,234 gallons.

Grain was used extensively in the manufacture of ardent spirits and malt liquors.

Geneva or gin was extensively consumed in this country, and distilleries of it, though but recently grown to any importance, were becoming of consequence, and required protection.

By an act of Congress of April 24, 1800, distillers who were unavoidably prevented from working their stills throughout the year were permitted to pay a monthly duty of 10 cents a gallon on the capacity of their stills in lieu of 54 cents yearly.

EARLY WHISKY MAKING IN KENTUCKY

Whisky as Money.—It is related in Hay and Nicolay's history of Abraham Lincoln that when his father, Thomas Lincoln, moved from Kentucky to Indiana he carried with him in addition to his family, a kit of carpenter's tools and several barrels of whisky. On reaching the Ohio River he constructed a raft in which he placed his earthly and spiritual belongings as above mentioned and started across. On nearing the Indiana side the raft parted, or capsized, or at least some accident happened to it, the kit of tools was precipitated into the River and the barrels of whisky were also set afloat. The legend says that he afterward recovered part of the whisky and also his kit of tools. This was the extent of his fortune on reaching the Indiana side. It is not at all likely that this whisky was taken along for consumption by his family, but as a means of trade and barter.

In the beginning of the last century it was a very common thing for the pioneers in Kentucky to have a small still on the farm. The settlers who came into Southern Indiana carried this practice with them so that there was a number of small stills erected and operated by farmers and used, not so much for home consumption as for trading for the necessities of life. The whisky was more easily transported than the grains from which it was made and as the roads were of the most primitive kind, and as it was difficult to bring the produce to the banks of the river, much freighting was saved by converting the corn into whisky and marketing it. The Ohio, Kentucky, Salt, Green, Cumberland, and the Tennessee

Rivers were used to a large extent as the means of carrying this article of trade. Whisky was convenient for barter because of the ease with which it could be divided into parts, it being thus possible to make change with despatch and convenience.

Whisky, even in those early days, was often stored because it was known that after ageing it would more than pay for the expense of holding it and for the loss by evaporation.

Origin of the Name "Bourbon."—Whisky made largely from Indian corn early became one of the principal products of Kentucky. The whole northeastern portion of Kentucky in the early days consisted of a single county which was called Bourbon county, and at that time it included practically all of the state in which whisky was made. In this way the term "Bourbon" came to be applied to the product of Kentucky, a name which it has since continuously borne. The term "Bourbon" has now the same relation to Kentucky whisky that the term "Cognac" has to brandy produced in the Charente region of France. The naming of the county "Bourbon" was doubtless due to the influence of the French settlers whose activities in the state of Kentucky are evidenced by many names of this kind, such as Paris, and Versailles. The names were doubtless a tribute to the home towns in France at the time they were applied.

Type of Distillery.—The early distilleries in Kentucky were of the most primitive type. They were built with logs in the cheapest possible way. No attempt was made at architectural display or perfection. No ornaments or decorations of any kind were ever thought of. No attention was paid to the style of the building, only so it would hold the grain and the still and protect them from the weather. Floors were usually of dirt or of split logs. The stills were made in the simplest kind of style, and in fact with the most simple arrangements that would produce the results. There is little difference between the old still of Kentucky and the still of the "moonshiner" at the present date, except that no attempt was made to conceal the locality of the old fashioned Kentucky distillery.

Locating the Stills.—The methods of preparing the grain for distillation in the early times were as primitive as the erection of the still itself. Usually, and perhaps always, the stills were built near water courses and the power for grinding the grain was derived from the water, which was usually impounded by dams. The grain was crushed in old-fashioned mill stones. Often the stills were placed on small water courses in which there was only a sufficient amount of water in the rainy seasons, that is the winter and spring, to keep the distillery in motion. In fact, the old still was not a business but an incident, hence the farmer did not care to operate it except during winter and spring when outdoor farming operations were unpleasant or impossible.

Mashing.—The mashing was done usually by hand in tubs prepared for that purpose. The mashes consisted of small quantities and they were mixed and manipulated with a paddle. In order to get a sufficient degree of lactic acid ferments each new mash was mixed with a considerable quantity of hot slop from the previous distillation, hence the term "sour mash" which still persists in many quarters. After scalding and cooling to a proper temperature, the conversion of the starch was obtained in the usual way by the addition of malt. Other grains, however, than barley were malted for this purpose and it was not unusual to malt the corn itself, at least partially, before the conversion process of the starch into sugar was accomplished. Initial fermentations of the mash were of course secured by the addition of ordinary yeast, the kind of yeast that was used for making bread being very commonly employed. After the distillery was once in operation, however, yeasts were skimmed from the preceding fermentation tanks and this form of yeast was commonly called "barm." After fermentation was completed in the fermentation tubs, the beer was carried by buckets and poured directly into the still. The copper still was often made in two pieces so that it could be taken apart and be filled up in this way by buckets. The stills were universally made of copper and were heated directly over a wood fire. They were not large and the rate of distillation was irregular. From the very first, however, it was found that the first distillation did not make a spirit of sufficient strength and therefore this spirit, after the first distillation, was put back into the same still and doubled. This doubling was supposed to have two effects, first, it brought the alcoholic content up to proof, or sometimes above, and in the second place it eliminated certain quantities of the gross materials which were carried over in the vapors of the first fermentation, thus improving the quality of the finished product. This doubling, however, did not appreciably diminish any of the important constituents which go to distinguish whisky from ordinary neutral spirits. After the distillation was completed the still was opened and cleaned. Naturally quite a quantity of the mashed grains not worked up into alcohol would find their way into the still. These deposits, settling on the bottom of the still, would protect the still from direct contact with the liquid contents and the result would be that often these deposits would be burned, giving in this case a scorched taste to the whisky.

Introduction of the Wood Still.—The danger of burning the whisky by introducing the beer directly into the still was so great that the early distillers, especially in Kentucky, conceived the idea of separating the first alcohol distilled from the grain by means of a current of steam. The very simplest possible apparatus was used for this purpose, namely, the splitting of a good straight log, the smoothing of the flat surfaces so they could be brought together again when necessary, the hollowing out of the interior of it so as to make a cylindrical center, the bringing of the two halves together and holding them with iron hoops. The mash was poured into the interior of this hollow log and a current of steam was blown into the bottom of it which gradually took the alcohol out without any danger of burning the grains which remained. This process rapidly gained in favor and has now become practically universal among the whisky distillers of the United States. The resulting product was placed back in the copper still and doubled in the usual way, thus securing the advantage of that form of distillation for the final product.

Changes Due to the Taxes laid upon Whisky.—From the time of the Whisky Rebellion in Pennsylvania up to the beginning of the Civil War the federal government laid no excise tax upon the production of distilled spirits. The great expense incident to the conduct of the Civil War made it necessary to increase the scope of taxable property. Distilled spirits were early selected as a proper product to bear some of the burdens of this increased taxation. Under the old system whisky was made in small lots and often sold fresh to the consumer, who if he wished to age it could do so himself. Very little whisky ever accumulated in the form of stocks and therefore very little of it was aged except incidentally. The collection of internal revenue made it necessary to keep a more accurate account of the production of distilled spirits and this led naturally to the establishment of the bonded warehouse where spirits could be kept until ready for consumption and the tax was paid only at the moment of withdrawal for consumption. The bonded warehouse, therefore, was for the mutual interests of the government and the distiller, making it much more convenient in both cases to meet the demands imposed upon this product by the high tax.

Manufacture of American Whiskies.—Whiskies in America are very rarely made from pure barley malt. There are some exceptions, and a few distilleries make pure malt whisky, barley being the material employed. Other grains are also malted and used alone for making whisky, but very seldom. Indian corn, for instance, may be the only grain used for making whisky, and rye is sometimes malted and used for making pure rye whisky. The great majority of American whiskies are made by a mixture of corn and unmalted or malted cereals.

Mashing.—The rye and Indian corn are ground into a coarse meal, mixed in the mash tub with sufficient water to make a paste of proper density and the temperature is raised to 160, 170, and even to 180°, by which the starch is reduced to a paste. The mash is then cooled to below 140° and the barley malt added. The mash is then thoroughly mixed until the diastase of the barley malt converts starch of the Indian corn and rye into sugar. The mash is then cooled to the temperature of about 60 to 65° or even lower, and the yeast is added. The mash after the addition of yeast is often called wort. There are two methods of mashing, called the sour mash and the sweet mash. In the sour mash a portion of the products of the preceding fermentation is added to the mash, and the acidity of the mash is thereby increased. Sometimes the spent lees from the stills are also added to the new mash, so that it becomes quite acid. On the other hand, the sweet mash is made directly by using only the mashed unmalted cereals with the malt. Each of the processes has its claims for superiority, but upon the whole the old-fashioned sour-mash whisky is regarded as superior to sweet-mash whisky. Sweet-mash whisky is a more modern invention.

The times of fermentation are limited for the different mashes. For the sour mash the fermentation should extend over 48 hours, and for the sweet mash not over 72 hours. During fermentation the temperature of the mash rises, so that the fermentation may begin at 65 rising to 85, or even higher, before the fermentation is finished.

The product of the fernmetation is called beer. The old-fashioned method of separating the alcohol from the beer in the United States is exactly the same as employed in Scotland, namely: in a Pot Still. Modern times have changed the character of the process of separating the alcohol from the beer. The alcohol is now usually separated from the beer by passing it through what is known as a beer still namely: a large cylindrical still with three departments.

When the distillation has proceeded to a certain extent in the first chamber, the contents of the top chamber are dropped into the second chamber, where again they come into contact with steam for a certain length of time. The contents of the middle chamber are then dropped into the bottom chamber into which the live steam is directly blown. When the still has once been charged it works continuously. The steam is blown into the bottom chamber and the vapors mix with the alcohol and pass to the middle chamber, boiling the mixture in the middle chamber; the volatile products of the middle chamber enter the top chamber and are driven over to the receiving tank. The stills are operated at such a rate that each time the spent beer is thrown off from the bottom chamber it is found to be practically free from alcohol. When the spent beer is thrown off from the bottom chamber the contents of the middle chamber are dropped into the bottom chamber; the contents of the top chamber dropped into the middle chamber, and the top chamber is filled with fresh beer. The objection to this process is, in the first place, that it is not a natural process of distillation, and in the second place, the live steam coming from an ordinary boiler gives it a bad flavor.

The distillate from the beer still is known as low wines. The low wines are dropped into a pot still, and it is then distilled either by building a fire underneath the still, the old fashioned, preferable way, or in modern times the distillation is carried on by means

20

of copper steam pipes in the bottom of the pot still. The distillate is collected in the receiving tank of the distillery under the supervision of the officials of the Internal Revenue, and are known in the distillery as high wines, usually having an alcoholic percentage of 60 by volume or more. When the high wines are collected they are drawn off by the gauger of the Revenue Service into casks, and before being placed in casks are diluted with pure water, to a proof of 50 percent alcoholic strength by volume. At the same time, under the law, the owner of the spirits is allowed to stamp upon the package the character of whisky he produces, namely: Rye Whisky, Bourbon Whisky, Corn Whisky, etc.

The last residues of the alcohol in the still are collected and thrown back into the beer for redistillation.

Selection of Raw Materials.—The raw materials of which whisky is made are of the greatest importance. The very best and finest grains of which the country is possessed should be used for making a whisky of high grade. Imperfect cereals are those which are moldy, frost bitten or decayed in any way. Such grains make poor spirits and never have the proper taste, that is, the proper flavor is not developed. The maker who values his reputation selects his Indian corn, rye, barley and malt with extreme care.

Bonding Whisky.—When whisky is withdrawn from the still and reduced with water, it is placed in a bonded warehouse in oak containers, which have been previously charred. This is different from the English, Scotch and Irish whisky, in which countries charred casks are not used. It is claimed that whisky kept in a burned barrel reaches the proper flavor at four years that would take the whisky in unburned barrels to reach in 8 years. The color extracted from the charred wood is very deep so that American whisky which has been four years in wood is very much deeper in color than Irish or Scotch whisky of the same age. Whether or not the whisky is improved by the charring of the barrel is a question left for experts to settle. Some claim that the character of the spirits is very much improved, while others claim that only age will give the proper flavor, color, etc.

Materials Extracted from the Wood Incident to the Storing of Whisky.—The alcoholic contents of a cask extract from the wood a certain quantity of matter soluble in alcohol of that strength. Among the substances extracted from the wood are tannin, and, to a considerable extent, the acids, which are contained in the wood. There is also a considerable quantity of coloring matter found in solution. The whisky in wood, during the process of ageing, gradually acquires, more or less, a red color. The matters extracted from the wood cannot be without effect upon the whisky, and a certain taste is given to the whisky, as well as flavor and aroma, from the extract of the wood itself. The matters extracted from the wood, although they may be pleasant to the nostrils and the palate are not the things valuable in whisky which give it its characteristic flavor. There are certain chemical changes which take place in the spirits, such as blending and mellowing, due to the chemical reactions of the constituents minute in quantity, and yet of great significance. The final result is an increase in aromatic ethers and acids.

Stamping of Whisky.—The stamping or taxing of whisky in the government warehouse is an interesting subject. When the container is put into bond a stamp is affixed to it showing the date on which the article was bonded, the number of the distillery, the name of the gauger, and the quantity of proof spirits that is contained therein. When this cask is withdrawn from bond for consumption another stamp is put on it, called the tax paid stamp. This is a stamp showing the date upon which the tax was paid, the number of gallons remaining in the barrel, and giving the serial number or identification of the container. Thus whisky of this kind is called double stamp whisky, one of the stamps naturally being old and the other fresh when the receptacle comes into the hands of the consumer. This double stamp assures the consumer that the product has not passed through the hands of a rectifier. The rectifier's package has but one stamp and that new.

Bottled in Bond Whisky and Rum.—Under the revenue laws of the United States when alcoholic spirits are distilled to make any kind of a product the manufacturer has the option either of paying a tax immediately or of storing the product in a bonded warehouse under the supervision of the United States until the article is sold, at which time the tax must be paid. When distilled spirits are put into a United States bonded warehouse they may remain not longer than 8 years, at the end of which time the tax must be paid, inasmuch as it is the common opinion, based upon reliable data and accurate observation, that distilled spirits which have in them the materials capable of developing into a palatable beverage improve in quality on storage in wood. While the rate of maturity varies according to temperature and whether or not the wooden containers are charred, and with the alcoholic strength of the distillate, there is a common agreement on the fact that in general a period of at least four years is necessary to make a really palatable and potable drink out of distilled spirits.

When the rectifying industry was thoroughly established in the country and doing practically all the business of manufacturing potable distilled spirits, it was found that there was great waste both of interest on the investment and also of leakage from the barrel in the storage of distilled spirits for any considerable length of time. The practice grew rapidly of making a perfectly neutral distillate which had only the character of ethyl alcohol, and which would lend itself upon reduction to proper proof to take in all colors and flavors necessary to imitate the old matured-in-the-wood beverage. Thus, by taking caramel, flavoring ethers, acids and extracts, and neutral spirits and water, a beverage could be manufactured in a few minutes which would have all the appearance of age and many of its characters, due to the added flavoring substances. It became increasingly difficult, for this reason, for anyone to be certain of getting an old matured beverage, whether it be whisky, brandy or rum.

In order to protect the consumer in this respect Congress passed what is known as the Bottling in Bond Act. The Bottling in Bond Act was signed by President Cleveland on March 3, 1897, his last official day in the White House. I do not know how he could have done a more salutary thing as his valedictory. The report of the committee favoring the passage of this measure through Congress said:

The obvious purpose of the measure is to allow the bottling of spirits under such circumstances and supervision as will give assurance to all purchasers of the purity of the article purchased, and the machinery devised for accomplishing this makes it apparent that this object will certainly be accomplished.

The interests of the Government and the revenue are carefully guarded, so that there will be no expense or loss to either.

The passage of the bill will enable American producers to supply a very large home and foreign demand which is now supplied, greatly to the injury of the home manufacturer, by Canadian producers, so that persons who desire to purchase bottled goods under a stamp must get the Canadian instead of the American article, and instead of having the option to obtain either.

We are not permitted to export to Canada in packages of less than 100 imperial gallons (equal to 120 of our gallons), while we permit the importation of bottled Canadian spirits, thus putting our manufacturers at a great disadvantage, especially as the bottled spirits thus imported are supposed to be guaranteed by the Government label and stamp placed thereon under a law somewhat similar to the pending measure.

It is believed that the enactment of the bill will give our home producers a large market for their goods which is now given to Canadian or other foreign spirits.

These considerations lead the committee to recommend the passage of the bill.

By the provisions of this act spirits cannot be bottled in bond until they have remained in wood at least four years, with this exception, that if the spirits are distilled in the autumn of any one year they may be bottled in the spring immediately preceding the four years of age. So it is possible that bottled in bond spirits may be not more than three years and a half old. The law prohibits the addition or subtraction of any substance or material or the application of any method or process to alter or change in any way the original condition or character of the product, except the reduction to proof with water. The transfer of this unmanipulated article from the barrels where it has been in the bonded warehouse for about four years into bottles is made under the most rigid government inspection, conducted under the custody and supervision of a detailed government official. The regulations provide that:

No material nor substance of any kind other than pure water can be added during the process of bottling or the preparation of the spirits for bottling, nor can any substance or material be subtracted, nor can any method or process be applied to alter or change in any way the original conditions or character of the product, except as authorized by the statute. The storekeeper or gauger will not, therefore, admit to the bottling house any material or substance capable of being incorporated with the spirits except pure water only.

The law also requires that the packages which contain bottled in bond spirits be of definite content and that the alcoholic strength

of the spirit must not be less than 50 percent by volume. The
green stamp over the cork which is used also gives the name of
the distiller, the location of the distillery, the quantity of spirits
in the bottle, and that it is at least 100° proof, that is, containing
50 percent by volume of alcohol. There is nothing mandatory
in the Bottling in Bond Act. It does not require the distiller
to bottle his spirits in bond unless he so elects, but if he does elect
to bottle his spirits in bond he can send them out from his distillery in bottles of various sizes, less than a gallon in quantity,
guaranteed by the United States itself as to purity, strength,
volume, age and character. The security which the consumer
gets by this Act has caused a continually increased sale of spirits
bottled in bond. The data, as taken from the records of the
Commissioner of Internal Revenue, show that the amount of
whisky bottled in bond during the fiscal year ended June 30, 1915,
was 9,741,639.2 gallons. The amount of rum bottled in bond
was 6,745.6 gallons, and the amount of gin bottled in bond was
594.1 gallons.

During the year ended June 30, 1916, the data are as follows:

Whisky...12,570,240   gallons
Rum..........8,212   gallons

The quantity of whisky bottled in bond for the year ended
June 30, 1917, was 16,495,203.7 gallons.

These data show an increasing amount of bottled in bond
spirits entering consumption. This indicates a growing confidence
on the part of the consumer in the little green stamp.

Only one rum distillery, viz., in the third Massachusetts collection district bottles its products in bond. For 1917 this quantity
amounted to 17,019.7 gallons.

SCOTCH WHISKY

Malting.—The Scotch distillers do their own malting. Home-
grown barley is preferred, but, inasmuch as that is not in sufficient quantity foreign barleys are purchased. These come
principally from Norway and from California. The malting is
done on cement floors, on which the grain is spread to a depth
of a few inches, properly moistened and allowed to sprout. The
season for malting for a Scotch distillery is from fall to spring,

the distillery usually not being in operation during the summer. Malting, therefore, takes place slowly, especially in winter. The more slowly the malting, as a rule, the better quality of malt is secured. From 10 to 15 days is supposed to be consumed in securing good malt, that is, in sprouting good grain so that the germ may be from a quarter to three-quarters of an inch in length.

During the early periods of malting before the sprout is developed, the grain is turned occasionally upon the floor so that all parts of it may be subjected to practically even conditions.

Drying Malt.—Malted barley is spread upon floors, with small orifices, to a depth of a foot or 15 inches, so that the hot air and smoke coming from the furnace below through the small openings in the floor pass up through the malt and escape through the chimney in the roof.

The malt is turned frequently during drying. Great care must be taken not to raise the temperature of the malt too high otherwise the activity of the ferment, that is, the *diastase* will be lessened.

The malt is dried until the percentage of water is reduced to a small amount, not to exceed 4 percent. Malt is considered better if it be kept stored for a time before it is subjected to manufacture.

Scotch whisky is made solely from barley malt; no other grain nor unmalted barley being used. It is therefore a type of a pure malt whisky.

Grinding the Malt.—The malt when ready for manufacture is ground to a coarse powder and placed in a mash tub. Here it is mixed with enough water to bring it to a proper consistency, and the temperature is raised so that the maltose and any remaining starch present is made into a jelly, and the remaining starch converted into sugar by the action of the diastase. Water which has come from peaty marshes is considered of superior quality for mixing with the ground malt to make the mash. The mash tub is usually quite large and supplied with a stirring apparatus driven by machinery whereby all parts of the mash are kept thoroughly mixed. The bottom of the mash tub is double; the first bottom being full of fine perforations, and the second a closed bottom. The object of this is to draw off the liquid portions of the mash, leaving the bran and undissolved portions on the upper bottom where they can be removed and used for cattle feed, or other purposes. The temperature of the mash tub usually is not allowed to go over 140°F., since at higher temperatures the diastase would not convert any remaining starch. After all the starch is converted the mash can be raised to a higher temperature, if desired. The liquid maltose and other sugars are now called the wort, and it is drawn off from the mash tubs, cooled and run into the fermentation vats. After the wort is cooled to a proper temperature, the yeast is added and the fermentation begins. It requires from 48 to 72 hours to complete the fermentation and convert all the sugar into alcohol. The fermented wort is now beer.

Distillation.—The beer is brought into a pot still, made of copper, and of a size, according to the dimensions of the distillery.

The first still, which is called the beer still, is usually double, that is, there are two of them, and between them is placed another still for the purposes of finishing the operation. The beer stills are provided with scrapers or chains which are dragged constantly over the bottom so that any solid matter coming from the remaining solid contents in the beer may not adhere to the bottom of the still and be burned.

The stills are heated, as a rule, by open fire. In some instances, steam coils are also found in the stills. The shape of the still is said to have a great deal of effect upon the quality of the whisky. The stills with big bellies and with rather long arm necks are preferred.

The product of distillation from the beer still is known as low wines, that is, containing alcohol less than 50 percent by volume of the whole mass. The distillation from the beer still is continued until the alcoholic proof of the distillate is too low to be placed in the low-wine tank. The residuary alcohol is then distilled and added to the next distillate. After the alcohol is all distilled the residue left in the still is called spent beer. It is rather of an unsavory character and is not allowed by the laws of the country to be run into the streams. It is therfore usually disposed of as a slop, or sometimes it is burned, or disposed of otherwise. It is not considered a very wholesome diet for domesticated animals. It is used, although not very extensively for that purpose. Instead of selling it for cattle food or burning it, it is sometimes run into tanks sunk into the earth until all the solid matter settles, and then it is used for manure. In some distilleries it is put into tanks and destroyed by bacterial action. The low wines, after they are collected from the beer stills, are put into a whisky still, which is smaller than the beer stills, and subjected to a second distillation for the purpose of bringing them to an alcoholic strength suitable for bonding, namely: full 50 percent alcohol.

When the distillate from the whisky still becomes too weak to be used for whisky, it is separated and used again with the beer in the next distillation. After all the alcohol is extracted the residue in the whisky still is known as lees, and the lees are disposed of like the spent beer above described. The spirit, after having been thus secured, is stored in oak casks. Those casks which formerly held sherry wine give a distinctly valuable character to the spirit, and, therefore are most highly prized. Such casks bring from $12 to $15 each.

There are in Scotland about 150 distilleries conducted according to the manner described above.

After the whisky is distilled and put into containers it is stored in bonded warehouses in the same manner as in this country.

Kinds of Scotch Whisky.—Of these distilleries 11 are equipped for the production of grain whisky only, 3 for both grain and malt, 98 for Highland malt, 23 for Lowland malt, 9 for Islay malt, and 21 for Campbeltown malt. The grain whisky production is in very few hands, viz.; The Distillers' Company, Limited, of Glasgow and Edinburgh, who have most of the grain distilleries and who produce the "Cambus" whisky, which is chiefly neutral spirit of which the name grain whisky is a synonym; the North British Distillers' Company, Limited, Edinburgh; and one or two private firms. The Distillers' Company, although the largest producers of grain whisky, also make Highland and Lowland malt whisky, but we are not aware of any other makers of malt whisky who also make grain whisky. While the production of grain whisky is in very few hands that of malt whisky is in the hands of quite a number of limited companies and private firms. Hence the large number of brands and of blends.

Smoke Flavor.—The whisky of Scotland is very famous, not only for excellence, but also for its peculiar flavor, which many people associate with creosote. The fact of the case is that the peculiar flavor of Scotch whisky is due to a remarkable character of the peat, which is produced in that country from aromatic heather. It may be a matter of some interest to know how the flavoring of peat gets into Scotch whisky, since peat is not a material which is used in the fermentation or in the mash in any way. The Scotch distiller makes his own malt, and in drying it after it has been sprouted he uses peat for a fuel. It is the burning of peat which gives rise to the aromatic substances, which are characteristic of Scotch whisky. These are absorbed by the moist malt during the process of drying, and thus pass into the mash, the fermenting tank and the still. After the fermentation, the flavor of the peat, being volatile, comes over with the whisky during the distillation.

Preparation of Peat.—Only peat of the character described above can be used for drying malt. Attempts have been made in Ireland to use the Irish peat, and in other countries to use native peats, but all without success. It is only the peat of the Scotch heather which has the quality necessary to produce the flavor.

The peat which is to be used for drying malt is cut into blocks and put away where it ferments and dries. During this process the aromatic properties of the peat are fully developed. The peat, as a rule, is not used until it is two years of age.

The malt is spread upon a perforated floor, below which is the furnace, in which the peat is burned. As the peat does not always give a very great degree of heat, it is a common practice to mix anthracite coal or coke with it, in order to aid its drying properties.

Irish Whisky.—Irish whisky should probably have been mentioned first in this list, as Ireland is supposed—I don't know how truly—to be the original home of whisky. Although Ireland is celebrated for its whisky, yet there are very few distilleries in that country at the present time and pot stills are used exclusively.

There are two kinds of Irish whisky. All malt, that is, made solely of barley malt, and a part malt and part unmalted cereals. The proportion of malt used in Ireland is very much larger than in the United States, ranging as it does from 100 percent for all malt whisky to 50 percent or slightly less, for the part malt whisky. Very little whisky is made in Ireland in which the proportion of malt runs as low as in this country. The Scotch distilleries are usually very small, while the Irish are very large. The following description of one of the largest plants, the Powers Distillery in Dublin, which I had the opportunity to study in detail, will prove of interest:

The distillery covers 8 acres on the south side of the Liffy and has an annual capacity of about 900,000 gallons. The corn (barley) screens and mills are of the most approved pattern. The elevators are of cast iron, so as to prevent the spread of fire, and being constructed of plates can be opened at any point.

The stores contain about 40,000 barrels and are used to their utmost capacity during the 9 months working season which extends from September to June.

The mash tubs are 33 feet in diameter and 7 feet 6 inches deep. The mash at each brewing being 500 barrels.

In the fermenting house there are four refrigerators and nine wash backs each holding one brewing, viz., 30,000 gallons.

The stills are all of the pot still type. There are six of these huge boiling vessels, working off 19,000 to 25,000 gallons each. The daily consumption of coal for the purpose being over 50 tons.

The walk through some of the bonded warehouses is an experience I shall never forget. There are nineteen upon the premises and two outlying warehouses, one under the South City Market, and the other under Westland Row Railway Station, the whole with a combined capacity of 50,000 casks, vessels ranging from the great 120 galloner down to the humble 28 gallon cask, piled one on top of another yet none touching its neighbor, there being planks between to allow a free circulation of the cool air kept always at one temperature; and with the knowledge that once the cask gets into its maturing position it does not leave it for at least five years, and in the majority of cases 7, 9, 12, and even 14 years.

To me the most interesting distillery in Ireland is the Old Bushmills near the Giant's causeway in the North of Ireland. The historic interest of its surroundings and the scenic beauty of the neighborhood will repay the sightseer for his trouble in visiting the locality.

"Moonshine" Whisky.—In all countries where an excise tax is laid on alcoholic beverages there are found persons who seek in every possible way to avoid the payment of the tax. The production of distilled liquors in a secret way to avoid the tax in this country is called "moonshining." The "moonshining" industry is confined particularly to the mountainous regions of Eastern Tennessee, Georgia, Virginia, North Carolina and South Carolina. In the mountain fastnesses this illicit industry is carried on by a large number of persons and in a large number of localities. The total number of these small illicit stills in existence of course is not known. We, however, do know, from the reports of the Commissioner of Internal Revenue how many are discovered and destroyed from year to year. During the fiscal year ended June 30, 1917, there has been a slight decrease in the number of illicit distilleries seized as compared with the number seized during the preceding fiscal years. No permanent abatement appears, and none can be expected until the officials of the United States receive more hearty coöperation on the part of local state officers in the various states and localities where local prohibition exists.

The following data from Report of the Commissioner of Internal Revenue show the activities of the Revenue agents since 1911:

ILLICIT DISTILLERIES SEIZED, CASUALTIES TO OFFICERS AND EMPLOYEES, AND PERSONS ARRESTED DURING THE LAST SEVEN YEARS

	1911	1912	1913	1914	1915	1916	1917	Total
Illicit distilleries................	2,471	2,466	2,375	2,677	3,832	3,286	2,232	19,339
Officers and employees killed.......	.....	.....	.....	.....	3	.....	.....	3
Officers and employees wounded....	2		2	1	.....	3	3	12
Persons arrested..................		494	459	504	893	1,314	1,008	5,201

The number of stills seized and destroyed in the various states is as follows: Georgia, 456; North Carolina, 411; Alabama, 342; Virginia, 160; South Carolina, 121; Tennessee, 106; Kentucky, 62; Florida, 13; West Virginia, 11; Mississippi, 10; Arkansas, 6; Pennsylvania and New York, each 1. The material used in the production of illicit whisky is mostly Indian corn. It is either malted directly by itself or else the starch is converted by barley malt. The stills are concealed in the mountain fastnesses in such a manner as to attract the least possible attention and in localities difficult of access. They are naturally of very small size, capable of making only a few gallons per day. This whisky is usually not stored at all, but is sent immediately into secret commerce by means of smugglers, bootleggers and others. It is, therefore, usually white in color and rather fiery in taste because of the lack of age. "Moonshine" whisky is produced solely in a pot still and hence if it had time to age would be a whisky of excellent properties and highest purity.

Excellence of Moonshine Whisky.—Morewood, in his work, on page 339, pays a compliment to illicit whisky, in the following words:

They (the Government regulations) bound down the legal distillers in such a manner by injurious restrictions, that it was not in their power to produce a spirit equal in flavor to that manufactured by the smugglers, who lie under none of those restrictions which bind down the ingenuity of the legal trader. This superiority induces a corresponding desire in the inhabitants of Scotland to possess themselves of smuggled whisky, even at a higher price than that for which they can purchase the same article from the licensed distillers. The smugglers, in consequence, are winked at, or rather encouraged, by a very considerable proportion of the inhabitants of the country. While this feeling existed, it was impossible to put an end to smuggling in Scotland. But Government of late has removed the restrictions by which the Scotch distillers were bound, as far as to allow them to distil from malt at nearly the same rate as they formerly did from raw grain. The consequence has been that the high reputation of smuggled whisky has gradually sunk, and smuggling has been nearly discontinued.

Composition of Whisky.—In addition to ethyl alcohol, whisky has acids, alcohols higher than ethyl (butyl, propyl, amyl, etc.) esters (ethers), aldehydes, furfurols and numerous compounds thereof. These are produced during fermentation (congeneric with ethyl alcohol) or are formed during ageing. Whisky also contains extractives from the wood (usually oak) in which it is aged. These extractives are tannin, acid and coloring matters. To these congeneric and synthesized bodies and extracts from the wood, whisky owes its flavor and aroma (organoleptic properties). For this reason whiskies which have long aged in wood are most highly esteemed and regarded as least harmful. Following is a table showing the chemical composition of some common whiskies.

APPROXIMATE COMPOSITION OF WHISKIES OF DIFFERENT KINDS

Parts per 100,000 of proof alcohol

Kind of whisky	Proof	Non-volatile matter	Acidity as acetic acid	Esters	Alde-hydes	Fur-furol	Higher alcohols	Total con-generic and secondary products
Old Rye...........	110	180	80	70	12	2.3	165	300
Old Bourbon.......	105	160	70	55	10	2.0	140	270
New Rye...........	100	8	9	30	5	1.0	135	174
New Bourbon......	100	8	10	30	3	0.5	130	170
Rye Bottled in Bond	100	190	90	75	18	2.0	150	335
Bourbon Bottled in Bond	100	170	74	60	25	1.5	125	275
Whisky Bottled in Bond...........	100	180	80	65	18	1.5	150	300
[1] Distillery Bottled..	80	650	80	45	12	1.4	115	250
[1] Bottled Bourbon..	85	480	45	24	7	1.5	85	150
[1] Bottled Rye.......	88	575	48	28	6	1.3	70	140
[1] Bottled Whisky...	85	600	55	34	8	1.3	75	150
[1] Bulk Whiskies.....	80	450	40	15	4	1.0	50	100
[1] Neutral Spirits....	190	00	1	15	0	0.0	5	25

The above table gives a fair average expression of the chemical constituents of different kinds of whiskies found upon the market. The remarkable chemical differences in new and old whiskies are shown by a comparison of the new Rye whisky and Bourbon whisky with old Rye and old Bourbon whisky. The principal difference, as is seen, is in the amount of extractive material, or solid matter. The extractive matter in old whiskies comes chiefly from the wood.

The Alcoholic Content of Whisky.—The content of alcohol by which its intoxicating power is measured varies greatly in different distillates. In Ireland and Scotland whisky is usually used, at a higher proof than in the United States. The legal strength or standard for whisky in the United States is 100 proof; that is, 50 percent of ethyl alcohol by volume. This has been fixed by what is known as the Bottling in Bond Act of 1897. This act provides that distilled spirits may be bottled in bond, provided their alcoholic proof is not less than standard. The Internal Revenue law also provides that any legal contracts in distilled spirits of which whisky is a type shall be based

[1] Commercial articles.

upon the standard proof gallon. In fact the greater part of the whisky delivered to the consumers of the United States is reduced, after the tax has been paid, by the addition of pure water until the alcoholic strength is about 90 proof, or 45 percent by volume or even less. This is done by consent of the revenue officials.

In certain forms of adulteration, where the cheapening of the product is the principal point to be taken into consideration, the proof of alcoholic spirits is diminished, often to a very great extent, falling as low as 35 percent in some instances. In general it may be said that an ordinary good strength whisky of the United States reaches the consumer, unless it is otherwise specified by him, at a proof strength of 90, or 45 percent per volume. The adulteration of whisky with water from the temperance point of view is regarded as a distinct advantage. It is nevertheless a typical fraud.

Rectified Whisky.—The term "rectified" in connection with distilled spirits at the present time is used in a double sense. Strictly speaking it means a spirit which has been purified in some way by redistillation or otherwise. In the process of distillation the character of the distillate varies greatly with the rapidity of the distillation and the shape of the still. Where stills are very large, very high, with large throats, and where the distillation is carried on at a reasonable speed, the distillate comes over in a very satisfactory condition. If, on the other hand, the still is small, with small throat, and the distillation is rapid, a great deal of material is carried over mechanically, as well as by the high temperatures, which give bad flavor and odor to the whisky. Such distillates are often purified either by the addition of a substance like sulphuric acid, which combines with the more undesirable impurities, and renders them non-volatile, so that they can be separated on a subsequent distillation, or the purification may take place by redistillation itself. To this end there was invented what is known as the chambered still. This still may be regarded as a large number of small stills constructed one upon the other. The spirit to be purified is placed in the bottom still or compartment and volatilized usually by means of steam coils. The spirit passes into the next still through a peculiar form of aperture whereby a portion of it when condensed must remain upon the floor of the still. As soon as the second compartment enters into activity a second distillation begins, and so on until all the still is filled. The steam is applied only in the lower compartment.

These stills often have from 10 to 40 chambers, and range in height from 20 to 70 feet.

At each distillation a more pure spirit is secured until finally at the top there is scarcely anything left but pure ethyl alcohol.

I will not stop here to describe the technique of the construction, but simply to illustrate its principle. It follows from the above that, as the spirit becomes more and more pure, that is, in essence consisting solely of ethyl alcohol, it becomes less and less like whisky. Finally, provided the spirit has previously been filtered through charcoal and further purified by a process of separation of the parts that come off first and the parts that come off last, a very pure spirit is secured, which is known to the trade by such names as "cologne spirit," "velvet spirit," "neutral spirit," "silent spirit," etc., already defined. This spirit has lost all the characteristics of a whisky and can in no way be considered as whisky in the beverage sense of that term. In the process of making whisky the spirits are subjected to two distillations, one to separate the spirits from the beer, and the other to double and concentrate the spirits so as to make its alcoholic strength greater. Then it is placed in casks and put away to age.

Control of Rectification.—The Internal Revenue provides that no process of rectification shall be carried on less than 600 feet distant from the distillery, and there shall be no underground or secret passage or pipe connections of any description between the rectifying house and the distillery.

When a spirit is taken to a "rectifying" house the tax is first paid thereon, as has already been described, and the purchaser is allowed, within certain restrictions of the Revenue Department, to do as he pleases with the product. He can mix together two or more spirits, forming what is known as a compound, or he can compound the spirit with other spirits or other substances, just as he sees fit. He may add coloring matter, sugar, spices, prune juice, sherry or any other substance he likes, provided always he is able, by the gauging of the finished product, to account for all the spirit for which he has been charged. Under the law this mixing or compounding is called "rectification."

Growth of the Rectifying Business.—At the present time the term "rectified" and "rectification" have come to mean in the trade the mixing or compounding of distilled spirits, either with each other, to form a compound liquor, or with some other substances to form a spurious, imitation or compound liquor. The legal definition of such a so-called "rectifier" is given by Act of Congress: "Every person who without rectifying, purifying or refining distilled spirits shall by mixing such spirits, wine or other liquor with any materials, manufacture any spurious, imitation, or compound liquors for sale under the name of whisky, brandy, gin, rum, wine, spirits, cordials, wine bitters, or any other name, is to be regarded as a rectifier and as being engaged in the business of rectifying."

The Economical Advantages of Rectification.—The advantages of rectification from an economical standpoint are easily understood when it is remembered that a genuine whisky, which has not been subjected to any rectification, except that incident to its manufacture, has to be stored in wood at least four years before it is deemed to be palatable or eligible to be put into bottles under government stamp. In point of fact, it is well known that four years is too short a time for properly ripening a whisky and securing its best qualities. For this reason whisky, that is 8, 10, 12, 16, and even 20 years old, is much more highly prized and esteemed, and brings a much greater price, than whisky only four years old.

One of the results of ageing whisky is that the harsh and unpleasant taste, which it at first has as it comes from the still, is softened and mellowed in such a way that the flavor is greatly improved. In addition thereto there are developed by oxidation ethereal and aromatic substances and acids, which give flavor, taste and value to the product. At the same time, incident to being stored in wood casks, the whisky extracts from the wood a certain amount of coloring matter, the depth of which becomes an index of age. It follows if it is possible to simulate a whisky of this kind without spending so much time for its ageing that the process would be a profitable one.

To this end it was soon discovered that the pure spirit, which has been described, such as cologne spirit, being of itself free of any characteristic odor or taste as a beverage, and being in point

21

of fact in this sense a "neutral" substance, becomes a vehicle for the practices of that legal rectification, which means mixing, compounding and adulterating. It was soon realized that, by mixing freshly made spirit, perfectly colorless, and which, therefore, had received no value from storage, with a small quantity of whisky, and coloring it by means of caramel to resemble old whisky and adding certain flavoring agents made by chemists, or such substances as prune juice, so-called blending sherry, etc., a liquor could be made resembling in color and flavor old whisky. The compounder could make up hundreds of barrels of this liquor in a day, and place it upon the market in such a guise that the unsuspecting consumer would be certain he was buying an old and valuable product.

The above so-called rectification, enabled mixers to undersell real whisky, which was driven practically out of the markets. So great was the extent of this adulteration that it has been estimated by some authorities that at the time of the passage of the food law in June, 1906, more than 90 percent of all the so-called whiskies sold upon the American markets were of the kind described above.

The total quantity of spirits "rectified" in the United States for the fiscal year ended June 30, 1917, was:

State or territory	Gallons	State or territory	Gallons
California	3,695,795.1	Nebraska	365,111.5
Connecticut	1,674,194.8	New Hampshire	107,732.7
Delaware	814,939.7	New Jersey	3,209,265.0
District of Columbia	324,032.2	New Mexico	21,882.4
Florida	1,234,790.8	New York	24,398,085.9
Hawaii	52,875.3	Ohio	14,009,547.6
Illinois	14,317,210.0	Pennsylvania	16,357,592.4
Indiana	1,565,614.7	Rhode Island	451,163.6
Iowa	27,494.8	Tennessee	1,356,014.9
Kentucky	7,815,805.2	Texas	451,218.6
Louisiana	1,177,697.9	Utah	113,198.3
Maryland	4,328,479.7	Virginia	482,349.5
Massachusetts	6,038,171.6	Wisconsin	2,585,297.8
Michigan	947,280.2	Wyoming	3,204.7
Minnesota	2,559,089.2	Total	114,596,201.7
Missouri	4,074,795.2		
Montana	36,270.4		

Magnitude of the Spirit Industry.—In the Annual Report of the Bureau of Internal Revenue for the year ended June 30, 1917, are found interesting data respecting the magnitude of the spirit industries. The quantity of distilled spirits produced for the fiscal years 1916 up to the end of the fiscal year 1917, is shown below:

STATEMENT SHOWING BY STATES AND TERRITORIES THE PRODUCTION OF DISTILLED SPIRITS DURING THE YEARS 1916 AND 1917, COMPARED

States and territories	Fiscal year 1917			Total production, fiscal year 1916
	Spirits produced from materials other than fruit	Fruit brandy	Total production	
	Gallons	Gallons	Gallons	Gallons
Arkansas..........	....	....	....	178.4
California........	9,979,723.4	7,871,759.0	17,851,482.4	11,845,251.1
Connecticut......	121,890.6	10,163.9	132,054.5	127,214.9
District of Columbia	608,812.2	....	608,812.2	1,664,389.3
Florida...........	....	....	....	2,715.1
Hawaii...........	14,015.5	....	14,015.5	13,671.9
Illinois...........	79,320,206.5	410.5	79,320,617.0	66,868,865.2
Indiana...........	43,332,771.0	28,504.9	43,361,275.9	51,108,395.3
Kentucky.........	36,407,614.9	34,162.9	36,441,777.8	33,254,129.4
Louisiana	26,545,832.8	....	26,545,832.8	23,291,661.1
Maryland.........	24,949,677.8	15,642.7	24,965,320.5	3,327,842.0
Massachusetts.....	12,511,179.8	58.5	12,511,238.3	11,609,189.1
Michigan..........	819,669.0	238.7	819,907.7	2,575,375.4
Missouri..........	283,972.8	5,687.7	289,660.5	194,171.9
Montana..........	244,772.5	....	244,772.5	52,385.8
Nebraska.........	2,938,594.1	....	2,938,594.1	2,476,219.1
New Jersey........	....	54,493.8	54,493.8	56,158.3
New Mexico.......	....	314.9	314.9	295.9
New York.........	13,817,034.5	39,019.1	13,856,053.6	13,802,024.2
Ohio..............	9,954,439.5	160,133.0	10,114,572.5	12,448,347.7
Pennsylvania......	12,177,573.5	13,190.1	12,190,763.6	14,408,130.3
Rhode Island......	224.2	....	224.2	236.7
South Carolina....	1,159,308.5	....	1,159,308.5	1,179,890.0
Texas.............	13,904.9	....	13,904.9	
Virginia..........	105,899.3	17,057.9	122,957.2	547,559.5
Washington.......	....	....	....	392.1
Wisconsin.........	2,527,249.3	....	2,527,249.3	2,428,480.0
Wyoming.........	....	259.7	259.7	103.0
Total..........	277,834,366.6	8,251,097.3	286,085,463.9	253,283,273.4

COMPARATIVE STATEMENT OF PRODUCTION OF DISTILLED SPIRITS DURING THE
LAST 6 FISCAL YEARS

	Gallons		Gallons
1912	187,571,808	1915	140,656,130
1913	193,606,258	1916	253,283,273
1914	181,919,542	1917	286,085,463

During the fiscal year ended June 30, 1917, 198 grain distilleries, 25 molasses distilleries and 284 fruit distilleries were operated in the United States. This of course does not include the moonshine stills.

Collections on Distilled Spirits.—The amount of tax paid on distilled spirits for the fiscal years ended June 30, 1916 and 1917, and the several sources thereof are shown in the following table:

Objects of taxation	Receipts during fiscal years ended June 30—		Increase	Decrease
	1916	1917		
Spirits				
Spirits distilled from fruit	$3,283,217.02	$4,035,536.42	$752,319.40	
Spirits distilled from other materials	146,565,963.45	177,096,234.20	30,530,270.75	
Rectifiers (special tax)	294,826.08	255,187.79	....	$39,638.29
Wine, etc., domestic and imported	2,631,529.98	5,164,075.03	2,532,545.05	
Retail liquor dealers (special tax)	4,309,656.02	3,967,270.45	....	342,385.57
Wholesale liquor dealers (special tax)	616,559.81	598,361.39	....	18,198.42
Manufacturers of stills, and stills and worms manufactured	3,400.86	2,416.69	....	984.17
Stamps for distilled spirits intended for export	45,839.40	49,867.55	4,028.15	
Case stamps for distilled spirits bottled in bond	440,244.00	558,180.40	117,936.40	
Grape brandy used in the fortification of sweet wines	491,202.91	384,188.89	....	107,014.02
Total	158,682,439.53	192,111,318.81	33,428,879.28	

FLAVORS OF DISTILLED SPIRITS

Natural Flavors.—Under each of the classes of distilled spirits statements have been made respecting the properties of the flavoring and aromatic constituents, which give to each one its character and value. Chemists and experts find great difficulty in detecting and isolating the particular individuals of the flavors and aromatic substances in general, as they exist often in such minute quantities. It is true that the general classes to which they belong could be determined, and it is known that they are members of those classes of bodies known as alcohols, ethers, esters, ethereal salts, essential oils, aldehydes, acids and terpenes. Each raw material, however, from which distilled spirits may be made gives a peculiar flavor of its own which cannot be isolated, nor can its chemical individuals be definitely distinguished.

Ethyl Alcohol not a Flavoring Agent.—The flavor of these bodies does not depend at all upon the quantity of ethyl alcohol which they possess. A distilled spirit may be reduced with water until it contains only 20 percent of alcohol, and still retain the flavor and aroma of the more highly concentrated article. Distilled spirits are found in the trade to contain alcohol ranging from 35 percent to 75 percent. Each spirit has its perfectly distinct characters irrespective of the amount of ethyl alcohol which it contains.

Fusel Oils.—That mixture of bodies known as fusel oil has already been described.

Distilled spirit contains notable proportions of fusel oil, probably three-tenths of 1 percent would represent the general average.

A distillate from potatoes contains less of these congeneric products, with the probable exception of amyl alcohol, than distillates from grains and fruits. For this reason a distilled spirit from potato has never become a popular drink, because it does not develop the aroma and taste which make a distilled spirit a desirable beverage.

Cherry as a Flavor.—Among the distilled spirits which are well known in Europe, but not very well known in the United States, may be mentioned cherry brandy, called in German, Kirsch or Kirsch Wasser, which is made from the fermented juice of the cherry. The wild cherry gives a distillate with the finest aroma. This distilled spirit is manufactured very extensively in Switzerland, in Baden and Alsace, and has a strong, fruity taste, with a most agreeable fragrance, is mild and easily drunk, although it contains a large percentage of alcohol. During the fermentation the alcohol which is formed dissolves the bitter principle known as amygdalin from the cherry stone, especially if the stones be broken. This amygdalin dissociates into a sugar, bitter-almond oil and a trace of prussic acid. Cherry brandy owes its distinctive taste to these bodies.

Prunes and Plums.—In Europe considerable quantities of brandies are made from prunes and plums. These are said to have a pure, pleasant fragrance and flavor, and also from the seeds of these fruits is derived a bitter principle similar to that in the cherry brandy.

In the Black Forest huckleberries and raspberries are fermented and a brandy derived therefrom which is highly praised on account of its aroma and taste.

In South Germany and Switzerland a distilled spirit is also made out of the root of the yellow gentian, which contains a kind of sugar capable of fermentation. This distilled spirit has a tart, aromatic flavor and is highly esteemed.

Influence of Wood.—The character of the wood which is employed in the storage of distilled spirits also has much to do with the aroma, and also to a certain extent with the flavor of an old spirit. The process of charring the barrels in which distilled spirits are stored gives to such spirits in the United States a distinct character not possessed by similar spirits stored in plain wood.

In the storage of brandy in the south of France great care is exercised in the selection of the wood and the manufacture of the barrels. A variety of oak known as Limousin is preferred. It is believed that this wood gives up very little bitter principle to the alcohol. The principles which are extracted from the oak are not only the tannin, but also quercin, which gives a distinct aroma, and quercetin, which contributes to the yellow coloration of the distilled spirits.

Flavors in Rectified Spirits.—In the artificial production of distilled beverages the flavors and aromas produced synthetically are very generally used. Small quantities of these concentrated oils, esters and essences serve to give a certain flavor to the mixture so that it may pass for a genuine whisky, brandy or rum. These are typical frauds and adulterations, but are permitted by our present revenue laws and the regulations made thereunder.

Adulteration of Whisky.—The adulteration of distilled liquors in all parts of the world has been continuous and extensive. Perhaps there is no other country where adulteration has reached such proportions as in our own. The discussion of the adulteration of whisky is typical of that which takes place with other distilled liquors, namely, brandy, rum and gin. All of them may be considered as coming under one head, and the statements which are made of whisky may be applied with equal force to the other forms of distilled liquors. The principal types of adulteration are the following:

First.—Selling raw spirits, that is, freshly distilled whiskies, under the guise of age by means of artificial coloring. When whiskies are stored in wood, especially in charred wood, they gradually take on a light amber, deep amber or reddish tint, according to the time of storage and the degree of temperature to which they are subjected. This process can be imitated at once by coloring to the desired tint by means of caramel (burnt sugar). Thus the spirit is mixed with the coloring matter, which gives it the appearance of age and deceives the customer. This method of adulteration may be practised in another way, namely, the spirit is stored in a warehouse which is artificially heated. The high temperature causes a more rapid absorption of the color from the wood than otherwise would take place, thus securing in a few months a depth of color which otherwise would have required three or four years to produce.

Second.—Substitution for whisky of a spirit rectified more or less completely and from which, therefore, the substances which give to whisky its character have been entirely or in part removed. This method of adulteration is by far the most extensive of all. A true whisky is a spirit which not only contains ethyl alcohol but a large number of other substances—alcohols, acids, esters, ethers, aldehydes, oils and other flavoring matters derived either directly from the grain, congeneric with alcohol in fermentation, produced by new combinations of ingredients during ageing, or extracted from the wood in which the spirit is aged. All of these ingredients act together to produce the flavor and character of the particular spirit. Substituting for this genuine whisky an article made from rectified spirit saves a great deal of time, expense and loss by storage. The true rectified spirit may be made from any source from which alcohol can be produced, but in point of fact is made principally in the United States from Indian corn. This rectified spirit is known as alcohol, neutral spirit, cologne spirit or velvet spirit, and other similar designations. The procedure is a perfectly simple one. All that is needed to make so-called whisky in this way are mixing vessels and a store of flavors and oils made by a chemist and having more or less resemblance to the natural flavors produced in the making of genuine whisky.

Third.—The third method of adulteration is the refilling of bottles and other packages which have contained genuine whisky with an adulterated article which is still sold under the same name. This has been a very common offense not only against the food law but the revenue laws, which punish with great severity an adulteration of this kind. The latest case of this kind of adulteration to which my attention has been called is a sentence passed on December 4, 1916, against certain offenders by Federal Judge Anderson of Indianapolis. The defendants were indicted for refilling "Bottled in Bond" bottles and selling their contents as "Bottled in Bond Whisky." This was done without destroying or removing the revenue tax stamp by which the "Bottled in Bond Whisky" is identified. Each of the defendants was fined a hundred dollars and costs and sentenced to serve 30 days in jail. The records of the Bureau of Internal Revenue show many offenses of this kind constantly occurring throughout the liquor trade.

Executive Indulgence to Adulterators.—During the time in which I was in charge of the execution of the Pure Food Law I made every effort to prevent the adulteration of distilled liquors by any of the above means. After a long drawn-out contest the definitions of whisky and other distilled spirits which I had established under the Pure Food Law were accepted by direction of President Roosevelt, and were put into the shape of regulations by the Bureau of Internal Revenue as a guide to the officials of that Bureau in properly marking and stamping packages of distilled spirits. This was done under the administration of President Roosevelt. When President Taft came into office he directed that these regulations be revoked and supplemented by another set, which permitted any distillate of grain, no matter of what color or character, to be designated as whisky. The two sets of regulations are as follows:

Roosevelt Regulations.—1. *High Wines.*—That which is practically the first product of distillation in which the substances congeneric with ethyl alcohol have not been transformed or their properties otherwise partially eliminated so as to convert them into any form of potable spirits, will be marked "high wines."

2. *Alcohol.*—(*a*) All forms of distilled spirits from which the substances congeneric with ethyl alcohol have been removed for practical purposes altogether, and which have been heretofore marked as "pure, neutral, or cologne spirits," will be marked "alcohol." (*b*) That product which has been commercially known as "alcohol" from which these congeneric substances have not been removed will be marked "commercial alcohol."

3. *Spirits,* "*as the Case may be.*"—Those products of distillation in which, by reason of the original material used and the methods of distillation employed, the characteristic substances congeneric with alcohol have been retained, which differentiate them into various forms of potable spirits—such as whisky, brandy, rum and gin—will be marked with the particular name of such potable spirit, as the case may be, without other description and without the addition of any adjective or descriptive word whatsoever; and the name of such particular spirit will be used even although when any of such spirits may be drawn from the receiving cisterns into casks certain congeneric products of distillation have not been changed by ageing or otherwise, so as to bring them to the potable form in which they are ultimately to be placed upon the market, provided such spirits have, before being drawn into the casks, been diluted to potable proof so as to then constitute a crude form of potable spirits.

Taft Regulations.—1. *High Wines.*—That which is the first product of distillation produced below 160° of proof and withdrawn from the cistern room above 110° of proof shall be marked and branded "high wines."

2. *Alcohol or Spirits.*—All distilled spirits produced at or above 160° of proof shall be marked and branded "cologne spirits," or "neutral spirits," or "alcohol," as the case may be.

3. *Whisky.*—All distillate from grain withdrawn from the cistern room at or below 100° of proof shall be marked or branded "whisky."

4. *Brandy, Rum, and Gin.*—Those products of distillation which, by reason of the material used and the method of distillation employed, are differentiated into various forms of potable spirits will be marked or branded with the name as known to

the trade, as "brandy," "rum," "gin," etc., and other truthful descriptive words may be used such as "apple," "peach," "grape," etc., provided that no distilled spirits produced from material other than grain shall be marked or branded whisky unless the word "whisky" is preceded by the word "imitation."

Nothing in these regulations shall be construed as prohibiting on the stamp end of the package the use of other truthful descriptive words or legends, such as "straight," "rye," "Bourbon," "corn," "aged in wood," "sour mash," "hand process," etc.

It will be seen by a study of the above regulations that in the first set the purity and identity of whisky was secured by the particular form of marking and branding required. In the second set of regulations the term "whisky" is made applicable to every form of spirit distilled from grain. Thus alcohol, velvet spirit, neutral spirit, cologne spirit, etc., when reduced to proof may all be marked "whisky." When it is remembered that the federal courts, especially those in Cincinnati and Peoria, upheld the first series of regulations as being in entire harmony with the Revenue and Pure Food Laws, and inasmuch as these laws have not been changed, it is easy to perceive that the present regulations established by executive order are in direct opposition to the rulings of the courts.

PART XI

BRANDY

Definition.—The term "brandy" used without any limitation is applied exclusively to the product of the distillation of wine in a pot still. If a fermented juice of any other fruit than the grape be employed in the making of a distillate of this kind, the name of the particular fruit is to be used in conjunction with the term brandy. For instance, if hard cider is distilled, the term "apple brandy" is used; if a fermented juice of the peach is distilled, the term "peach brandy" is used; if fermented cherry juice is distilled, the term is "cherry brandy;" or if the blackberry, blackberry brandy, and so on.

It is fair to presume that brandy was the first form of distilled spirit known, since the method of converting starch into sugar was not known at the time of the discovery of distillation. Brandy, therefore, ought to be regarded as a type of all distilled beverages, having ethyl alcohol as the principal constituent.

Cognac.—The term "Cognac" is applied to brandy distilled in the southwestern portion of France, in the departments of the Charente, and of the Charente Inferieure. These brandies have the highest reputation of any in the world, and the term "Cognac," therefore, is guarded with great jealousy by the French as the particular designation of their product. The term is derived from the city of Cognac, which is the Capitol of the Department of the Charente.

Large quantities of brandy are made in France outside of the two departments mentioned. Under the French law it is not permitted to be called Cognac, but is known by the term "Eau de Vie," followed by the name of the department, or locality where it is manufactured.

It is a very common practice outside of France to call all brandies in general "Cognac" on the supposition that the name is no longer distinctive of a locality. This is an untenable position, since the French have delimited with the greatest care the area in

331

Fig. 38.—View of the vineyards at Charroux.

which the brandy can be made, legally bearing the name, as will
be shown further on.

Character of the Wine from which the Brandy is Made.—The
wine from which the brandy is made in the Cognac regions is not

FIG. 39.—Superposed vats.

of itself very palatable. It is somewhat thin in body, acid in
taste, and unpleasant as a beverage. In fact, the wine which
makes the finished brandy is regarded with little favor as a beverage in its own home. It is true that, in the brandy region, the
peasants often drink of the wine of the character used in making

brandy, but this is only a local consumption. The wine has little value as a commercial beverage. On the other hand it has the qualities which give to the distillate a character which is superior to that produced in any other country.

Kinds of Vines Grown.—The kinds of vines cultivated in the Charentes for the production of brandy are not specially found in that region, but they are the same as are found in other places, even far removed from the Charentes. None of them perhaps, or at least very few, can be regarded as indigenous.

Among the most important of the various kinds of grapes cultivated are the following:

Folle Blanche.—The origin of this name is difficult to explain. There is nothing about the plant itself which suggests either one of the names by which it is known. It does not have a very exuberant vegetation, but only of the ordinary kind. It is, very productive; its fruit is thickly borne upon the bunch and the individual grapes are only of moderate size and not very much different from those of other varieties in this respect. Moreover, the Folle Blanche is known under other names in different parts of the country and even in the Charente. It will be unnecessary to give here a list of the various names by which this variety of grape is known. It is supposed by some authorities that the Folle Blanche is practically identical with a variety of grapes which grow upon the banks of the Rhine and also in Wurtemburg. There are marked differences between these fruits. The color of the grape when ripe is of a yellow green, the fruit being almost round, of moderate size, thickly gathered on the bunch and quite sweet in taste. There are many variations from these characteristic qualities, but they all more or less correspond to the original French form. As the grapes may be more light in color in one instance than another they are sometimes called Folle jaune, and when more green Folle verte.

Under some of the different names referred to this grape is widely extended throughout the vineyards of France, and from the west of France even to the Pyrenees and as far as Bretagne. It covers thus, more or less, a surface which is limited on the west by the ocean and on the east by a line which passes by Toulouse, Cahors and Orleans. The Folle Blanche develops early in spring.

Composition.—In a bunch of grapes from the Folle Blanche vine it is found as a rule that the stem totals about 4.7 percent of the weight and the grapes about 95.3 percent. The composition of the grapes themselves is as follows:

Pulp............................87.22 percent
Skin............................9.9 percent
Seeds............................2.9 percent

The average composition of the juice of the Folle Blanche is as follows:

Specific gravity......................... 1.077
Water................................ 78.93
Sugar................................ 16.95
Tartrate of lime......................... 0.35
Free organic acids........................ 0.61
Crude protein............................ 0.09
Crude fiber............................. 2.74
Ash.................................... 0.05
Insoluble fiber.......................... 0.28

From the above it is seen that the grapes of the Folle Blanche in the region of the Charente if used for a wine would have rather a low alcoholic content. Seventeen percent of sugar, the average which they contain, would not produce more than 8½ percent of alcohol by weight in the wine, and this would be regarded as a light wine insofar as its alcoholic content is concerned.

Other varieties of grapes which are used in the Charente for brandy making may be mentioned, only by name:

Colombard, Saint-Emilion, Le Jurancon, Blanc-Rame, Balzac Blanc, Chalosse, Saint-Pierre, Bouilleaud, Saint-Rabier, Balzac Noir and Petit Noir.

Antiquity.—Grapes have been cultivated in the Cognac region of France since the earliest historical times. The cultivation appears to have been introduced concurrently with that of wheat. All common methods are in use in the preparation of the field for the vineyard. In some cases the work is done by hand, in others by the plow, and in still others by a steam engine. In each case it is important that the ground be deeply stirred before the vines are planted, in order that the hard layers underneath the surface may be broken up to permit an entrance of the roots. The vine, however, is a hardy plant and will seek its nourishment even in the most unpropitious localities, sending its roots deep in between the rocks, if necessary, to get the required nourishment. It is desirable that the vine shall yield a crop as soon as possible after planting, and this is usually secured in three years; but the greatest care must be taken in the setting of the vines, in cultivation and in fertilizing. The soil is often broken up to a depth of half a meter, or even more. The surface is brought into a fine condition of tilth by harrows and other apparatus. Where the vine is planted in stony land, of course, the preparation is more laborious.

Manure.—The vines have scarcely ever been manured in the Cognac region. What manure is produced is usually reserved for the cereals or meadows. This is not considered an economical process, since it is proved by experiment that generous fertilization produces much more abundant growth of grapes. Not only is stable manure valuable in the vineyard, but also chemical fertilizers. Organic manures are preferred to the mineral manures, when they can be procured, especially for furnishing nitrogen to the growing plant. These organic compounds decompose more slowly and thus produce their plant food as it is needed for the growth of the grape.

The Vintage Season.—The harvest of the grapes begins about the first of October and is usually over by the middle or end of November. The wine has by the end of the year undergone its first fermentations and the sugar which the juice contains has been practically converted into alcohol. The distillation of wine begins in the early winter and is continued up to the spring months. The distilling season usually ends in March or May.

Manufacture.—The production of the wine from which the brandy is made is not very different in the Cognac region from that of other parts of the world. The red wines which are made are produced by the fermentation of the crushed red grape, while the white wines are made from the green-yellow grapes or by expressing the grape juice and fermenting the expressed juice.

Distillation.—Distillation of the wine is extremely simple. It is accomplished in small pot stills, holding from 600 to 1,000 liters, and operated by a slow fire gauged in such a way that about 8 hours is required to make one complete distillation. If the stills are run night and day, each still will have about three charges in the 24 hours. The first products of the distillation are collected together and are subsequently subjected to a second distillation in order to secure the necessary alcoholic strength. The stills are ranged in batteries, often 6, 12, or even a large number of stills being placed in one line. The stills are usually furnished with a heater, into which the succeeding charge of wine is placed and through which the tube bearing the distilled vapors passes. This arrangement has a double advantage. In the first place, it condenses and throws back into the still the more watery portions of the alcoholic vapor, thus permitting a stronger alcoholic liquor

FIG. 40.—BRANDY DISTILLERY AT SAINTES, SENT BY SENATOR A. CALVET.

The stills are embedded in the masonry. The necks of the still are connected with the top-shaped vessels (chauffe vins) which contain the wine intended for the next charge. The cylindrical vessels to the extreme left and right contain the condensing worms. The raw brandy flows directly into the casks.

to be made on the first distillation; and in the second place, it warms the wine which is to be used for the next distillation, so that when it is placed in the still it may at once be brought to ebullition.

The stills are not completely filled with the wine, but usually only about one-half filled, making about 300 liters in the ordinary charge. The product obtained by the first distillation is known in the still as eau-de-vie brute.

22

Form of Still.—Many different forms of stills have been used, among them those known as column or continuous stills, but these were found to give a brandy without character. Although more economical in operation, and making a spirit at less expense, it was found undesirable to use a still of this kind because of the failure of the product to mature in the manner which would entitle it to represent the old-fashioned and high-grade brandies of the district. A type of still in general use is shown in the figure.

Storage of Brandy.—After the second or third distillation the brandy is found to have an alcoholic content of about 65 percent to 70 percent. After distillation, it is immediately placed in a cask containing about 500 liters (over 100 gallons). It is kept in these casks until it is ready for sale. These casks are made of oak from Trieste, or Burgundy, or from Limousin, and are usually prepared by the distiller himself. Almost every distiller has his own coopershop where these oak casks are made from the imported oak. The casks are kept for a long while; usually the brandy is sold and the casks are returned to the owners. Old casks are preferred, especially if they have been used for making very old brandy. Care must be taken in all cases, however, not to use a cask which is moldy or decayed in any way. The brandy is extremely susceptible to the influence of foreign odors and may easily lose its good flavor if kept in casks of this kind.

The casks are placed in storehouses called "chais." These storehouses are all above ground and the casks are kept until by age their contents acquire the qualities which are associated with mature brandy.

Change of Alcoholic Content.—While in the United States it is well known that whisky in our storehouses undergoes certain changes in alcoholic strength of a character which produces a finished product much richer in alcohol than at first, the contrary operation takes place with brandy in France. This may be due in particular to the fact that the brandy is stored at an alcoholic strength of 65 percent or 70 percent, while in America the whisky is stored at an alcoholic content of only 50 percent. At any rate, the effect of age upon the brandy is to diminish to a certain extent the percentage of alcohol therein. Water also passes off with the alcohol, but as a rule with residual contents of the cask of old brandy, while diminished largely in volume, contain also a much less content of alcohol than they did at first. While the greater part of the alcohol which is lost is, doubtless, evaporated through the pores of the wood, some of it is changed into ethers and other aromatic substances, which give to old brandy some of its most desirable characters. At any rate, the brandy becomes more and more fragrant, and more delicious to the taste, as well as to the nostril, by keeping in wood. Very old brandies; that is, those that have been kept for 20 or 30 years, or more, in wood, have a character which is entirely distinct from the new brandies, and often brandies which are kept only four or five years ameliorate in their qualities and become much more potable and pleasant, both to the taste and the nostril.

Time of Maturity.—In the Cognac region the best brandies rest for 20 or 25 years, in wood. This, of course, makes them much more expensive, since the interest on the cost of producing the brandy must be reckoned and the great loss in volume taken into account. The result of this is that most of the brandies which go into commerce are kept only a few years, and thus do not have the superb qualities which the French themselves attach to the brandies consumed at home.

Composition.—The following table shows the average composition of the various grades of brandy produced in the Cognac region —at least all of the grades which are entitled to be known by a special name. Of these grades the Grande fine Champagne is the best and the Bois à terroir is the least desirable.

	Grande fine champagne	Grande champagne	Petite champagne	Borderies	Bons bois	Fins bois
Percent of alcohol..........	69.3	67.9	25.4	66.9	50.85	65.3
Percent of extract per liter..	0.16	0.25	2.08*	0.05	2.58*	0.16
Percent of fixed acids.......	0.048	0.024	0.0252	0.024	0.0348	0.024
Grams per hecto-liter of pure alcohol. Acids	36.36	35.59	141.98	30.49	205.31	40.43
Aldehydes	9.26	14.81	35.86	13.08	49.99	9.75
Furfurol	2.10	2.74	1.59	1.73	1.26	1.76
Ethers	160.13	121.39	154.50	116.26	190.36	130.55
Higher alcohols	124.72	123.72	140.42	143.23	186.94	165.31
Total constituents other than alcohol	312.57	296.25	474.35	304.79	633.86	347.80

*Mostly added sugar.

Method of Sale.—New or old brandies are seldom sold directly to the consumer; they go to an intermediary. This intermediary is the house of the merchant. Each one of these great houses, at least the most important of them, in each locality has a special agent or representative, who knows where the good brandies are found, and also is acquainted with the needs of the proprietor. These agents secure samples of the different brandies which are for sale and submit them to the merchant, or the merchant on the other hand often depends upon the judgment of the agent in regard to the quality of the goods offered. All brandies are tested by taste and smell by experts, in order to judge of their qualities, before sale.

Blending.—Brandy is rarely given to the consumer in the same condition in which it is bought from the producer. There are several reasons for this. In the first case, the brandy as it is often found by lack of proper care in its keeping, is not entirely limpid; thus the first step in preparing for sale to the consumer is to carefully filter it. The greater reason is that each of the great houses dealing in brandy supplies a clientele which is accustomed to brandy of a particular kind. This kind is usually the product of a mixture of a great many brandies of the region. Brandies differ from year to year in character, and it is the art of the great houses to make mixtures which produce a beverage varying little from year to year in quality and character. Thus, by selling these brandies under a certain name they secure a constant demand for them, and the persons who use them expect the same quality from year to year. Great art and skill are required to take from hundreds of different producers different kinds of brandy and by mixing make a uniform variety.

History of the Cognac Vineyards.—The vineyards of Cognac are of very ancient date. Like those in the vicinity of Bordeaux, they probably were established even before the conquest of the country by Ceasar.

Authentic documents prove that as early as 1323 wines of these departments were exported from the port of La Rochelle to northern countries, particularly England, Scotland and Scandinavia. Later on the Dutch, who were the greatest voyagers in those days, were accustomed to sail up the river Charente and buy from the farmers the wines on the banks of the Charente, called Borderies.

As a result of this trade, the natives of Cognac and the surrounding district began to plant vines in greater volume and soon there was an overproduction in the neighborhood; huge stocks of wines could find no purchasers. About 1630 it occurred to a few of these producers, who had large quantities of wine on hand, to turn their wines into brandy by the process of distillation, which at that time was supposed to have been invented by the Greeks, but which had seldom been used outside of the apothecary's laboratory for any commercial product. This new idea was found to be not only practicable but profitable. Indeed, the business was so profitable that what at one time had been a surplus in the stock of wine became a deficit, and a tremendous impulse was given to the cultivation of the vine, which rapidly spread over that part of the country suitable for grape growing.

The Phylloxera.—The industry continued to flourish, and especially under the rule of Napoleon the Third the Cognac district reached a degree of prosperity such as had never before been known. This prosperity continued even during the period of the Franco-German war, which did not affect the region inasmuch as the operations of the war did not extend to this part of the country. It was not until 1875 that any serious check was given to the prosperity of the country. At that time, the dread disease, the phylloxera, an insect pest which attacks vines and which had already created ravages in France appeared in the Cognac vineyards and in a few years they were destroyed. In fact, the whole industry seemed to be on the verge of absolute extermination. There was one remarkable phenomenon connected with the attack of the phylloxera, namely, low-lying vineyards subject to overflow were not affected. It was found that the annual overflow of the water destroyed the life of such of the phylloxera as might have found a lodgment and thus protected the vines. The only original vineyards which still exist in the Cognac region are those which were subject to periodic overflow. Some of these vineyards have vines of remarkable size and age probably exceeding 100 years, and which are still bearing vigorously.

American Vines.—Scientific men were called in to combat the dread disease. The attempts to kill the phylloxera were abandoned as impracticable. Happily, it was discovered that if the vines were grafted on American roots, largely taken from the southern part of the country, especially Texas, they produced a plant which the phylloxera did not attack. This discovery lead to the renovation of the vineyards, not only of the Cognac region, but of all parts of France. In the last 15 or 20 years the vineyards have been entirely re-established and even extended beyond their original boundaries, so that the prosperity of the Cognac region has now been renewed.

It is interesting, in this connection, to call attention to the fact that the celebrated American scientist Dr. C. V. Riley, did very much of the scientific work which resulted in the restoration of the French vineyards.

Magnitude of the Industry.—From 1801 to 1815 the highest number of hectoliters of brandy sent out of the Charentes in one year was 99,157, in 1807; and the lowest quantity 16,562, in 1812. In 1815, the last year of the First Empire, the amount sent out was 38,188 hectoliters. The 100,000 mark was first passed in 1823, with the exportation of 122,424 hectoliters. The 200,000 mark was first passed in 1849, with the exportation of 213,139 hectoliters. The 300,000 mark was first passed in 1864, with the exportation of 320,621 hectoliters. During the time of the ravages of the phylloxera there was a remarkable decrease in the exportation of brandy, in 1874 the amount sent out of the country being only 160,310 hectoliters. In 1878, however, the vineyards were largely restored and 433,660 hectoliters were exported. In the last 40 years the magnitude of the industry has fluctuated with the abundance of the harvests. The widespread imitations of Cognac brandy has exerted a very depressing influence in the trade and caused a great accumulation of stocks. The closing of the ports of the United States to the importation of distilled spirits for beverage purposes has still further restricted the trade. It is not likely that the future promises any expansion of the industry beyond the palmy days that followed the recovery from the devastation of the phylloxera.

Designations of Cognac.—The quality of brandy produced in the Cognac region, as has already been intimated, is not uniform. Some of it is of an exceptionally high grade, while others form a brandy which compared with the finest may be designated as low grade; but even the low-grade brandies of the Cognac region rank very well with the best of the brandies of other localities.

The different kinds of brandy which are produced in the Charente have received, by common consent, certain names or appelations which distinguish them from each other, and all other brandies. These names are strictly regional and they are guarded with great jealousy by those who are entitled to use them. These designations have become so firmly fixed that they have all the force and effect of law. A merchant or producer who, for instance, is not entitled by the location of his vineyard to use the term "Grande fine Champagne" upon his product, would be subject to a process by law if he should employ it, even if his brandy might be made just on the edge of the region entitled to bear that particular name, and would be, as far as ordinary methods of detection are concerned, equally as good. Following are the divisions of the vintage brandies in the Charentes:

1st.   Grande fine Champagne.
2d.   Grande Champagne.
3d.   Petite Champagne.
4th.   Borderies.
5th.   Bois.

The variety known as Bois are further sub-divided as follows:
(a)   Fins Bois.
(b)   Bons Bois.
(c)   Bois ordinaires.
(d)   Bois à terroir.

The proprietors of the vineyards which produce the Fins Bois have an ambition to be permitted to enter into the region of the "Champagne," on which they, moreover, border. Already to some of these brandies the term Champagne Boisée has been applied.

The differences in the qualities of the brandies are sometimes not perceptible to the uninitiated, and experts even may often be deceived by the taste and flavor of the brandy in regard to the class in which it should be placed. The factors which make these variations are numerous, and competent authorities do not always agree respecting their number and importance. Among those which may be mentioned as effective are, first, the nature of the soil, particularly with regard to the amount of chalk it contains; second, the character of the climate; third, the method of cultivation of the vine; fourth, the character or kind of vine; fifth, the distribution of the rainfall; sixth, the character, quality and extent of fertilization.

The Limits of the Cognac District.—Long before any legal action was taken by the French Government it was recognized that there was a certain geographical area to which the term "Cognac" as applied to the production of brandy should be delimited. In the beginning of the enforcement of the law against the importation into the United States of adulterated and misbranded foods, I made a ruling that only brandy made in the Cognac region of France could be imported as Cognac.

FRENCH DELIMITATIONS

When the French Government by law provided for the delimitation of the Cognac region they established almost exactly the limits suggested in this earlier decision.

The exact limits as established by the French Government are as follows:

The two Departments which comprise the principal regions of the Cognac are the Charente and Charente Inferieure. The Department of the Charente contains the following divisions: First; Angoulême, the most important division of the Charente and also the name of the most important city. It contains about 37,000 inhabitants and is beautifully situated. It is principally placed upon the summit of a hill, which is about 250 feet above the level of the plain. The city is constructed in a most substantial manner out of the famous stone of Angoulême, which is very soft when first mined and which hardens on exposure. This stone is famous for its building purposes and is exported to great distances. There are a number of ancient monuments in the city and the cathedral is constructed according to the style of the 12th century of the Roman Byzantine architecture. Angoulême was a city which existed at the time of the invasion of Gaul by Caesar. At that time, however, it was only a small village.

The next most important place in the Charente is Cognac, the city which gives its name to the celebrated product of the country. It is a beautiful city, containing about 20,000 people, and is the center of the commerce in brandy. It is situated on the banks of the Charente, upon an undulating plain, and is of very large extent considering the number of its people. It is not so well constructed as Angoulême, and has more the aspect of a large village than of a small city. One of the most important personages who have lived in Cognac was Francis I, who, before he became king of France, was the Lord of Cognac. His castle has now become a house engaged in the brandy industry.

The next most important point in the Charente is the town of Jarnac, which is situated on the border of the region producing the best brandies of the country. It is also situated upon the Charente river, and is celebrated, especially in history, by the victory which the Duke of Anjou gained there in 1569 against the Prince of Condé. It is not now a warlike region, but has a long time been the center of an agricultural and commercial population.

Another important region of the Charente is the town of Segonzac, which is situated directly in the center of the region producing the most famous of the Cognac brandies, *i.e.*, Grande Fine Champagne. This kind of champagne must not be confounded with the same word which is used in another part of France to indicate a sparkling wine. It is used here simply to denote the brandy which is produced upon this almost level plain, known as the Champagne.

Chateauneuf is another port of some importance and is the center of the region which produces the second grade of wine called the Petite Champagne.

Charente Inferieure.—The lower part of the brandy region lying near the ocean is known as the Department of the Charente Inferieure. The chief city of this Department is La Rochelle, a town of 29,000 inhabitants. It was at first built upon a small rock, whence its name. On account of its maritime importance it early assumed a position of considerable value, both in regard to commerce and the protection of the coast. It naturally became the scene of many bloody and long-continued sieges and conflicts. It was a region which was principally Protestant in its religious views, and upon the revocation of the Edict of Nantes a large part of the population moved to Canada. It is a town which preserves to a large extent its aspect of former years, its forts and ditches being well preserved. The importance of La Rochelle as a maritime port has been somewhat diminished with the growth of more suitably situated cities, such as Nantes and Bordeaux, which are better able to handle a large volume of commerce. There has recently been an improvement in the maritime activities of La Rochelle by the building of a new bridge to La Pallice which is joined with La Rochelle by a tramway. La Pallice has an admirable harbor and the construction of this port will doubtless do much to restore the maritime importance of the city.

The next most important place in the Charente Inferieure is Saintes. This town is situated on the river Charentes and has many remains of the Roman invasion. The amphitheater which the Romans built there was capable of holding from 20,000 to 22,000 spectators. At the present time Saintes is the center of the agricultural region of great prosperity and is also the market for large quantities of brandy.

The brandies which are produced in the Charente Inferieure as is well known, are not so highly prized as those which are produced in the Charente.

The largest city of the Charente Inferieure is Rochefort. It is one of the five great maritime ports of France. It is a town of 34,000 inhabitants and has a large commercial and industrial population.

Another important town of the Charente Inferieure in regard to the production of brandy is Jonzac. It is a picturesque although small city and is engaged almost exclusively in the production of wine, brandy, and building stone.

Bouquet.—The bouquet of a brandy is the name given to the odor and cannot be detected from the chemical point of view. It is certain that the bouquet owes its properties chiefly to the ethers, esters and volatile oils.

It plays a very important part physiologically and organoleptically in the brandy itself. The medicinal properties of a brandy, if there be any, are considered to be to some extent due to the bouquet, and the impressions which it makes upon the olfactory nerves, and its exciting effects upon the digestive glands. The bouquet of a brandy cannot be ascertained in the ordinary method of drinking from a small glass, nor from the odor alone which emanates from the bottle when the cork is withdrawn. The true way of ascertaining the bouquet of brandy is to use a very large glass, holding at least half a liter, and of sufficient size so that the mouth and the nostrils and the greater part of the face can be included within the circumference of the glass when the brandy is sipped. A few drops of brandy, when placed in a glass of this kind, and at the proper temperature, which should be between 70 and 80°F., develops very rapidly its bouquet, so that when the glass is raised to the lips the full impression of the bouquet upon the receptive nerves is secured. These minute components of the bouquet, are difficult to separate. Many of them are found in the wine itself from which the brandy is made, yet, as has already been stated, distillation of brandy is usually from the young wine, and when of this age it has not developed the bouquet which wines of older age possess. These odoriferous substances which exist in the wine are carried over by the distillation and there increased and modified by the processes of ageing to which the brandy is subsequently subjected. It is very difficult to describe, so evanescent a thing as the bouquet of fine brandy. It is only the connoisseur who, by long years of practice has developed his abilities to discriminate, can distinguish in a proper way the delicacy of the bouquet.

Brandy from Marc.—The pomace which is left from the pressing of grapes is called "marc." In many instances the marc is fermented and distilled, forming a pomace brandy, or as it is called in French "eau-de-vie de marc." This brandy has no resemblance to the true Cognac in flavor, taste, or other properties, except the common property of alcohol. It is extremely strong and acrid in taste and can hardly be swallowed by those who are not used to its comsumption. It contains a very high percentage of aldehydes, ethers and acids. Nevertheless, in spite of its somewhat revolting character it is consumed in considerable quantities by many people. In Burgundy the brandy which is made from the marc is very extensively used and is highly thought of by those accustomed to its use.

From the seeds of the grapes an oil is made which is devoted to many industrial uses, but there is no industry of this kind in the Charente.

Marcs are also made use of for the extraction of essences with which to flavor neutral spirit in imitation of brandy. It is needless to say that such an imitation is but a poor substitute for the genuine article.

The residues which are left after the distillation of brandy are called in French "vinasses." The residue of the distillation is still largely liquid, and it consists of all those parts of the wine, including some water, which are not volatile at the temperatures at which distillation takes place. The vinasses contain practically all of the tartaric, malic and succinic acid, and gums which are found in the original wine. Also the whole of the tartrate of potash, the bitartrate of lime, alumina, phosphates of all kinds, and any mineral acids that may have been in the wine or produced by the use of sulphur in its manufacture, notably sulphuric acid. In addition to these, the nitrogenous matters of the wine are also found in the vinasses, as well as those of a pectic character and the tannin derived from the wine, and the glycerin produced during fermentation.

Tartrate of Lime.—The residues of the distillation are frequently used for the manufacture of bitartrate of potash, being converted for this purpose first into tartrate of lime. The hot vinasses, or after it has been cooled, is treated with milk of lime until it is neutralized, which can be recognized by the color of the liquid, or with litmus paper. The tartrate of lime which is formed is precipitated, the liquor decanted, and the precipitate cooled and dried. During the precipitation of the tartrate of lime many other bodies foreign to this substance are thrown down, which renders the whole mass impure. It is necessary to add a sufficient quantity of lime to perfectly neutralize all acids present; otherwise the precipitation of the tartrate of lime is not complete. If care is taken to carry on the precipitation in a solution which is just the least bit acid a much better tartrate of lime is secured, even being almost white. In this way the tartaric acid is fixed and separated from the other bodies which remain in the residue of the distillation. The tartrate of lime is well dried and reduced to powder or coarse consistency and preserved in bins which are kept perfectly dry. In a humid atmosphere it may undergo fermentations which practically destroy the tartaric acid which it contains. From this tartrate of lime much of the tartaric acid and bitartrate of potash of commerce are made.

Official Definitions.—Brandy in the Pharmacopœia according to the Eighth Edition of the United States Pharmacopœia brandy should have the following properties: The Pharmacopœial name is Spiritus Vini Gallici. Definition: An alcoholic liquid obtained by the distillation of the fermented, unmodified juice of fresh grapes. The following remarks are then made in the Pharmacopœia:

A pale amber-colored liquid, having a distinctive odor and taste, and a slightly acid reaction.

Brandy should be at least four years old.

Its specific gravity should be not more than 0.941, nor less than 0.925 at 15.6°C. (60°F.), corresponding, approximately, to an alcoholic strength of 39 to 47 per cent. by weight, or 46 to 55 per cent. by volume, of absolute alcohol (see Appendix, Alcohol Tables, page 604).

"If 100 c.c. of Brandy be slowly evaporated in a tared dish on a water-bath, the last portions volatilized should have an agreeable odor free from harshness (absence of fusel oil from grain or potato spirit); and the residue, when dried at 100°C. (212°F.), should not weigh more than 0.5 Gm. This residue should have no sweet or distinctly spicy taste (absence of added sugar, glycerin, and aromatic substances), and it should almost completely dissolve in 10 c.c. of cold water, forming a solution which is colored not deeper than light green by a few drops of ferric chloride T. S. diluted with 10 volumes of water (absence of more than traces of oak tannin from casks).

If 50 c.c. of Brandy be shaken vigorously in a stoppered flask with 25 Gm. of kaolin, and, after standing half an hour, be filtered, the color of the filtrate should not be much lighter than that of the Brandy before treatment (absence of caramel coloring).

To render 100 c.c. of Brandy distinctly alkaline to litmus, not more than 1 c.c. of normal potassium hydroxide V. S. should be required (limit of free acid).

Both brandy and whisky were omitted from the Ninth Revision of the United States Pharmacopœia issued Sept. 1, 1916. This omission was due to the fact that a great majority of physicians testified that they were no longer sufficiently important as medicines to warrant retaining them in the Pharmacopœia. The supposed medicinal virtues of brandy and whisky are found to be illusionary by modern medical experience.

Official Definition.—The definition of brandy adopted by the Association of Official Agricultural Chemists and the Association of State and National Food and Dairy Departments upon the recommendation of the Joint Committee on Food Standards is as follows:

New brandy is a properly distilled spirit made from wine, and contains in one hundred (100) liters of proof spirit not less than one hundred (100) grams of the volatile flavors, oils, and other substances derived from the material from which it is made, and of the substances congeneric with ethyl alcohol produced during fermentation and carried over at the ordinary temperature of distillation, the principal part of which consists of the higher alcohols estmated as amylic.

Brandy (potable brandy) is new brandy stored in wood for not less than four (4) years without any artificial heat save that which may be imparted by warming the storehouse to the usual temperature, and contains in one hundred (100) liters of proof spirit not less than one hundred and fifty (150) grams of the substances found in new brandy save as they are changed or eliminated by storage, and of those produced as secondary bodies during aging; and, in addition thereto, the substances extracted from the casks in which it has been stored. It contains, when prepared for consumption as permitted by the regulations of the Bureau of Internal Revenue, not less than forty-five (45) percent by volume of ethyl alcohol, and, if no statement is made concerning its alcoholic strength, it contains not less than fifty (50) percent by volume of ethyl alcohol as prescribed by law.

Cognac, cognac brandy is brandy produced in the departments of the Charente and Charente Inferieure, France, from wine produced in those departments.

Esters and Ethers.—The following compounds have been obtained from brandy 25 years old by Ordonneau, a French chemist:

> Normal propyl alcohol;
> Normal butyl alcohol;
> Amyl alcohol;
> Hexyl alcohol;
> Heptyl alcohol;
> Ethyl acetate;
> Ethyl propionate, butyrate, and caproate;
> Oenanthic ether;
> Aldehyde;
> Acetal;
> Amines.

Manufacture of Brandy in the United States.—Very little true brandy is made in the United States. There is a good deal of spirit distilled from grapes, or the wine made from grapes, but little of it has any claim to the term "brandy." Much of the so-called brandy made in this country is the product of rectifying stills.

The principal part of the brandy in the United States is made in California. The greater part of the grapes, which grow in California, are so radically different from those grown in the Charente that it is not to be expected that the products will compare in quality. In addition to this the manufacture of brandy in California is mostly an incident of the winery, instead of its object. In point of fact, a very large part of the wine set aside for distillation for the making of brandy is wine that has gone bad during fermentation, and is so sick that it cannot be kept for development. Wines which are commencing to turn acid or to go wrong are sent at once to the distillery, while the sound wines are preserved for beverage purposes. In addition to that a considerable amount of sugar is added to the pumace and refermented, but this brandy is of a bad character. A single sip of this so-called brandy would give a far better description of its character than words could convey.

The brandy industry of the United States may be regarded as practically not established upon an ethical basis. The result is that no one drinks or cares for the brandy of this country. There is no reason why many regions of California and other wine-producing states may not make a good brandy if proper precautions are taken. I have personally tested brandy made in California of good quality and bouquet.

Regions where Brandy may be Produced.—There are large areas of our country producing grapes in great profusion, and which are of a character, to some extent, resembling that of the product of Charente. The wine which is made from such grapes usually is thin of body, acid in taste, and of a general character like those produced in Charente. If these grapes, which grow so abundantly in may parts of the country, especially in New York and Ohio, were properly distilled into a brandy and kept for a sufficient length of time, they would in time, in my opinion make a brandy of most excellent character, and one which would commend itself with great favor to the people in general.

Uses of American Brandy.—While American brandy is drunk to some extent by the Californians and others who make it, it has no distinct place on the American market. One of the chief uses of so-called American brandy is for the fortification of a so-called sweet wine, so that a sufficient amount of alcohol may be present in the wine to prevent further fermentation.

FRUIT BRANDIES

Apple Brandy.—When the juice of apples is fermented, the product is known as hard cider. When this is subjected to distillation, apple brandy is produced, very commonly known by the phrase "Apple Jack." This is a true brandy, and has the characteristic properties imparted to it by the congeneric substances produced during fermentation of the ethyl alcohol and developed on ageing. It does not take a connoisseur to distinguish at once between apple and grape brandy, for both have characteristics exclusively their own. Apple brandy has been very extensively made in the United States, especially by the smaller farmers in New England, New York, New Jersey, North Carolina, and other parts of the country, where apples grow in great abundance. Unfortunately it is usually consumed in rather a fresh state, and thus does not have the proper age which it should have, and which would convert it into a very delicious liquor. Freshly made brandy is a rather rank tasting article and, like other freshly distilled spirits, it is very much more injurious than those distillates which have been softened and modified by age. When an apple brandy has been long matured in wood, it acquires a deliciousness and fragrance of character which is not much inferior to that of grape brandy.

Peach Brandy.—The product of the fermentation of the juice of the peach is known as "peach brandy." Generally this brandy has the properties of the distillates from fermented fruits, but has characteristic flavors and odors, which distinguish it from other fruit brandies. It has a special flavor due to benzaldehyde which is quite abundant in the peach kernel.

In the United States peach brandy has not been made so long and so extensively as apple brandy, but it is produced at the present time in considerable quantities. As in the case of apples, the peaches may be previously dried and subsequently subjected to fermentation, but this impairs the quality of the distilled product.

The wide expansion of the peach industry in the United States, in the South, especially in Georgia before the advent of prohibition, increases the problem of the utilization of very large and abundant crops. Often the transportation facilities are not sufficient to move such perishable properties as peaches with sufficient rapidity to save the whole crop. In this case distilleries would be extremely advantageous for using up the parts which could not be shipped, or those portions of the crop which are too ripe at the time of gathering to bear transportation. These very ripe peaches would be better suited for the production of industrial brandy than those less mature and suitable for transportation.

Ordinarily the market value of good peaches is so far superior to the market value of brandy that there would be no necessity of distillation. It is only in the cases which have been mentioned of superabundance, or in the production of fruit too ripe for transportation that a distillation of the product would be advisable. A distillery which could be utilized for other purposes during part of the year, in this case, would prove profitable for peaches, but, if the distillery could only be utilized during the ordinary time of the ripening of the peach crop, say from two to two and one-half months, it would be hardly profitable to invest very much in the erection of a factory. Since prohibition bids fair in the near future to become nation wide, it is better to seek an outlet for superabundant peaches in canning factories or drying kilns, or for denatured alcohol.

Ageing.—Peach brandy must be aged in order to secure its best properties, as in the case of other distillates. The principal trouble with the consumption of peach brandy is the same as with apple brandy, that is, its lack of maturity. It is expensive to keep a product 4 or 6 years before it is offered to the market, and hence, in the case of smaller farmers, no attempt is made to store the product in a government bonded warehouse. It is sold as soon as possible after manufacture.

Blackberry Brandy.—Another brandy, which is found abundant upon the market in the United States, is that made from blackberries. True blackberry brandy is made only from the blackberry which has been crushed and fermented and the fermented liquor distilled. There is thus produced a true blackberry brandy, which not only has excellent properties as a beverage but is highly prized for medicinal purposes. Blackberry brandy is a home remedy of great efficiency in almost all forms of diarrhœa. In general the blackberry brandy of commerce is produced by expressing the juice of the blackberry, and mixing with sugar before fermentation. This makes in no sense of the term a brandy, but probably it might pass as a cordial.

This liquor is also highly esteemed for its medicinal properties for the same reason stated above. It cannot, however, be regarded

23

in the same light as a true brandy, and should not, of course, be sold under the name of blackberry brandy.

Immense quantities of blackberries grow wild in the United States, and if this industry were properly fostered and placed upon a scientific basis and the product properly aged it is believed it would become a profitable source of income to the farmer. All these things depend upon the installing of small stills for agricultural uses.

The passages of the denaturated alcohol act has not encouraged the erection of such stills through the agricultural regions, and the state and national restrictions on the manufacture and sale of fruit brandies, keep the industry from growing. Of course, it would be necessary that there should be trained men to conduct both the fermentations and the distillations.

Cherry Brandy.—Another source of brandy, which is highly esteemed, is that of the fermented cherry. The crushed fruit, with the stone, is fermented in the usual way, and the distillate forms a brandy which is highly regarded both for its odor and flavor. Cherry brandy is subject to the same form of adulteration as that just described for blackberry brandy, namely, the mixing of the juice of the cherry with sugar and even with alcohol. This, of course, as in the above case, is not a cherry brandy, but a cherry "bounce" or a cherry cordial. Cherry brandy has a distinctive flavor, due to the bitter principle extracted from the seed during fermentation. The seed of the cherry possesses a particular bitter flavoring material, which is known as amygdalin. There is only enough of it present in a cherry brandy to produce a very delightful bitter taste, quite characteristic of the product.

Cherry brandy, to be potable, should be aged and cared for as other brandies. Very little of it is ever produced in this country and kept as long as four years in wood before consumption.

Cherry brandy is produced very largely in foreign countries, especially in Switzerland, Eastern France and Southern Germany. It is usually sold without aging, and therefore appears in commerce as a water white product. It is prized chiefly for its medicinal qualities, and is drunk in very small portions after dinner, usually not more than a tablespoonful of it is consumed at any one time.

Other Forms of Fruit Brandy.—All fruits, containing, as they do, the mother substance of alcohol, may become the raw materials from which brandies are made, and there is scarcely any kind of berry or common fruit of any description which is not at some time or other utilized for the manufacture of alcohol.

Blueberries, huckleberries, raspberries, strawberries, etc., have all been utilized in this way. These brandies are not made in any commercial quantities, and they are mentioned simply to indicate what their true composition should be.

General Conclusions.—From the above summary it is seen that a potable brandy is an old distillate from the fermented juice of fruits, and each kind contains a characteristic flavor or aromatic substances which distinguishes it as coming from one particular kind of fruit. A true brandy is never mixed with another kind of liquor, and in order to be of the proper palatability and wholesomeness should be aged at least four, or better, six or eight years in oak wood.

Adulteration of Brandy. Methods Employed.—The adulterations and misbranding of Brandy are similar to the same offenses in the case of whisky. The use of sugar for masking the taste of young brandy and of caramel to suggest age, although quite common are none the less for that reason objectionable. The use of artificial flavors such as œnanthic ether serves to impart to neutral spirit an odor similar to that of true brandy. Applying any term such as Cognac, or any of the names of the regional products of the Charente, such as Fine Champagne, Borderie, etc., which gives a false idea of origin is typically deceptive. Flavoring and coloring any distilled spirit which has been rectified by any means to pure alcohol is a typical method of making an adulterated brandy.

Artificial Coloring.—After the blending of the brandies, and filteration which have already been described, the brandies are left for a long while before they are placed in bottles.

Unfortunately, it has become a very common practice to imitate great age by adding burnt sugar to brandies comparatively young. There is, of course, no objection whatever to the addition of a small quantity of burnt sugar to a brandy on account of health. There is a very serious objection to the deception which it practises upon the eye of the consumer.

Many of the brandies are sold while still young, or at least some of the constituents of the blend are young brandies having a harsh taste. In order to modify this taste, more or less sugar is added to the brandy. The same remark may be made in regard to the sugar as is made respecting the caramel or burnt sugar—it is not objectionable on account of itself, but only because it may conceal inferiority by masking the taste of a brandy which is too young.

Brandy not a Proper Name.—The term brandy cannot be properly applied to any artificial mixture of alcohol, sugar, fruit juice and flavoring extracts. These are all simply artificial preparations, and should be called liqueurs, cordials, or some such name. The consumer should insist upon it that there should not be sold to him as a brandy any alcoholic spirit which has not been obtained directly and solely from the distillation of the fermented juice of fruits.

Early Adulteration of Brandy.—An interesting description of French Brandy and its adulteration is found in a book entitled, "The Grocer and Distiller's Useful Guide," by William Beastall, Chemist, published by the author in New York in 1832. On page 208 of this work brandy is defined as "a spiritous and inflammable liquor, obtained by distillation from wine."

The author says:

French brandy acquires by age a great degree of softness, and at the same time a yellowish brown color, which our distillers have imitated in their artificial preparations. But this color being found only in such brandies as have become mellow by long keeping, it follows that the ingredient from which it is extracted, is the wood of the cask, and that the brandy in reality has received a tincture from the oak. The peculiar flavor which French brandies possess, is supposed to be derived from an essential oil of wine, mixed with the spirit; but more probably, it originates from the very nature of the grape, or the wine lees. The color, however, is usually given by burnt sugar. It deserves to be remarked, that our distillers frequently make use of the spirit of nitrous ether, commonly called dulcified spirit of niter, a very small proportion of which, added to pure whisky, or a liquor obtained by the distillation of malt, imparts to it a flavor not unlike that of French brandy.

Imitation brandy is made by running malt spirits through fresh charcoal, adding to each 10 gallons, a pint of the tincture of bitter almonds, and a sufficient quantity of coloring, and one-fourth of good French brandy.

The author also gives a recipe for making imitation brandy, which reads as follows:

Take a hogshead of rectified spirits, to which add four pounds of salt of tartar, as an additional rectifier. Then take three pounds of orris root, three pounds of bitter almonds, ten pounds of prunes, two gallons of wine vinegar, three and a half pounds of salt-petre, three and a half pounds of figs; add these together, and distill with a strong heat until the feints rise, and color with highly burnt brown sugar.

Another recipe for making imitation brandy is found on page 213 of the work, which reads as follows:

The best, and indeed the only method of imitating French brandy to perfection, is by an essential oil of wine, this being the very thing that gives the French brandies their flavor; it must, however, be remembered, that in order to use even this ingredient to advantage, a pure tasteless spirit must be first procured; for it is not likely that this oil should give the flavor of French brandies to any one of our foul malt spirits. The best spirit to convert into French brandies are these; cider spirit, raisin spirit, or crab spirit.

Also a spirit distilled from natural grapes of this country, will prove preferable to any, providing they can be procured in sufficient quantities to make it an object to the distiller.

PART XII

RUM

Definition.—Rum is an alcoholic beverage distilled from the unrefined fermented products of the sugar cane. The term "rum" is often given as synonymous with all distilled liquors, much as brandy and whisky are used in the same sense. Distinctively speaking, rum is applied only to a spirit distilled from molasses, or from sugar cane products. The sugar cane juice may be fermented directly, or the products of manufacture, notably the molasses, after the separation of a crop of sugar, are used particularly for this purpose. In fact, for all practical descriptions it may be said that rum is an alcoholic distillate derived from the fermented molasses of sugar cane. Rum may be made wherever sugar cane is produced, but experience has shown that there are certain localities, as in almost every other instance of this kind, in which the product is of greater value than in other places.

Rum is one of the oldest and most widely known of distilled alcoholic liquors. Its particularly peculiar flavor and aroma come from the aromatic volatile bodies naturally present in cane juices, produced in the course of manufacture, or formed during the distillation and aging of the product. It is evident that very little volatile aromatic substances can remain in molasses, by reason of the fact that in the making of molasses a very high temperature is reached, especially if the sugar cane juices be boiled in an open kettle. If a vacuum pan be used the heat of boiling is very much lower but the volatility of the bodies therein is correspondingly increased by the diminished pressure. Nevertheless, all molasses has that fragrant aromatic odor peculiar to this sugar cane product and this fragrant odor is preserved to a large extent to the distillate. Rum, as is the case with other distilled liquors, improves greatly on keeping in wood, and both for beverage and medicinal purposes it is highly important that

358

the rum be well aged. The Act of Congress prescribing a term of four years of storage for distilled spirits bottled in bond is based largely on the fact that during the first four years after manufacture the improvement in the quality of distilled spirits is extremely rapid. In a country where the temperatures in summer are equal to those of the United States, the ripening of the distilled spirits makes great progress in this time, though it is by no means complete. The term "old" probably should be applied to a rum much older than four years, although at the end of four years the rum has assumed quite a fragrant and attractive character. Among the localities which produce rum of the highest character may be mentioned the islands of the West Indies, especially Jamaica.

Jamaican rum is probably the most famous of the rums of commerce. Rum is made in almost every country where sugar cane grows, and very largely in countries where it does not grow, as for instance, New England, where the rum industry was established more than a century ago and where it has flourished up to the present time.

Character of the Raw Material.—The molasses which is produced in Louisiana has not been used very extensively there for the manufacture of rum, because it is not sufficiently aromatic, or because the refinements of manufacture have extracted from the molasses too much of its saccharine contents to make it a suitable source for the manufacture of rum. Moreover, it may be said that the use of the fumes of burning sulphur in treating the juices of the expressed sugar cane tends to render the molasses unfit for the manufacture of rum. It would be impossible to make a rum, which would have any character at all from black strap, a low-grade molasses, or molasses in which the content of sulphur dioxid had been increased to a very large amount by the successive concentrations and extractions of the sugar therein contained.

Manufacture.—The manufacture of rum does not differ in any essential particular and principle from the manufacture of other distilled liquors. In the case of rum there is one difference which is quite marked between that industry and the whisky or spirit industry in general. There is no starch which must previously be converted into sugar before the fermentation takes place. Inasmuch, as the cane sugar which remains in the molasses is not acted upon directly by the ferments, it is important that it should be changed into invert sugar before or during the process of fermentation. Fortunately, the yeasts which are used in the fermentation secrete a diastase which is very active in converting cane sugar into invert sugar. Hence, it is usually not found necessary to convert the cane sugar into invert sugar by treatment with an acid or otherwise before the fermentation begins. As a rule, the invertase secreted by the yeasts is quite sufficient for the purpose mentioned.

Time of Fermentation. The period for the fermentation of rum is longer than that for the manufacture of whisky, and this is recognized in the regulations, which allow a longer period for fermentation in a distillery surveyed for the manufacture of rum than when surveyed for the manufacture of whisky or alcohol.

While, as is said above, molasses may be considered the base supply for rum, other materials derived from sugar cane are used in many countries in its preparation, wholly or in part. The skimmings which are taken from the boilers of the old open-kettle processes of manufacture are often mixed with the molasses, and there is also found employed in the manufacture of rum the counterpart of the sour mash process of making whisky. In other words, a portion of the residue from the previous distillation, known in some countries as "dunder," is added to the mash. The rum which is produced wholly from refuse molasses is of a very inferior character, and the same term is applied to it in many countries as was applied to the low-grade whisky made in this country, namely, "nigger rum."

Dunder.—The nature of "dunder" may be described as follows: When the fermented mash from which the rum is to be made is placed in the still, it contains practically all of the yeast cells, living and dead, which aided in or were produced during fermentation. The warming of the fermented material in the still produces a rapid extraction of the soluble materials from the yeast cells. These soluble materials, as is well known, are of the nature of diastases or enzymes. This whole mass, after the removal of the alcohol by distillation, is naturally concentrated and the extracts are in a more usable form than they were before. This material, consisting of numerous mineral matters and other substances, as well as the remains of the yeast cells, forms an excellent food for the nourishment of the new yeast cells of the succeeding fermentation. Naturally, the "dunder" itself does not afford any alcohol, but it stimulates the growth of the yeast, so as to produce a larger yield and of a finer character, provided the quantity of "dunder" employed is not too great.

It must not be forgotten that not only does this residue of the fermentation of the rum contain valuable qualities, but it also has some disadvantages. The "dunder" is very apt to be infected with bacteria, some of which are not killed during the process of distillation. Especially is this true of the bacteria which are produced from spores. Inasmuch as the constituents of this residue are an excellent food for yeasts, they also become likewise a most excellent food for bacteria.

The development of bacteria may interfere very seriously with the succeeding fermentations and introduce elements of activity which tend, or may tend, to produce a product of an inferior quality. Hence, as in the case of using sour mash, attention must be paid to the process of fermentation, in order that no undesirable strain of bacterial or yeast life may be produced.

Manufacture of Jamaican Rum.—Attention has already been called to the fact that rum of most excellent quality, perhaps the most famous of all rums, St. Croix alone excepted is made in Jamaica. Various grades of rum are made in this island, according to the nature of raw materials used and the processes of fermentation and distillation employed.

Varieties of Rums.—In Jamaica a distinction is made between the ordinary "clean" or Jamaican rum, and the very highly flavored product, which by way of distinction is known as "German Rum." Various theories have been advanced to account for the difference between the "clean" rum or the ordinary rum, and the highly flavored rums which are produced in this island. The common, belief, as has been expressed, is that the flavoring qualities which distinguish rum from other distilled spirits are peculiar to the sugar cane. They either exist naturally in the products of the cane, or they are produced from pre-existing sources during the processes of fermentation. The essential chemical difference between the ordinary rum of the Jamaican product, and the highly flavored rum, as might be inferred, is in the quantities of esters, or ethers, which they contain. The highly flavored rum contains considerably larger quantities of these ethers than that of the ordinary character; in fact, the quantity is almost twice as great.

Manufacture.—The common rum of Jamaica is made in a very simple way, which may be described as follows:

Molasses is used as the base raw material, and in the regular operation of the distillery "dunder" is used in the fermenting tanks, as has already been described. The skimmings which come from the open kettle used in boiling the product are also added to the fermenting tank. The skimmings are supposed to be particularly valuable by reason of increasing the acidity of the fermented mash. Some manufacturers allow the skimmings to stand in tanks until they become sour, while others allow them to trickle through cisterns over cane trash, which produces a rapid oxidizing effect.

In the manufacture of higher flavored rums an attempt is made to produce a greater etherification, and, consequently, a larger production of aromatic substances due to the action of microbes. These additional flavors cannot be regarded as pre-existing in the sugar cane, but they are the product of bacterial activity exercised on the original materials. The two organisms which are most active in this respect are the *bacillus butyricus* and the *bacillus amylobacter*, and other forms allied thereto. These bacteria are very common and exist frequently in soil and are not difficult to introduce into fermenting solutions. They are mostly produced from spores and are difficult to destroy by the ordinary processes of sterilization.

The bacillus butyricus is an anaerobic organism and it will not develop well unless it is grown out of contact with oxygen. For this reason, in pure cane juices its action is not at all vigorous. Where the cane sugar has been more or less inverted, this bacillus acts with much greater vigor, especially if some albuminous matter be present. The extract of yeast cells adds very much to the activity of these organisms, and in this we see a scientific reason for the use of "dunder." The exclusion of the air from the fermenting tank presents practical difficulties in the production of the maximum activity of these anaerobic organisms. Before the aid of scientific investigation was placed at the disposal of the rum maker, he found that to produce very highly flavored rum he would have to add some material to the fermenting mass, which contained, although he did not know it, a considerable quantity of nitrogenous matter, and this was applied from the "dunder" of the preceding fermented mass. The utilization of these special ferments is another reason why the period of fermentation for rum must be longer than that for the manufacture of whisky or spirits. These bacteria are not of quick action; they move slowly and they require some time to produce their full effect. Hence, the period of the rum production may be well above 72 hours without going too far.

There is a popular impression in Jamaica that good rum cannot be produced in a modern sugarhouse, using modern machinery. While some authors doubt the truth of this belief, it must be borne in mind that modern methods, which take away increasingly large quantities of sugar, must, of necessity, diminish the fermenting value of the molasses, and add thereto, proportionately, very much larger quantities of foreign matters than in the style of molasses formerly made. It is perfectly reasonable to suppose that improved machinery would result in a depreciation in the character of the product.

Further Divisions of Rum.—Jamaican rums are further divided from a commercial point of view, into three classes, namely:

1. Rums for home consumption.
2. Rums for export to Engand.
3. Rums for export to the continent of Europe.

Although rum is one of the principal products of Jamaica, fortunately it is not consumed in very large quantities at home. The statistics show that the local consumption of rum does not much exceed a gallon per head per annum. The citizen of Jamaica cannot be regarded as an inveterate rum drinker. While this is regretted from the point of view of inland revenue, it is a matter of congratulation from the point of view of temperance in the use of all things.

In Jamaica, as in other countries, the introduction of the manufacture of pure spirit has opened the doors for mixing native rums with neutral spirit, and as this is cheaper than making rum in the old-fashioned way, it is not at all surprising that the consumption of these mixed articles has practically driven the consumption of old-fashioned straight rum out of the market. As has been stated by Mr. H. H. Cousins, Government Analyst for Jamaica.

The high-class trade in old rums of delicate softened flavor, which were formerly so highly thought of by the planters and moneyed classes, has largely disappeared, and it would probably be most difficult to obtain a choice mark of an old rum, which has not been blended, from any spirit merchant in Jamaica today.

In regard to the second class, which is intended for shipment to England, it may be said that the same manipulations during the last few years were tried with this class. To such an extent was the manipulation carried, especially after reaching England, that the Jamaican Government sent a special representative, Mr. Nolan, to London to protest against the adulteration of Jamaican rum and see if he could not re-establish the trade in the genuine article. Mr. Cousins refers to the rums of class two, as follows:

The rums of the class to which I now refer, and which constitute the bulk of the rum exported from Jamaica, represent the type of spirit which Mr. Nolan is seeking to advertise, and to protect from fradulent adulteration, and from the competition of spurious Jamaica rum in the United Kingdom.

The rums of this class are produced by a slower type of fermentation than those intended for the local trade. Some of the best varieties are produced by fermentations in ground cisterns, slightly flavored by the addition of some soured skimmings to the fermented materials. These rums are very rich in ethers, being hardly less than 300 parts of ethers per 100,000 parts of alcohol, and sometimes a great deal more.

Class three rums are intended for consumption on the continent of Europe. The trade in Jamaican rum has been long established on the European continent. This trade has largely declined in recent years, and is, doubtless, due to the adulteration of the article with molasses spirit and other substances, which so depress its character as to make the beverage unpopular. Another reason which has tended to diminish the consumption of Jamaican rum is the extremely heavy duty imposed upon it in Jamaica. This, in connection with the lower rates of duty on alcohol, has rendered the competition of the artificial with the imported article, even with its fine flavor, very keen.

The rums exported to Europe are commonly those already described as German-flavored rums. Not only are these rums treated as has already been described, with "dunder" and under special circumstances, but the period of fermentation is very much prolonged, reaching sometimes as high as 15 or 20 days. The fermentation takes place slowly, under very acid conditions, the acidity being produced by the addition of soured skimmings, or by the slow process of fermentation to which the mass is subjected.

If the ordinary rum shipped to England may be said to contain about 300 parts of ethers to 100,000 parts of alcohol, the German-flavored rums will contain practically double that amount, namely, from 600 to 700 parts. It is readily seen that they are very aromatic and are considered the very highest flavored rums of commerce. Some of these very fine rums have been found to contain as high as 1,500 parts of ethers per 100,000 parts of alcohol. These ethers are those of the coördinate alcohols, namely, acetic ether, derived from ethyl alcohol, and the ethers derived chiefly from the higher alcohols, such as butyl, propyl, and amyl.

Differences in Flavor.—All of these three classes above mentioned vary in flavor. It is said that there are no two plantations in Jamaica which produce rum of the same flavor. The output of rum from each place is influenced by the peculiar bacterial flora of that place, and by the methods employed in fermenting and distilling.

Rum is, of all kinds of distilled alcoholic beverages, the most fragrant, and makes the greatest mass effect upon the olfactory nerves. Only connoisseurs of the highest character can pick out the shades of flavor to a certainty, but no one would be misled in respect to the character of the goods, as a rule, even if he were not a connoisseur.

It is a material of this character, so highly flavored, which is so valuable in the stretching of rums; that is, by mixing them with spirits made from any given source, and thus using the genuine article merely as a flavor to the whole mass. The legitimate trade in rum has almost been destroyed, it may be said, by this system of admixture and adulteration.

Average Composition of Jamaica Rums

	"Clean" rums	Flavored rums
Alcohol, volume, percent.....................................	79.1	77.3
Total solids, grams per 100 c.c...........................	0.43	0.31
Total acids as acetic, grams per 100 liters of alcohol..........	78.5	102.5
Volatile acids, grams per 100 liters of alcohol.................	61.0	95.5
Esters as ethyl acetate, grams per 100 liters of alcohol........	366.5	768.5
Higher alcohols as amylic, grams per 100 liters of alcohol...,,	98.5	107.0
Furfurol, grams per 100 liters of alcohol...................	4.5	5.2
Aldehydes, grams per 100 liters of alcohol..................	15.3	20.7

The above data show in striking contrast the difference between the "common" or "clean" rums of Jamaica, and the flavored rums or those made carefully with "dunder." There is little difference, as is seen, in the alcoholic strength, the salts, the total acids, the higher alcohols, the furfurol, and the aldehydes; the great difference is in the volatile acids and the esters. It is easy to see that the flavoring matters must belong chiefly to those two classes.

Rum in Guadeloupe.—The French name "Rhum," and also the name "Taffea" is applied to the alcoholic products obtained by the fermentation and distillation of the juice of the sugar cane, and of the molasses produced in the factories making sugar cane, in Guadeloupe. The name "rhum" in this island is particularly applied to the product obtained by the distillation of the juice itself, and the name "taffea" to the product having molasses as its origin. These different usages of the term are not absolute, and especially in France the term "rhum" is applied to "taffea" which has been stored some years in the cask. Unhappily, in France it is usually mixed with industrial alcohol before it reaches the consumer, or worse still, some artificial product is sold under its name, a product made with artificial essences and with industrial alcohol and the whole colored with caramel.

The true "rhum," that is, the beverage made from the juice of the sugar cane in Guadeloupe, never reaches France, although its manufacture is a very important item in the island. It is practically consumed in the home of its production. The "rhum" made from the sugar cane juice, when kept for several years in wood without the addition of anything whatever, acquires a bouquet which approaches in excellence that which is acquired by old brandies.

The total production of "rhum" and "taffea" in Guadeloupe is not very great, and as has already been indicated very little of it, if any, is exported. More perhaps of the variety known as "taffea" is exported than of the true "rhum." The quantity of "taffea" produced is also larger than that of "rhum."

Other varieties of rum are also made in Guadeloupe, for instance rum made from the boiled cane juice. In the manufacture of this drink the juice of the cane is heated in the boiler in such a way as to boil violently for some minutes. Its density is then considerably increased, the scum is carefully removed and it is allowed to cool. It is diluted with water in such a way as to bring it back to its original density. Finally, this product is taken to the cisterns of fermentation and treated in the ordinary way. Evidently the object of this boiling is to sterilize the juice, thus destroying the adventitious ferments, and at the same time purifying it to a certain extent by removing certain quantities of the matter coagulable by heat and commonly known as "skimmings."

There is still another variety, made in Martinique, from what is known in that country as "gros sirop." This molasses is obtained in the manner of the old-fashioned open-kettle molasses, being the dripping from the sugar cane juice boiled over a naked fire in an open vessel until it reaches a crystallizing density. The quantity of sugar made in the island in this way at the present time is almost nil, and hence the amount of rum thus produced is inconsiderable.

In the manufacture in Guadeloupe "dunder" is also used in the fresh fermentations, but the name of it in this island is "vinasse," namely, the residue of the distillation of the preceding fermentation, in order to produce the rum. This vinasse is very strongly acid and contains usually only traces of sugar, but has a dry residue amounting from 35 to 40 grams per liter. The ash is rich in phosphoric acid and potash. It serves, as has already been stated, for the nourishment of the yeasts, especially in the addition of certain quantities of nitrogenous matter and of mineral substances such as phosphoric acid and potash, which the yeasts require for their proper nourishment.

Distillation in Guadeloupe.—The distillation in Guadeloupe is carried on very much in the same manner as that of Cognac in France. The apparatus consists of practically three parts: the heater in which the fermented mash is raised to a high temperature by the vapor escaping from the still; the still itself, which is the ordinary pot still; and the cooler, which is the ordinary worm surrounded with cold water.

Rum in Demerara.—Rum is also made to a considerable extent in Demerara. Inasmuch, however, as the principal product of the sugar cane is sugar, the raw materials do not have such a high character as those used in Jamaica and in other islands where the process of extraction of the juice for sugar-making purposes is less perfect. The rum made in Demerara is distinctly inferior to that produced in Jamaica and Guadeloupe, St. Croix, and other West Indian countries. Very little, if any, dunder is used in Demerara. The fermentation is produced solely by the added yeasts and is carried on much more rapidly than in Jamaica and Guadeloupe. Moreover, the Demerara rum is often, or largely, produced in chamber stills, or rectifying columns, which is never the case in Jamaica, where only pot stills are employed. This is another reason why the Demerara rums average more nearly the character of alcohols in proportion as they depart from the true character of rums.

Early History of Rum in New England.—The early history of rum in New England is set forth in a work by Alice Morse Earle, entitled "Customs and Fashions in Old New England," published by Charles Scribner's Sons in 1893. On page 174 the author says:

Aqua-vitæ, a general name for strong waters, was brought over in large quantities during the seventeenth century, and sold for about three shillings per gallon. Cider was distilled into cider brandy, or apple-jack; and when, by 1670, molasses had come into port in considerable quantities through the West India trade, the forests of New England supplied plentiful and cheap fuel to convert it into "rhum, a strong water drawn from the sugar cane." In a manuscript description of Barbadoes, written in 1651, we read: "The chief fudling they make in this island is Rumbullion alias Kill-Divil—a hot hellish and terrible liquor." It was called in some localities Barbadoes liquor, and by the Indians "ahcoobee" or "ockuby," a word of the Norridgewock tongue. John Elliot spelled it "rumb," and Josselyn called it plainly "that cussed liquor, Rhum, rumbullion, or kill-devil." It went by the latter

name and rumbooze everywhere, and was soon cheap enough. Increase
Mather said, in 1686, "It is an unhappy thing that in later years a kind of
drink called Rum has been common among us. They that are poor, and
wicked too, can for a penny or two-pence make themselves drunk." Burke
said, at a later date, "The quantity of spirits which they distill in Boston from
the molasses they import is as surprising as the cheapness at which they
sell it, which is under two shillings a gallon; but they are more famous for the
quantity and cheapness than for the excellency of their rum." In 1719, and
fifty years later, New England rum was worth but three shillings a gallon,
while West India rum was worth but two-pence more. New England distilleries quickly found a more lucrative way of disposing of their kill-devil
than by selling it at such cheap rates. Ships laden with barrels of rum were
sent to the African coast, and from thence they returned with a most
valuable lading—negro slaves. Along the coast of Africa New England
rum quite drove out French brandy.

The Irish and Scotch settlers knew how to make whiskey from rye and
wheat, and they soon learned to manufacture it from barley and potatoes,
and even from the despised Indian corn.

The drinking of compounded liquors was also practised in
old New England, as shown by the following extract from page
178 of this work, beginning:

Flip was a vastly popular drink, and continued to be so for a century and
a half. I find it spoken of as early as 1690. It was made of home-brewed
beer, sweetened with sugar, molasses, or dried pumpkin, and flavored with
a liberal dash of rum, then stirred in a great mug or pitcher with a red-hot
loggerhead or bottle or flip-dog, whch made the liquor foam and gave it a
burnt bitter flavor.

Landlord May, of Canton, Mass., made a famous brew thus: he mixed
four pounds of sugar, four eggs, and one pint of cream and let it stand for
two days. When a mug of flip was called for, he filled a quart mug two-thirds
full of beer, placed in it four great spoonfuls of the compound, then thrust in
the seething loggerhead, and added a gill of rum to the creamy mixture. If
a fresh egg were beaten into the flip the drink was called "bellowstop,"
and the froth rose over the top of the mug. "Stonewall" was a most intoxicating mixture of cider and rum. "Calibogus," or "bogus" was cold rum and
beer unsweetened. "Black-strap" was a mixture of rum and molasses.
Casks of it stood in every country store, a salted and dried codfish slyly hung
alongside—a free lunch to be stripped off and eaten, and thus tempt, through
thirst, the purchase of another draught of black-strap.

A terrible drink is said to have been made popular in Salem—a drink
with a terrible name—whistle-belly-vengeance. It consisted of sour
household beer simmered in a kettle, sweetened with molasses, filled with
brown-bread crumbs and drunk piping hot.

24

In a work published in 1890 entitled "Economic and Social History of New England," by William B. Weeden, it appears that distillation began in Salem as early as 1648. On page 186 we find the following:

Emanual Downing writes that Leader has cast the iron pans to be used in the process. Downing began distilling in Salem this year. Frequent commerce with the West Indies carried out unmerchantable fish to be exchanged for molasses.

On page 188 it is stated:

Rum was much used by the common people, and malt liquors were the favorite drink of the English colonists. The native New England beverage was cider and the presses began to work about 1650. Much barley had been raised in Plymouth. The many malthouses were not so common after this.

In 1686 the Southern part of the Colonies had commenced to buy rum from New England, as stated on page 376. In 1690 it is stated (page 416), "Cider and vinegar corrected the West Indian sugar and molasses always coming in; that is, when the molasses did not evolve itself into the fiery rum. Rum was beginning to be the important commercial factor which it came to be later in the century."

By 1670, it is stated on page 459:

The West Indies afforded the great demand for negroes; they also furnished the raw material supplying the manufacture of the main merchandise which the thirsty Gold Coast drank up in barter for its poor, banished children. Governor Hopkins stated that for more than thirty years prior to 1764, Rhode Island sent to the Coast annually eighteen vessels carrying 1,800 hhds. rum . . . . . Newport had 22 still houses; Boston had the best example, owned by a Mr. Childs . . . . . The quantity of rum distilled was enormous, and in 1750 it was estimated that Massachusetts alone consumed more than 15,000 hhds. molasses for this purpose . . . . . The consumption of rum in the fisheries and lumbering and ship-building industries was large; the export demand to Africa was immense.

The adulteration of rum was an early practice. Captain Potter, in 1768, gave directions for the trade on the African Coast, as follows:

Make yr Cheaf Trade with The Blacks and Little or none with the white people if possible to be avoided. Worter yr Rum as much as possible and sell as much by the short mesuer as you can.

Order them in the Bots to worter thear Rum, as the proof will Rise by the Rum Standing in ye Son.

Evidently the Captain was provided with a rectifier's license.

In 1740, it is stated, page 501:

The most important change in the manufacture of this period was in the introduction of distilleries for rum. Massachusetts and Connecticut undertook the business, but Rhode Island surpassed both in proportion . . . . . . The trade in Negroes from Africa absorbed quantities of rum.

The 18th century brought in the manufacture of New England rum with far-reaching consequences, social as well as economical. It was found that the molasses could be transferred here and converted into alcohol more cheaply than in the lazy atmosphere of the West Indian seas.

No Rectification.—In all the references which are made to the manufacture of rum in New England not a single intimation is made that the spirit was ever rectified or mixed with a neutral spirit color and flavor to make a beverage. In fact, the neutral spirit was unknown as a commercial proposition. The stills which were used were the old-fashioned pot stills. It is stated, on page 502, that Mr. Thomas Armory

built a "still-house" in 1722, bringing pine logs 28 feet long, 18 inches in diameter, from Portsmouth for his pumps. In 1726 he orders a copper still of 500 gallons capacity from Bristol, England. The head was to be large in proportion, the gooseneck to be of fine pewter and two feet long, with a barrel in proportion to the whole still.

This shows the character of the still and the nature of the spirit which must have been made therefrom. In the early history of its production there is nothing known of the modern process of rectifying, mixing, adulterating, compounding, coloring and flavoring. In 1659 a hogshead of rum was quoted at 12 pounds, 12 shillings. In 1670 it had fallen to 7 pounds. In 1671 it was quoted at five shillings per gallon. Insofar as can be judged by the early history of these substances, the contention that has been made, that distilled beverages were always rectified, colored, adulterated, mixed and flavored before consumption, does not appear to have any basis in fact.

Morewood, on page 334, of his work, writes of New England rum in the following language:

The rums of New England are considered of good quality, and some deem them not inferior to the best that are produced in the West Indies. In 1810, they distilled in this State 2,472,000 gallons of rum; from grain, 63,730 gallons; from cider, 316,480 gallons, while the breweries yielded 716,800 gallons. Besides this extensive manufacture, much is imported. Geneva is successfully imitated, particularly since the tide of emigration has brought many intelligent men from Holland, who possess sufficient knowledge of this branch of trade, to render the American article equal to that manufactured in the Netherlands. Many of the Irish emigrants distill, in genuine purity, that description of spirits commonly called *Innishowen* or *Potheen*, which is no less a favorite on the other side of the Atlantic, than on the shores of Magilligan, or the banks of the Shannon. The following mode of making it at an early period, is thus described by an eyewitness: To a bushel and a half of rye, four quarts of malt, and a handful of hops, were added fifteen gallons of boiling water, which were allowed to stand for four hours. These being increased by sixteen gallons more, two quarts of home-made yeast were thrown in, and in this proportion either a large or small quantity of worts was prepared, which, after being allowed ample time to ferment, was distilled in a simple apparatus. One bushel of rye produced about eleven quarts of weak and inferior spirit, and sold at the rate of 4s. 6d. per gallon. The refuse of these small stills was used in feeding swine.

Description of Rum Making in the Early Part of the Last Century.—An interesting description of the manufacture of rum is found in book written by John Bell and published in 1831 in Calcutta. Mr. Bell describes the manufacture of rum as a part of an article on the manufacture of sugar on a West India plantation. He says his work would not be complete without going into the details of the disposition of the feculencies which form the base of that highly esteemed spirit usually sold in Great Britain under the title of Jamaica Rum. He gives directions for adapting the capacity of the distillery to that of the sugar house, and especially provides that there must be at least one large cistern equal in size to the contents of four fermenting vats, to receive the lees and the dunder. He regards the lees as indispensable in the distillation of rum, and the want of them is seriously felt at the beginning of the season. He advises the following mixture as a proper one for fermented skimmings:

Skimmings....................40 percent.
Water....................40 percent
Lees....................20 percent

When molasses is procurable, he recommends one-third each of skimmings, lees and water, and 5 percent of molasses.

After the crop, however, has been harvested and there is a scarcity or total want of skimmings, the distiller must have recourse to his molasses, and the proportion of lees must be increased with regard to the increased tenacity of the sweets.

In this condition he recommends equal proportions of lees and water 40 percent, with 20 percent of molasses. The fermentation, it is stated, is not finished before the end of 10 days, but after a good supply of lees is obtainable it may be finished in five days.

The still used in these days was a very old fashioned one and the top was luted on instead of being secured by clamps or packing. Great attention must be paid to the luting of the head of the still, he says, which is done with clay, and that unless great care is used the alcohol may break out through the crevices and take fire, to the imminent danger of the persons in attendance. The overseer he advises never to leave the still, unless relieved by equally competent assistants. The first distillation in these days was called "low wines." The strength of the spirit was proved by the bubble, or in the absence of bubble by olive oil, and the spirit was to be made of such a density that olive oil would sink to the bottom of it.

Disposition of the Rum.—A great deal of the rum made in the United States is entered for consumption. Other parts are bought by rectifiers, mixed with neutral spirits, made from corn or molasses, artificially colored, and sold as rum. A very considerable portion of the rum made in the United States, is sent to Africa, and disposed of to the semi-civilized and savage tribes of that continent. The well-known love of the natives of Africa for alcoholic drinks indicates that the markets of that country are particularly favorable to the disposal of large quantities of rum. What the effect is upon the welfare of the natives is a matter which, insofar as I know, has not been taken into consideration in the preparation and shipment of these products.

Quantity Produced, 1917.—Practically all the rum produced in 1917 was made in the Third Massachusetts District (2,706,414 gallons) and Sixth Kentucky District. The total quantity was

2,842,834.3 gallons. The quantity remaining in bond was 906,-
042.5 gallons.

The total quantity withdrawn from bond and tax paid was
642,798 gallons and the quantity bottled in bond 17,019.7.

The quantity of rum exported was 1,030,249.9. Of this there
was sent:

To Africa	778,057.3 Gallons
To England	122,081.5 Gallons
To China....................	42,490.3 Gallons
To Holland..................	30,730 Gallons
To Canada	24,232.6 Gallons

The dearth of ships has greatly decreased the amount of
exported rum to Africa over that of previous years. In 1916,
1,196,905 gallons of rum were exported to Africa alone.

Adulteration of Rum.—As in the case of whisky and brandy
rum has been subjected to all kinds of adulteration. So great
had become the adulteration of rum shipped to England from
Jamaica that for every barrel of rum sent 6 barrels were sold.
It is not difficult to see how seriously the industries of the island
of Jamaica have been crippled. Thus by the better execution of
the English food law and the application thereto of the merchandise marks act, a great deal of the adulteration has been eliminated.
There is still enough practised to excite grave concern in the minds
of those who make the pure product, and depend upon England
for its market. There is no doubt that similar frauds are perpe-
trated by the rectifiers in this country.

PART XIII

GIN

Definition.—Gin is a distilled alcoholic beverage having approximately the same alcohol content as whisky, brandy and rum, and owes its character to the fact that it is flavored with the juniper berry. The juniper tree belongs to the family Coniferæ, and is therefore related to the pines, hemlocks and balsams. The juniper tree which yields the seeds employed for flavoring gin is known as the *Juniperus communis*. Oil of juniper is a product distilled from the berries of the *Juniperus communis*. It has the flavor and aroma which are characteristic of gin. The French term for the juniper berry or juniper is "genèvre." "Genèvre" therefore became the name of the beverage flavored with juniper, and this in Holland became "Geneva," the name which was for many years given to gin. This was afterward shortened to "gin."

Holland Gin.—Many persons suppose that "Geneva" was first made in Switzerland and took the name of the town where it was manufactured. In point of fact gin was invented in Holland, by Professor Sylvius according to Emerson.[1] Sylvius lived in Leyden. He was born in 1614 and died in 1672. His proper name was Francis de le Boë. He was the father of medical chemistry and was professor in medicine at Leyden. In England the drink was known as Royal Poverty, because when beggars got drunk on gin they imagined themselves kings. It was the idea of Professor Sylvius that the gin which he invented was to be used only as a medicine, and it was first dispensed by apothecaries in Holland, at Leyden.

Definitions in Other Countries.—The official definition of gin in France is found in the Public administration ruling of the 5th of September, 1907. It is defined as an alcoholic drink obtained under the conditions set forth in Article 15 of the law of March 30, 1902, by the simple distillation in the presence of the seeds

[1] Beverages, Past and Present, Vol. II, p. 65.

of the juniper tree of the fermented must of rye, of wheat, of barley, or of oats.

The Congress of the White Cross at Geneva adopted the following definition of gin: "Gin is the product of simple distillation in the presence of the seeds of the juniper tree of the fermented must of cereals."

The definition of gin in Belgium is more flexible, but it is of such a character that it defines nothing by defining too much. The definition says: "Gin is the alcoholic liquor prepared either by the distillation of added fermented materials or not, of the seeds of the juniper tree, or by simple mixture with alcohol and water in determinate proportions to obtain the alcoholic strength required.

The definitions in Switzerland and Italy are more restricted, and according to these, gin is obtained only from the fermentation of wheat and the seeds of the juniper tree.

In the north of France gin is the distilled alcoholic liquor which is most employed. Its distillation and consumption are much anterior to the production of industrial alcohols of high strength. One of the most ancient distilleries of gin in the north of France near Lille was erected in 1817. There are 11 distilleries in the north of France, producing gin; 8 in the Department of the North, and 3 in the Pas-de-Calais, making together nearly 38,000 hectoliters.

Gin is often called the "Cognac" of the North. In the north of France gin is often sold under the name of "Schiedam, Wambrechies." The method employed in the manufacture in the north of France is described as follows: "100 hectoliters of ground grain, of which rye is the dominant component, are mixed with from 10 to 30 hectoliters of barley malt, and suspended in 250 liters of water of a temperature of 50°. This temperature is maintained for an hour and afterward it is raised for one and one-half hours to 63° in order to permit the inversion of the starch. At the end of this time cold water is added until the temperature falls to 28 or 30°, at which temperature it is treated with the yeast. The fermentation lasts about 30 hours, and the alcoholic mass thus obtained is then submitted to distillation in a column still. The alcohol obtained in this way is mixed with variable quantities of juniper berries and redistilled from a simple pot still. When the work has been completed there is obtained about 30 percent of pure alcohol, or about 60 liters of gin for each 100 kilos of the grain employed. The quantities of juniper berries employed are very variable. In the region of Lille they usually use about 100 grams per hectoliter of gin. The berries contain at least 1 percent of the essence so it is seen that there is about one gram of essence of the juniper berry per hectoliter or one-hundreth of a gram per liter. The gin called "Schiedam" made after the Holland style is somewhat more strongly flavored.

Juniper Berries.—The berries of the juniper tree usually require two years to mature. Many methods are employed for flavoring. The rawest and what is regarded as adulterated gin, is simply neutral spirit flavored with oil of juniper. A very close imitation of the real liquor is made in this way. As gin is not usually aged, being made of neutral spirit, the white, colorless liquor made artificaly is easily passed off for the genuine. In the original way of making gin the berries were ground and mixed with the malt with the mash at the time of fermentation. In this way the flavoring essence was introduced at the very time of manufacture and the volatile products were secured during the subsequent distillations. This is probably the best method of manufacture.

Distillation.—If a very pure spirit is not used in the manufacture of gin, or if it be made by original fermentation in presence of the flavoring matter, more than one distillation is required to produce a fine article. Gin should not have any other flavoring matters than those derived from the juniper berry. In the mash originally fermented, however, all the congeners which give to a distilled liquor its character are present in more or less abundance. To remove these a second and even a third rectification is often necessary. Gin presumably is made of barley. Other cereals are introduced, however, but barley malt is the usual material by which the starch is converted. Wheat is considered a fine grain for producing some of the highest grades of gin. Rye is also used quite extensively. In this country the neutral spirit from which gin is made is derived chiefly from Indian corn.

Varieties of Gin.—There are many other materials which are used for flavoring spirits, making a kind of gin. The most common substitute for the juniper is the sloe berry. The sloe berry is the fruit of the *Prunus spinosa*, or blackthorn. It is used very extensively in England and is regarded as making a very satisfactory gin. Its methods of manufacture are practically the same as for gin. Unless the term "sloe" is used in speaking of gin the juniper gin is always understood. Buchu is also used as a flavor for a beverage known as gin, and likewise coriander and other seeds. Whatever substance is used other than juniper, the name of the material must be prefixed to the gin in order to have it properly named.

The Blind Tiger.—The whole list of substitutes used for juniper is a rather long one: coriander, almond, powdered angelica and cayenne pepper are mentioned by Emerson as among the hosts of substances that are found to be cheaper and easier to use than the juniper berry. The kind of gin which is made in England is commonly called "Old Tom." The story which explains the origin of this term also discloses the first account of the "Blind Tiger" which I have been able to find in literature. It is quoted by Emerson from "The Life and Uncommon Adventures of Captain Dudley Bradstreet." When the Captain decided to start the "Blind Tiger" he laid in a stock of 13 pounds worth ($65) of good gin. He had the cargo sent to his house, at the back of which there was a way to go in or out. He had a person to inform a few of the mob that gin would be sold by the old cat at the window next day provided they put the money in his mouth, from which there was a hole which conveyed it to the proper authority. It was over three hours before anybody called, which made the Captain almost despair of his project. At last, however, he heard the clink of money and a voice say, "Puss! give me two pennyworth of gin!" The Captain instantly put his mouth to the tube and bid the buyer receive it from the pipe under her paw, and then measured and poured it out. The first day he took in 6 shillings, the next day 30 shillings, and afterward three or four pounds a day. From all parts of London people came in such numbers that his neighbors could scarcely get in and out of their houses. The Old Tom Cat became an institution and since that time gin has been known as Old Tom.

Center of the Gin Industry.—Holland is still the center of the gin industry and probably the best and most genuine products are made in that country. The town of Schiedam is the principal city of manufacture. It is a common practice in this country to represent gins as Hollands or Old Toms, and the courts unfortunately have permitted manufacturers and dealers to use this form of deception provided they will plainly state on the label that the article is manufactured in the United States. This is only condoning misbranding, since it is not at all necessary for the purchaser to look at the bottle. He is therefore deceived regarding the origin of the beverage which he buys.

Magnitude of the Industry.—The quantity of gin produced in the United States in 1916 was 4,118,064 gallons, made chiefly in the Fifth Illinois District—viz.: 1,002,995 gallons and in the First Ohio District viz.: 506,562 gallons. The quantity of gin imported in 1915 was 711,669 gallons valued at $696,636 (tax free).

The total product of the United States in 1917 was 5,756,666.8 gallons. The importation of gin into the United States is forbidden by the war legislation.

PART XIV

CORDIALS AND LIQUEURS

Character.—A very important class of beverages is embraced under the general name of cordials and liqueurs. Cordials and liqueurs are distinguished from the beverages which have been considered before in two important characteristics. In the first place they are compounded articles, and are not produced in any natural way, nor do they represent any natural product. In the second place they are all characterized by the presence of a certain quantity of sugar and of alcohol. They differ among themselves principally in the character of the flavoring materials which are used in their composition. It is easy to see that there is no end to the variety of cordials and liqueurs that may be produced. It would not be possible in the confines of a book of this kind to attempt to describe all the beverages belonging to this class. New ones are offered constantly, while many of the old that do not possess any very great merit, die and disappear. There is a large number of liqueurs and cordials which have become established both in name and character, and are widely known and consumed. It will suffice our purpose to describe these more important members of the family after giving a little further insight into the character of the class as a whole.

Classification.—First of all the cordials may be divided into two distinct classes, namely: those that contain a deleterious drug or ingredient, other than acohol, the use of which develops into a habit, and exercises remarkably injurious effects upon the consumer, and those whose components are regarded as harmless.

Cordials of the first class naturally would contain a habit-forming drug, the use of which would enslave the consumer to a certain degree, and greatly undermine his moral and physical strength. A type of liqueurs of this kind is represented by absinthe. In the second class may be found all the other liqueurs and cordials which contain fragrant, odoriferous and harmless, in so far as

380

known, flavoring matters. At least these cordials do not contain any alkaloid or resin or other compound capable of forming a habit and inducing the injurious effects spoken of above. They are sweet, fragrant, and have a flavor peculiar to their composition.

Some of the cordials and liqueurs have well-established reputations, are made in the most careful manner, and usually according to secret formulæ, and have secured a lasting popularity. As a type of this kind of cordial that made by the Carthusian Monks may be cited namely, Chartreuse.

Character of the Alcohol Used.—The kind of alcohol used in the manufacture of liqueurs and cordials is not always the same. Usually pure neutral spirit or cologne spirit is selected for this purpose. In the United States spirit of this kind, as has already been stated, is made from Indian corn; in Germany it is made from the potato; in France it may be made from the grape or from the sugar beet; in Spain it may be made from the grape or sometimes from the sweet potato. If neutral spirit is not used then the next best thing to employ, in the way of spirit, is brandy. Neutral spirit, being very much cheaper is usually preferred for this purpose. It is highly important that this neutral spirit be as pure as technical skill can make it, since it is not aged in wood before it is employed in the making of this liqueur. It should be as nearly pure ethyl alcohol as can be secured.

Character of the Sugar Employed.—At the present time, as a rule, only pure white refined sugar is used in the manufacture of liqueurs and cordials. This sugar may be indifferently derived, either from the sugar cane, or the sugar beet. On the Continent of Europe the sugar derived from the sugar beet is principally employed. In the United States the sugar derived from the sugar cane is the most common.

Flavoring Matters.—The flavoring matters used may be derived from almost any spicy or aromatic herb or shrub. Among those which are most common are blackberry, cherry, anise, mint, and various aromatic herbs, growing especially in the Alps and in the Pyrenees. There is no limitation placed upon the selection of these materials, except that they should be free of any harmful drug, alkaloid or resin.

Drinking of Cordials.—Cordials should be consumed in extremely small quantities, and this, as a rule, is the method in which they are employed. A small conical glass, containing not much over a tablespoonful, is the proper size for the cordial. They are to be used rather as a flavoring after the dinner, and especially after the coffee than at any other time or between meals. It is a very common custom at dinner parties, among the richer classes, to have a small glass of cordial after the coffee. The French name generally applied to cordials of this class is *pousse café*. The drinking of cordials and liqueurs of a habit-forming character, such as absinthe, has been productive of most serious injury.

Chartreuse.—The first distillations of Chartreuse were made by St. Bruno in 1084. In 1656 after the lapse of 6 centuries the profits were so great that the monks erected a million dollar monastery at Fourvoire. The maximum annual Fourvoire production reached the huge figures of 80 million liters. Twelve different kinds of herbs were used. When gathered they were dried in well aired cellars underneath the monastery and then macerated in water. It was this aqueous extract which mixed with alcohol and distilled gave the desired flavor to the product.

Use of the Term.—Since the expulsion of the religious orders from France and the consequent emigration of the Carthusian Monks from Grenoble, considerable confusion has arisen in different parts of the world respecting the use of the term Chartreuse. It was claimed by the French and this claim was sustained by their courts, that the administrator of the estate of the monks, appointed by the Republic to conduct the operations for the making of Chartreuse, was authorized to use the old name upon the product which he advertised. It is true this product was not made according to the secret formula of the monks, because no one knew exactly what that formula is. It was possible with the skilled labor which the administrator could secure, to produce a liqueur which resembles in many of its important respects the genuine.

The New Product.—On the other hand, the Carthusian Monks when they established themselves in Tarragona in Spain, continued the manufacture of the liqueur after the recipe which they had used at Grenoble, gathering the same kinds of herbs they had used at Grenoble from the Pyrenees, and in every other respect imitating the liqueur formerly made at Grenoble. The natural difference in the aromatics employed, and the change in the environment of manufacture resulted, as might have been expected, in the production of a liqueur which was distinctly inferior to that previously made near Grenoble. They called the new product *Liqueur Pères Chartreux*.

Thus the world was a double sufferer, since the French with the same materials at Grenoble could not make the Chartreuse the monks made before, and the monks with the same materials at Tarragona could only make an article inferior to their former product.

Decision of the Courts.—The question as to who had the right to use the word "Chartreuse" has been decided in the United States Courts. Judge Coxe, called attention in an interesting way to the essential points of contention:

The courts of France by a decision on March 31, 1903, dissolved the order of the Carthusian Monks at Grenoble and sequestered the entire property and appointed a receiver therefor. The court also held that all the business of the monks, including their good-will, clientage, trade marks, commercial names, models of bottles, flagons, cases, furniture, machinery, raw material, manufactured goods and the exclusive right to the industrial name L. Garnier, was the property of the monks and as such passed to the receiver to be liquidated. Thus it appears that every right and title which belonged to the monks, whether corporeal, or incorporeal, tangible or intangible, was, so far as the laws and courts of France are concerned, vested in the receiver appointed by the French Government.

The federal court also summarized the situation as to the monks and found that had they chosen to do so they could, with some necessary changes, have used the old label and trade marks in Spain; but they have seen fit not to do so probably because the labels would have been prohibited in France and they would thus have lost the French market, which, of course, is the most important. They could not have used the trade mark in the form registered in France, for it would have been a falsehood and a fraud on the public to assert that liqueur made at Tarragona, Spain, was manufactured at the convent of the Grand Chartreuse in France. This especially would have been a false statement, since the monks even had claimed that the peculiar excellence of their product came from the plants and herbs grown in the Alps in the vicinity of their Monastery.

It appears, therefore, that on their establishment in Spain, the monks of their own accord abandoned the use of their former labels and trade marks and put on their bottles an entirely different label, calling their product Liqueur Pères Chartreux.

The U. S. court also held that the use of the old labels by the French liquidator, or the parties to whom he sold the right, would prove deceptive to the customer, who would not only think that the liqueur was made as before at the Grand Chartreuse at Grenoble, but unless he was familiar with the processes of the courts in France, would think also it was made by the monks themselves. Hence, any label used by the liquidator, or any one authorized by him, which would convey such an idea; that is, any label which was exactly similar to the old label used by the monks, must, of necessity, be deceptive. Thus any liqueur made subsequent to 1903, cannot be legally called Chartreuse in the United States.

Benedictine.—An aromatic liqueur resembling Chartreuse made originally at Fécamp, France by the Benedictine Monks. It is much used in France as a substitute for Chartreuse but does not have the exceptionally fine flavor of yellow Chartreuse formerly made by the Carthusian Monks at Grenoble.

Curaçao.—This is a liqueur flavored with the rind of a bitter orange originally found in the Island of Curaçao, Dutch East Indies. It possesses a bitter taste and a pleasing aroma. It has been made chiefly at Amsterdam in Holland.

Dubonnet.—Dubonnet is a type of the many so-called *aperitifs* which are so common in France and take the place of the cock-tail in this country. It is mildly bitter, decidedly alcoholic, and generally agreeable to the taste. The best appetizer in the world is hard work and reasonable abstention from too generous dining. Those who sit down to the table and are unable to eat until they have stimulated their stomachs by alcohols, bitters and spices, will do far better to fast than to feast.

ABSINTH

Definition.—Absinth is an alcoholic liqueur which takes its name from the botanical name of wormwood, *Artemisia Absinthium*. This is a perennial plant which is indigenous to Europe and Northern Africa. It is cultivated to some extent in the

United States in limited localities in New York, Michigan, Nebraska and Wisconsin. The dried leaves and flowering tops of the plant are used for the extraction of the drug. The active principle is a volatile oil, of which the leaves and dried flowering tops contain about 0.5 of 1 percent. The oil is of a dark blue or greenish color, has a bitter lasting taste, and consists essentially of a substance known as absinthol in various forms of combination. There is also a bitter principle, probably a glucoside, which is known as absinthin. Absinth is of a most seductive character, and its victims are as helpless almost as those addicted to the alcohol or opium habit. The continued drinking of considerable quantities of absinth seems to break down the morale, sometimes paralyzing, or deranging digestion and general health, and reducing the victim to complete subjugation. The victims of absinth are more to be pitied, as a rule, than those of alcohol, opium, or cocain.

Kinds of Absinth.—As liqueurs, the extracts of absinth are divided into four classes, namely, ordinary absinths, demi-fine absinths, fine absinths, and Swiss absinths. The latter are sub-divided into absinths of Pontarlier, or Montpelier and of Lyon.

Ordinary absinths are the kind known only in Paris and other large cities. They are generally manufactured, as are the demi-fine absinths and the fine absinths, by the rectifiers. Ordinary absinthe is represented by the following formula:

        Clean and dry ground absinth . . . . 2.5 kilograms
        Dry hyssop. . . . . 500.0 grams
        Dry citronated balm-mint. . . . . 500.0 grams
        Green anise . . . . 2.0 kilograms
        85 percent alcohol. . . . . 16.0 liters

These ingredients are infused for 24 hours in a suitable vessel, 15 liters of water are added, and distilled until 15 liters of the product are secured. To this 15 liters are added 40 liters of 85 percent alcohol and 45 liters of water. The product amounts to 100 liters with an alcoholic strength of 46°. It is mixed and allowed to stand until clear. Filtration is not necessary in the manufacture of absinths. After standing for 48 hours they clarify themselves.

25

Extracts of Swiss absinth do not necessarily imply that they are made in Switzerland but they are made in that part of France to the south and east near Switzerland. They are manufactured particularly in large quantities in the cities of Pontarlier, Montpelier and Lyon.

Different flavoring and coloring materials are used in different liquids, but the absinth remains the principal ingredient.

Adulterations.—Absinths of bad quality are often made, some of them manufactured without distillation and with essences to replace the plants and seeds which are used in the genuine process. There are others which are distilled with crude alcohol of beets, the taste of which leaves much to be desired. Other absinths are made by adding aromatic resins after distillation.

Absinth and the War.—On the breaking out of hostilities in Europe all the nations engaged in the conflict sought to increase the efficiency of their citizens in every possible way. The French Republic realized that the use of absinth was undermining the character and endurance of her citizens. Instead of any longer trying to regulate the sale of absinth, it was decided by the authorities of the republic to ask for its entire prohibition. This prohibition was established by the law which became effective on the sixteenth of March, 1915. This law has the following provision: "The manufacture, the sale, wholesale and retail, as well as the commerce in absinth and liqueurs similar thereto are forbidden." This law has been sustained by the courts, which have held that not only is the sale interdicted, but that the person who does sell is penally responsible for all damages arising therefrom. In like manner the prohibition of the sale of distilled beverages at retail in the saloons where beverages are dispensed has been upheld by the courts, maintaining that such a regulation, made with a view of securing and maintaining good public order, is legally admitted to the circle of those exercises of authority on the part of the prefects as well as by the military authority in places where martial law has been declared. The manufacture and sale of absinth has long been prohibited in Switzerland and Belgium. In Switzerland the question of prohibiting absinth was submitted to a plebecite and was adopted by a large majority.

Absinth Poisoning.—The symptoms which attend the excessive or prolonged drinking of absinth are extremely revolting. The effects which the patient undergoes are perhaps more pronounced than in the case of addiction to the use of alcohol, opium, or cocain. The physical and mental deterioration goes on much more quickly in the case of absinth poisoning than in the case of the use of alcohol, and the disease maintains its characteristic symptoms, which are marked principally by epileptoid attacks, and by delerium and hallucinations, followed by specific paralysis of the lower extremities. Absinth is regarded as the most injurious and dangerous of the alcoholic concoctions which are commonly used as a drink among the people.

Action of French Academy of Medicine.—The whole subject of essences was studied by the French Academy of Medicine, and they were divided into two chief classes.

First, these essences which present a character particularly noxious; these being subjects of the absolute prescription.

Second, those of a relatively inferior degree of toxicity, of which the use may become dangerous, and, therefore, they are subjects of special regulation.

Among the essences which should be excluded from beverages entirely, according to the report of the Academy of Medicine, are the following:

Grand absinth.
Small absinth.
Badiane.
Angustura.
Queen of the Meadow.
Salicylic aldehyde.
Oil of Wintergreen.
Mythl salicylate.
Stones of fruit containing benzoic aldehyde and prussic acid, such as bitter almonds.
Rue.

The French Academy of Medicine makes the following comment on the first category:

"This category comprises the principal essences entering into the fundamental composition of liqueurs, especially those which are so commonly advertised as *aperitijs* (appetizers); they constitute the most dangerous class of beverages, and at the same time are those which predominate in habitual consumption. It is sufficient to place at the head of the list of these essences, the one that merits the position in every respect, but especially by its superior toxicity and characteristics, namely absinth."

Composition.—Chemically, essence of absinth is a complex product, composed, first, of a hydrocarbon analogous to the essence of terebinthine; second, of a bluish oil, and third, of a hydrocarbon absinthol isomeric with camphor. These three elements vary in quantity with the place of origin and the period of the harvest of the plant.

With its varying composition the toxicity of absinth also varies. In some cases the distinctive features of absinth poisoning seem almost absent. Animals on which experiments have been made, most frequently die after a series of attacks, or they die subsequent to the symptoms altogether, from lung troubles, which probably are produced by the irritating action of the essences, or they sometimes die as a result of embolism in the capillaries of the lungs.

As has been intimated before, however, it is not the essence of absinth alone to which the bad effects, especially of the epileptoid form, can be attributed. The other essences which are combined with absinthe in the drink are also culpable.

Cautions in Prescribing Absinth Extracts for Invalids.— Physicians should be careful to avoid prescribing absinth and its allied essences in cases of persons whose nervous excitability is already near the breaking point, and who possess decided irritability. These are the kinds of drinkers who too often serve as subjects of observation, and upon whom the epileptic attack has been noticed after the drinking of the absinth liqueur. To such fragile beings it is necessary for the physician to forbid absolutely not only the use of absinth, but of all the substances which belong to the epileptogenic group, which are more widely distributed and more important than has generally been supposed.

Among this group may be mentioned Hyssop, Rosemary, Fennel, and a variety of Sage. The essence of sage is often more toxic than essence of absinth, according to some authors.

The following table shows the quantity of the different substances which are required to produce convulsions in animals, the quantity being for each kilogram of weight:

Essence of sage........................... 10 milligrams.
Essence of absinth .... 25 milligrams.
Essence of hyssop......................... 31 milligrams.
Essence of Rosemary...................... 62 milligrams.
Essence of fennel.......................... 85 milligrams.

Some of the Results of the Habitual Use of Absinth.—The habitual use of absinth determines a series of disorders in addition to those already described, of which some are like those of ordinary alcoholism, while others differ therefrom either in degree or nature.

All the symptoms of absinthism are correlated directly or indirectly with the nervous system. The character of suffering from absinth poisoning is distinguished by its impressionability and by a succession or mixture of irritability and sadness. The dreams of those suffering from absinthism are similar to those afflicted with alcoholism, but more terrible, if that is possible. The hallucinations of vision and of hearing are much more frequent in absinthism than in alcoholism. Absinth delerium does not differ greatly from alcohol delerium, since it is produced from a similar cause. The most characteristic symptoms of absinthism, in addition to the epileptiform convulsions, are the phenomena of depression and sadness, either spontaneous or provoked.

More often than in alcoholism, absinthism produces a feebleness of the members of the body, approaching almost veritable paralysis. General convulsions are symptomatic of acute absinthism. They are not, however, observed so generally in chronic absinthism, unless there should be developed some lesion of the brain.

The digestive troubles in absinthism are analogous to those of alcoholism, but less pronounced.

Those who are afflicted with absinth poisoning are peculiarly susceptible to the ravages of tuberculosis, and invariably die if the disease is contracted. It is very unusual that any one addicted to the use of absinthe lives to be 60 years of age. There seems to be in absinthism, as in alcoholism, a premature condition of old age established; that is, the precipitation of the albuminous contents of the original cells. In other words, a hardening and segregation of the protoplasm.

This array of symptoms is certainly one that is large enough to deter any one from becoming a victim of chronic absinthism.

Fortunately, the absinth habit has not gained any great vogue in the United States, and it is hoped that before such an unfortunate state of affairs as existed in France arrives, there may be such a rigid control of traffic in absinth as to amount to prohibition.

Cocktails and Mixed Drinks.—A grand passion prevails among drinkers of alcoholic beverages for new, strange and mysterious mixtures of various kinds. There is one typical compound of this character which has gained a great vogue and which is known as a class by the name "cocktail." A cocktail is a mixture of an alcoholic beverage, usually brandy, whisky or gin, with various bitters and flavors. It is served, usually before dinner with a piece of lemon peel or a dash of lime juice as an additional spice. These beverages are employed usually before a meal, though by no means confined to consumption at the table. The name of these preparations is legion and as a rule gives no indication of composition.

Common cocktails are known under such names as Bronx, Manhattan, Martini, Saratoga, Rob Roy, Sunshine, Sazarack, etc. One prime condition of the cocktail is that it should be be chilled to as low a degree as possible.

Punch.—Another variety of mixed drinks is that listed under the name of punch. The term "punch" is applied to any alcoholic mixture in which fruits, aromatics and other substances of a similar character are combined with alcohol. The common idea of a punch is one made with wine as a basis, together with plain or carbonated water, sugar and fruit juices, mostly of citrous fruits or fragments of fruits of various kinds, such as pineapple and orange. Another punch is made using sparkling wine as the basis, commonly known as champagne punch. A plain fruit juice punch is now coming into vogue, which contains no alcohol, made chiefly of grape juice, pineapple juice and lemon juice, with sugar and water.

Gin Fizz.—A fizz, or gin fizz, as it is mostly made of gin, is a beverage containing some carbonated liquid, usually soda water, together with gin or other alcoholic beverage and sugar. Often milk or cream is used with beverages of this kind. A celebrated gin fizz is made in New Orleans, and is known by the name of its originator as the Ramos. This drink is made of gin, sugar, lemon juice, white of egg, with vanilla or other flavoring essence and some seltzer water. These are all shaken together, and afterward cream or milk added and shaken until an emulsion is produced. This is a very popular drink in New Orleans and has spread to many other parts of the country.

Appetizers, or, as known in France, Aperitifs, as, for instance, Amer Picon and Dubonnet, are drinks which are somewhat of the nature of a cocktail, though usually they come ready mixed and are supposed, like the cocktail, to be taken before the meal to stimulate the appetite. These appetizers are not known so well in this country under their own name, but are included under the cocktails, but in France they are very common.

Toddy, if true to its original signification, is an alcoholic beverage made in connection with a roast apple, though now the name "toddy" has been extended to a variety of drinks very much of the quality of a cocktail. In looking through the list of toddies in one of the new bartenders' guides I do not find the old-fashioned toddy described at all. The mixtures which are sold under the name of "toddy" might just as well be called "cocktails."

Formulæ.—There is no set formula that the barkeeper follows in making the mixed drinks. Each one may vary it to suit himself or the taste of his customer. I think that if the number of these mixed drinks which are sold in different parts of the United States could be brought together, there would be more than 500 of them, yet practically all are built upon the same plan. The list of the more common ones follows:

Brandy and Soda.—This beverage is true to name and consists of about half brandy and half plain soda water, seltzer or other carbonated mineral water.

Wine Cobbler.—Cobblers are made with various kinds of wines, sweetened and flavored with pineapple, orange, lemon peel, various fruits, Angostura bitters, etc. It is not to be understood that all these are used in any one cobbler, but some of them are contained in every mixed drink that bears that name.

Applejack Cocktail.—Apple brandy is commonly known as applejack, and this gives the name to the cocktail in which applejack is the alcoholic constituent.

Bronx Cocktail.—This drink contains orange as one of its chief flavors, and also French or Italian vermouth. The alcoholic constituent is gin.

Manhattan Cocktail.—This is one of the most common cocktails served. It is made in many different ways. The original Manhattan used whisky as the alcoholic base, flavored with Angostura bitters and vermouth. Usually a so-called red cherry is served with this cocktail. There are various modifications of this drink, but generally built upon this plan.

Martini Cocktail.—This is built up much in the same style as the Manhattan, using gin as the alcoholic ingredient and flavoring with vermouth and orange bitters. These cocktails may be dry or sweetened. When ordered dry no sugar is used in their manufacture.

Saratoga Cocktail.—The alcoholic constituents of the Saratoga cocktail are whisky and brandy in equal parts. Angostura bitters are the chief flavoring agents but vermouth is also employed.

Sazarack Cocktail.—The alcoholic constituent of the Sazarack cocktail is Bourbon whisky. Angostura bitters and vermouth are the flavoring agents. All apparatus employed, as well as the drink itself, is to be chilled.

Egg Nogg.—Egg nogg is one of the most common of the mixed drinks, especially during the holidays. There are dozens of recipes for making egg nogg. The alcoholic basis of egg nogg is usually whisky, brandy and rum. These should all be old and mellowed in order to make a good article. In the making of egg nogg the eggs are separated into whites and yolks, and the yolks of the eggs are beaten up with a sufficient quantity of sugar to add the required sweetness to the beverage. Various flavoring re-agents are used, as, for instance, sherry and Madeire wine, nutmeg and cinnamon. The whites of the eggs are beaten until firm, cream is added to the mixture of egg and alcoholic beverages, and after all is thoroughly mixed and cooled the white of the egg is placed on the top and dusted over with cinnamon. To make two gallons of egg nogg requires about one and three-quarter pounds of powdered sugar and a dozen fresh eggs, a quart of brandy, half a quart of rum, and half a quart of whisky. When all the ingredients are mixed there should be a sufficient amount of cream or rich milk added to bring the volume up to that desired.

Highballs.—Another class of mixed drinks, and one which is now very common, is known as highball. The highball in its simplest conception is not a mixed drink at all, except in the mixing of water with the alcoholic beverage. The highball is served in a tall glass, the whisky, brandy or gin being first poured into the glass with, usually, a lump of ice, and the glass is then filled with soda water or seltzer or some carbonated beverage. There are certain forms of highballs which are made on the same principle as cocktails, but these are not the real article.

Horse's Neck.—A very common mixed drink is known as Horse's Neck. This is also made in a glass similar to that of a highball. In a Horse's Neck the lemon is peeled usually in a manner to keep the lemon in one piece. The glass is then filled with ginger ale. This form of Horse's Neck is not an alcoholic beverage, but a soft drink.

Mint Julep.—A beverage, especially common in Kentucky, is the mint julep. The mint julep is made with Bourbon whisky, sugar and some water, and the glass which contains it is partly filled with fresh mint. Mint juleps are also made with other beverages, for instance, brandy, gin or rum.

Mamie Taylor is a mixed drink of considerable vogue. This drink is made, usually, of Scotch whisky, lime juice and ginger ale.

Milk Shake.—Milk shake is a mixed drink that is often served, but when made in the simplest way it is not an alcoholic mixture. A true milk shake is composed of sugar, egg and milk, with usually a little nutmeg grated on top after it has been well shaken.

Gin Rickey.—Gin Rickey is a drink which has a great vogue. A gin rickey is made with gin, lime juice or lemon juice, and soda, seltzer or other carbonated waters.

Shandy Gaff.—Shandy Gaff is a common mixed drink which is composed of equal parts of beer or ale and ginger ale.

Brandy Smash.—Brandy smash is a kind of mint julep, composed of brandy, sugar and mint.

Brandy or Whisky Sour.—Brandy or Whisky Sour is a mixture made with brandy, whisky or rum, and lime or lemon juice. Usually a carbonated water is used in the manufacture.

Tom Collins is made of Old-Tom gin, sugar, lime juice, and plain soda or seltzer.

Tom and Jerry.—This is a kind of egg nogg. The whites of the eggs are beaten separately, and the yolks with sugar and a little carbonate of soda. The brandy or whisky is mixed with other ingredients, except the white of the egg, and made up to the required strength with hot water, and the whites of the eggs placed on top and sprinkled with a little nutmeg.

After-dinner Drinks.—There are various after-dinner drinks which are grouped under the name of *pousse café*. The simplest *pousse café* is a small glass of brandy. Various liqueurs of different colors are used in making the compound *pousse café*, which is very attractive to the sight and which is built up in the following way: A few drops of a heavy syrup is placed in the bottom of a conical glass. This is followed by liqueurs of different characters, arranged in the order of density. These liqueurs also have different colors, so that in pouring them into the glass each one remains as a separate layer. The top layer usually consists of Kirschwasser or Maraschino, that is, brandy distilled from cherries. The liqueurs which are used are usually some of the following: Benedictine, Chartreuse, Curaçoa, and cherry brandy. Any other liqueurs may be employed, which have the required density, taste and flavor.

Mixed Drinks Containing Rum.—Rum is a very common constituent in many mixed beverages. Among the beverages which are formed with rum may be enumerated:

Hot Rum with Lemon and Sugar.

Rum and Rock Candy.

Rum and Milk.

Cherry Rum.—Prepared with old rum and ripe wild cherries. In the preparation of cherry rum the cherries are allowed to stay in the rum for a week or longer. At the time the maceration begins the cherries are pressed, but the force used is not sufficient to break the stones. At the end of a week a small quantity of sugar is added and the mixture thoroughly stirred and strained through filter paper or closely woven cloth.

Rum and Molasses.—Rum and molasses is an old family remedy which has been used for many years, especially in New England. The preparations of molasses and rum should be about one tablespoonful of molasses to two ounces of rum; well mixed.

Old Rum Punch.—Made of sugar, lemon, water and rum.

Rum Mint Julep.—Made in the same manner as the ordinary mint julep except that rum is used instead of whisky.

Rum Cordial.—Made of one-third rum, one-third rock candy syrup, one-third Chartreuse.

Rum Punch.—Made of one wine glass of hot water, juice of half a lemon, two wine glasses of hot rum and a slice of lemon.

Among the other preparations in which rum is used may be mentioned:

RomanPunch.
Rum Cream.
Frozen Pudding.
Rum and Grape Fruit.
Rum and Tea.
Rum and Cheese.
Rum Omelette.
Rum Milk Punch.
Rum and Ginger Ale.
Rum Highball.
Brandy and Rum Punch.
Rum Daisy.

Rum and Soda.
Rum Flip.
Rum-Egg Mllk Punch.
Hot Spiced Rum.
Egg Nogg.—Rum is an important constituent of egg nogg.
Rum Cocktail.
Tom and Jerry.
Rum Sling.
Rum Sour.
Rum Smash.
Rum Crusta.

PART XV

ALCOHOLIC REMEDIES

Beverages under the Guise of Medicinal Preparations.—Many preparations which contain alcohol also lay claim to being remedial agents. The classification of beverages of this kind is a source of much trouble to the Revenue Department, where they are called upon to examine these preparations and determine whether or not they are medicinal and therefore can be sold without a special license, or whether they are of an alcoholic character as a beverage and require a special license the same as ordinary alcoholic beverages. According to the ruling of the Treasury Department[1] the following articles cannot be sold except on payment of a special tax as a retail liquor dealer.

Alcoholic medicinal preparations

List of alcoholic medicinal preparations for the sale of which special tax is required. Revision of T. D. 2222

TREASURY DEPARTMENT,
OFFICE OF COMMISSIONER OF INTERNAL REVENUE,
Washington, D. C., October 19, 1917.

To Collectors of Internal Revenue, Revenue Agents, and Others:

The accompanying list of alcoholic medicinal preparations which have been examined by this office and held to be insufficiently medicated to render them unfit for use as a beverage is published for the information of all concerned.

Special tax will be required for the sale of any of the preparations herein named, even though such sales are for medicinal use. The liabilities of dealers for sales *for medicinal use* of any of the preparations marked with an asterisk (*) will, however, be held to date from and after January 1, 1918.

The names of most of the preparations heretofore given on the various lists which have been published will be found included in

[1] T. D. 2544, October 19, 1917

this list, the only exceptions being those the manufacturers of which have revised their formulæ to meet the requirements of this office or which are no longer on the market. Special tax should not, therefore, be required for the sale for medicinal use of any alleged medicinal compound not on this list until this office has been communicated with and specific instructions received.

The preceding paragraph does not, however, apply to the class of compounds usually described by the term "cocktail bitters," which are suitable for and usually used as beverages.

It having been found in various instances that there are several preparations of the same name on the market, the names of the manufacturers of the preparations examined by this office are here given, and it should be understood that only the preparations as compounded by the manufacturer whose name is given are embraced in this list.

Special tax will be required for the manufacture and sale of beef, wine, and iron, unless it contains at least the percentages of beef and iron given in the formula on page 1821 of the nineteenth edition of the United States Dispensatory or is otherwise sufficiently medicated to be unsuitable for use as a beverage. Special tax will also be required for the sale of compounds ordinarily sold under the name of rock, rye, and glycerin, and ginger brandy.

Collectors and revenue agents should continue to secure and forward to this office samples of preparations which they have reason to believe are or may be used as a beverage.

List of alcoholic medicinal preparations

Ale and Beef, Ale & Beef Co., Dayton, Ohio.
Allen's Restorative Tonic, Faxon & Gallagher Drug Co., Kansas City, Mo.
Alps Bitters, Peter Rostekowski, Chicago, Ill.
Amer Picon, G. Picon (imported).
Angostura Aromatic Tincture Bitters, E. R. Behlers, St. Louis, Mo.
Arbaugh's Newport Bitters, Daniel Stewart Co., Indianapolis, Ind.
Aroma Bitters, V. Gautier, 287 Hudson Street, New York.
Aromatic Bitters, Hanigan Bros., Denver, Colo.
Atlas Life Tonic, Atlas Medicine Co., St. Louis, Mo.
Atwood's La Grippe Specific, Excelsior Medicine Co., Chicago, Ill.
Angauer Bitters, Angauer Bitters Co., Chicago, Ill.
Angauer Kidney-Aid, Angauer Bitters Co., Chicago, Ill.
Augustiner Health and Stomach Bitters, A. M. August, Milwaukee, Wis.

Beef, Iron and Wine, The Jarmuth Co., Providence, R. I.

Beef, Iron and Wine, Lion Drug Co., Buffalo, N. Y.

Beef, Wine and Iron, Chas. C. Miller, Chicago, Ill.

Beimer's Walnut Beverage, Winona Liquor Co. (Inc.), Winona, Minn.

Beimer's Walnut Bitters, Winona Liquor Co. (Inc.), Winona, Minn.

Belvedere Stomach Bitters, Loewy Drug Co., Baltimore, Md.

Bentrovato Blood Bitters and Alterative Tonic, Lyons Bitters Co., New Haven, Conn.

Best Bitters, A. J. Lukwinski, Cleveland, Ohio.

Bismark Laxative Bitters, C. Lange & Co., Chicago, Ill.

Bitter Wine, Struzynski Bros., Chicago, Ill.

Bitters, The Atlantic Vineyard & Vine Co., Philadelphia, Pa.

Blackberry, Karles Medicine Co., Aberdeen, S. Dak.

Blackberry Cordial, International Extract Co., Philadelphia, Pa.

Blackberry Cordial, Irondequoit Wine Co., Rochester, N. Y.

Blackberry Cordial, Strother Drug Co., Lynchburg, Va.

Blackberry and Ginger Cordial, Standard Chemical Co., Fort Smith, Ark.

Black Hawk Bitters, Meyer Bros. Drug Co., St. Louis, Mo.

Bon Campo Bitters, Dr. A. H. Doty, St. Paul, Minn.

Bonekamp Bitters, J. S. Smith & Co., Burlington, Wis.

Bonekamp of Maagen Bitters, Teuscher & Co., St. Louis, Mo.

Bonus Elixir of Bitter Wine, Bonus Drug Co., Duquesne, Pa.

Bracer Bitters, Bracer Bitters Co., Chicago, Ill.

Bradenberger's Colocynthis, Standard Chemical Co., Fort Smith, Ark.

Brod's Celery Pepsin Bitters, Jno. Brod Chemical Co., Chicago, Ill.

Brown Gin, H. Obernauer & Co., Pittsburgh, Pa.

Brown's Aromatic Cordial Bitters, Chas. Leich & Co., sole agents, Evansville, Ind.

Brown's Utryme Tonic, A. E. & E. U. Brown Co., Mobile, Ala.

Buckeye Bitters, Geo. Albert, Milwaukee, Wis.

Buhrers Bitters, Weideman-Fries Co., Cleveland, Ohio.

Cardinal Stomach Bitters, P. J. Bowlin & Son, St. Paul, Minn.

Carmeliter Bitters—Dark—Elixir of Life, Burhenne & Dorn, 347 Hamburg Avenue, Brooklyn, N. Y.

Carmeliter Bitters E-Z Laxative, Burhenne & Dorn, 347 Hamburg Avenue, Brooklyn, N. Y.

Carmeliter Bitters—Light—Tonic and Appetizer, Burhenne & Dorn, 347 Hamburg Avenue, Brooklyn, N. Y.

Carmeliter Ginger Brandy, Burhenne & Dorn, 347 Hamburg Avenue, Brooklyn, N. Y.

Cascara Roots, American Bitter Wine Co., Chicago, Ill.

Cauffman's Ginger Brandy, E. Cauffman & Co., Philadelphia, Pa.

Celebrated Baja California Damiana Bitters, Naber, Alfs & Brune, San Francisco, Cal.

Celery Bitters and Angostura, Frank J. Maus, Kalamazoo, Mich.

Clarke's Rock Candy Cordial, Colburn, Birks & Co., Peoria, Ill.

Clayton & Russell's Stomach Bitters, Adams & Co., New York City.
Clifford's Cherry Cure, Standard Chemical Co., Fort Smith, Ark.
Clifford's Peruvian Elixir, Standard Chemical Co., Fort Smith, Ark.
Cocktail Bitters, Milburn & Co., Baltimore, Md.
Columbo Elixir, Columbo Elixir Co., Philadelphia, Pa.
Columbo Peptic Bitters, L. E. Jung & Co., New Orleans, La.
Columbo Tonic Bitters, Iler & Co., Omaha, Nebr.
Cooper's Nerve Tonic, Muller & Co., Baltimore, Md.
Cordial Panna, The Cordial Panna Co., Cleveland, Ohio.
Cossack Stomach Bitters, D. Vandewart & Son, New York, N. Y.
Cross Bitter Wine, Eugene Parisek Co., Chicago, Ill.
Damana Gentian Bitters, Milburn & Co., Baltimore, Md.
Dandy Bracer, Dandy Bracer Co., Philadelphia, Pa.
Der Doktor, Schooemer & Stoppenbach, Milwaukee, Wis.
Dr. Bergelt's Magen Bitters, Imported.
Dr. Bouvier's Buchu Gin, Dr. Bouvier's Specialty Co., Louisville, Ky.
Dr. Gray's Tonic Bitters, Central Botanical Co., Cherry Creek, N. Y.
Dr. Hoffman's Golden Bitters, F. Trandt, St. Louis, Mo.
Dr. Hopkins Union Stomach Bitters, F. S. Amidon, Hartford, Conn.
Dr. Hortenbach's Stomach Bitters, Minneapolis Drug Co., Minneapolis,
Minn.
Dr. Munro's Stomach Bitters, A. Du Chateau Co., Green Bay, Wis.
Dr. Rattinger's Bitters, Rattinger's Medical Co., Sappington, Mo.
Dr. Sherman's Peruvian Tonic and Systematizer, Des Moines Pharmacal
Co., Des Moines, Iowa.
Dr. Theodore Hartwig's Stomach Tonic, Jno. Behrendt, successor to Dr.
Theodore Hartwig, Grafton, Wis.
Dozier's Apple Bitters, Bitter Apple Bitters Co., Hattiesburg, Miss.
Dubonnet Wine, imported.
Dubonnet, imported.
Ducro's Alimentary Elixir, imported.
Duffy's Pure Malt Whiskey, Duffy's Malt Whiskey Co., Rochester, N. Y.
Elderberry Tonic, M. P. Kappel & Co., Chicago, Ill.
Elixir of Bitter Wine, Pleasant Tonic Bitters Co., Chicago, Ill.
Elixir of Bitter Wine, V. Bokr, Chicago, Ill.
Eureka Stomach Bitters, Iowa Drug Co., Des Moines, Iowa.
Excelsior Bitters, Des Moines Drug Co., Des Moines, Iowa.
E. Z. Laxative Bitters, Carmeliter Bitters Co., New York, N. Y.
Fabiani's Marsala Chinato, Fabiani's Pharmacy, Philadelphia, Pa.
Famous Wiener Bitters, Foxman Bros., Rock Island, Ill.
Faxon's Beef, Iron and Wine, Faxon, Williams & Faxon, Buffalo, N. Y.
Fernet-Branoa, L. Grandolphi & Co., New York City (imp.)
Fernet-Carlisi Fernet Bitters, C. Carlisi Co., New York City.
Ferri Rheumatic Cure, Luis Ferri, Butte, Mont.
Ferro-China Bascal, Basilea & Calandra, New York City.
Ferro-China Berna, W. P. Bernagozzi, New York City.

Ferro-China Bissleri (Felice Bissler), imported.

Ferro-China-Blotto, Vittorio Blotto, New York City.

Ferro-China Carlisi Tonic Bitters, C. Carlisi Co., New York City.

Ferro-China-Columbia, Columbia Distilling Co., Albany, N. Y.

Ferro-China-Derna, C. Matalone, Chicago, Ill.

Ferro-China-Florentino, Commercial Wine & Bottling Co., 182 Commercial Street, Boston, Mass.

Ferro-China-Salus, Italo-American Liquor Mfg. Co., New York.

Ferro-China-Trionfo, Basilea & Calandra, New York City.

Ferro-China Universale, Imported.

Ferro-Quina Bitters, D. P. Rossi, San Francisco, Cal.

Fine Old Bitter Wine, Struzynski Bros., Chicago, Ill.

F. Miller & Co.'s Stomach Bitters.

Fort Henry Ginger Compound, Reed, Robb & Breiding, Wheeling, W. Va.

Franz Urban Bonekamp of Maag Bitters, Wm. Straube, 1497 Twenty-fourth Street, Detroit, Mich.

Gastrophan, Edward Rimsa, Chicago, Ill.

Genuine Bohemian Malted Bitter Wine Tonic, Edward Rimas, Chicago, Ill.

Genuine Herb and Root Bitters, Schloemer & Stoppenbach, Milwaukee, Wis.

Germania Herb, Root, and Fruit Tonic Bitters, Dr. F. G. Nordman, Chicago, Ill.

German Stomach Bitters, Geo. Kuevers, Granite City, Ill.

German Stomach Bitters, Wm. W. Torge, Waukesha, Wis.

Ginger Tonic, Loewy Drug Co., Baltimore, Md.

Graham's Brand Orange Bitters, Chas. Jacquin, New York City.

Green's Chill Tonic, M. V. Green, Selma, N. C.

Gross Bros. Blood and Liver Tonic, Gross Bros., Illinois.

Harrison's Quinine Tonic, I. X. L. Chemical Co., Chicago, Ill.

Health Bitters, H. Bitzegeio, Chicago, Ill.

Herb Bitters, Otto F. Lenzt, Petersburg, Ill.

Heublein's Calisaya Bitters, G. F. Heublein & Bro., New York City.

Himmalia Tonic, C. O. F. Burkstrom & Co., Chicago, Ill.

*Holland Type Bitters, John Bardenheier Wine & Liquor Co., St. Louis, Mo.

Holtzermann's Bitters, Ahrendt & Sons Co., Toledo, Ohio.

Hop Bitters, Hop Bitters Mfg. Co., Rochester, N. Y.

Horke Vino, H. Obernauer & Co., Pittsburgh, Pa.

Humbolt Stomach Bitters, M. Koenigsberger, Kansas City, Mo.

I. X. L. Bitters, I. X. L. Chemical Co., Chicago, Ill.

Jack Pot Laxative Bitter Tonic, J. B. Scheuer Co., Chicago, Ill.

Jamaica Ginger, Yough Chemical Co., Connellsville, Pa.

Jamaica Type Ginger Drops Compound, V. Gautier & Co., New York City.

Jensen's Celebrated Kidney and Liver Bitters, Hans Jensen Co., Chicago, Ill.

*Jensen Laxative Bitters, Hans Jensen Co., Chicago, Ill.

Jones Stomach Bitters, Natchez Drug Co., Natchez, Miss.

Juniper Kidney Cure, Juniper Kidney Cure Co., Fort Smith, Ark.

*Kaiser Wilhelm Bitters, B. P. Sexton Co., Sandusky, Ohio.

Kapuziner Kloster Bitters, Union Wholesale Liquor Co., Chicago, Ill.

Karle's German Stomach Bitters, Karle German Bitters Co., Aberdeen, S. Dak.

Katarno, Katarno Co., New York City.

Kernel Stomach Bitters, Meyer Bros. Drug Co., St. Louis, Mo.

Koehler's Stomach Bitters, Koehler Bitters Co., New York City.

Kennedy's East India Bitters, Iler & Co., Omaha, Nebr.

Kidniwell, Brown Drug Co., Sioux Falls, S. Dak.

Kil-A-Kol, Pond's Bitters Co., Chicago, Ill.

Kobolo Tonic Stomach Bitters, Kobolo Medicine Co., R. D. Weisskopf & Co., proprietors, 1714 South Ashland Avenue, Chicago, Ill.

Ko-Ca-Ama, The Wm. Brooks Medicine Co., Russellville, Ark.

Kola and Celery Bitters, Milburn & Co., Baltimore, Md.

*Kratos Wine Bitters, Rochester Distilling Co., Rochester, N. Y.

Kreuzberger's Stomach Bitters, H. H. Shufeldt, Peoria, Ill.

Krummel's Bonekamp Maag Bitters, Hry, Krummel, New York City.

Kudros, A. M. Hellmann & Co., St. Louis, Mo.

Laxa Bark Tonic, Natchez Drug Co., Natchez, Miss.

Lee's Celebrated Stomach Bitters, Lee's Anti-Trust Medicine Co., Joplin, Mo.

*Leipziger Burgunder Wein Bitters, Hochstadter Co., New York City.

Lekko Stomach Bitters, Struzynski Bros., Chicago, Ill.

Liverine, T. S. Mitchell Co., Providence, R. I.

Lutz Stomach Bitters, Chas. M. Lutz, Reading, Pa.

Lyons Stomach Bitters, Lyons Bitters Co., Chicago, Ill.

Magador Bitters, E. J. Rose & Co., Tacoma, Wash.

Magen Bitters, A. J. Waberksy, Chicago, Ill.

Magen Bitters, Mrs. Ingeborg Rosmer, Milwaukee, Wis.

Marks' Famous Stomach Bitters, R. Marks, Milwaukee, Wis.

Marvelous Sweeping Model Wine Tonic, Marvelous Sweeping Model Wine Tonic Co., Chicago, Ill.

Mexican Stomach Bitters, Iler & Co., Omaha, Nebr.

Milburn's Kola & Celery Bitters, Milburn & Co., Baltimore, Md.

Miller Brand Bitters, Pure Food Cordial Co., New York City.

Miod Honey Wine, Struzynski Bros., Chicago, Ill.

Nature's Remedy for Kidney Troubles and Blood Poisoning, Dr. J. T. Sumpter, Bowling Green, Ky.

Neuropin, J. B. Scheuer Co., Chicago, Ill.

New Tonic Bitters, Chas. C. Miller, Chicago, Ill.

Newton's Nutritive Elixir, Parker-Blake Co., New Orleans.

26

Nibol Laxative Kidney and Liver Bitters, Lobin Distilling Co., St. Louis, Mo.

Novak's Stomach Elixir, Jno. Novak, Chicago, Ill.

*Old Country Bitters, Hans Jensen Co., Chicago, Ill.

Oro Kidney and Liver Tonic, J. B. Scheuer Co., Chicago, Ill.

Our Ginger Brandy, Rex Bitters Co., Chicago, Ill.

Ozark Stomach Bitters, Lee's Anti-Trust Medicine Co., Joplin, Mo.

Pale Orange Bitters, Field, Son & Co., London, England.

Panama Bitters, Richardson Drug Co., Omaha, Nebr.

Panama Bitters, W. R. Reeve, Dorchester, Mass.

Parker's Bitters, Louisiana Distillery Co. (Ltd.), New Orleans, La.

Pater Emanuel's Herb Wine, The Ambrose Co., Bridgeport, Conn.

Peppermint Drops Compound, V. Gautier & Co., New York City.

Pepsin Stomach Bitters (E. L. Arp), Imported.

Peptonic Stomach Bitters, Ross, Flowers & Co., Chicago and New York.

Peruvian Bitters, Reed, Robb & Breiding, Wheeling, W. Va.

Peter Paul Stomach Bitters, Paul P. Fasbender, Detroit, Mich.

Peychaud's Bitter Wine Cordial, L. E. Jung & Co., New Orleans, La.

Pilsener Bitter Wine, Prenstat Bitters Co., West, Tex.

Pioneer Ginger Bitters, Dr. Koehler Medicine Co., Appleton, Wis.

Pond's Ginger Brandy, Pond's Bitters Co., Chicago, Ill.

Pond's Rock and Rye, Pond's Bitters Co., Chicago, Ill.

Quinquina Dubonnet, Imported.

Red Jacket Bitters, Monheimer & Co., Chicago, Ill.

Rex Elixir of Bitter Wine, Red Bitters Co., Chicago, Ill.

Rex Ginger and Brandy Tonic, Rex Bitters Co., Chicago, Ill.

Rex Ginger, Rex Bitters Co., Chicago, Ill.

Rex Horehound Tonic, Rex Bitters Co., Chicago, Ill.

Rheinstrom's Stomach Bitters, Rheinstrom Bros., Cincinnati, Ohio.

Richard's Celebrated Tonic Bitters, Minneapolis Drug Co., Minneapolis, Minn.

Riley's Kidney Cure, Jas. S. Riley, Hayne, N. C.

Rimsovo Malto-Sovo Vino Chino, Ed. Rimsa, Chicago, Ill.

Rockandy Cough Cure.

Root Plant Medicinal Gin, Lobin Distilling Co., St. Louis, Mo.

Rosolio, The Cordial Tanna Co., Cleveland, Ohio.

Royal Pepsin Tonic, L. & A. Scharff, St. Louis, Mo.

Royal Pepsin Stomach Bitters, L. & A. Scharff, St. Louis, Mo.

S. B. C. Essence of Peppermint, Star Bitters Co., Sacramento, Cal.

S. B. C. Extract of Jamaica Ginger, Star Bitters Co., Sacramento, Cal.

S. B. C. Ginger and Brandy Compound, Star Bitters Co., Sacramento, Cal.

S. B. C. Wild Cherry Tonic, Star Bitters Co., Sacramento, Cal.

Sanitas Stomach Bitters, Sanitas Tonic Medicine Co., Chicago, Ill.

Sarasina Stomach Bitters, Wm. Blech, New York City.

St. Rafael Quinquina, Imported-Scheetz.

Schier's Famous Bitters, Wendelin Schier, Alexandria, Ind.

Schmit's Celebrated Strengthening Bitters, Schmit Pharmacal Co., Evansville, Ind.

Schroeder's German Bitters, Milburn & Co., Baltimore, Md.

Schuster's Bitters with Pepsin, The Schuster Co., Cleveland, Ohio.

Serravallo's Tonic, J. Serravallo's Pharmacy, Trieste, Austria.

Simon's Aromatic Stomach Bitters, Samuel B. Schein, St. Paul, Minn.

Sirena Tonic, Sirena Manufacturing Co., New York City.

Smart Weed, Francis Cropper Co., Chicago, Ill.

Smith's Bitters, Van Natta Drug Co., St. Joseph, Mo.

Smith's Vitalizing Bitters, Ben Smith, Scranton, Pa.

Smyrna Bitters, Smyrna Bitters Co., Dayton, Ohio.

Steinkonig's Stomach Bitters, Adam Steinkonig, Cincinnati, Ohio.

Stomach Bitters, Imported by J. G. & J. Boker, New York City.

Strauss Exhilarator, Wm. H. Strauss, Reading, Pa.

Sure Thing Tonic, Furst Bros., Cincinnati, Ohio.

Tatra, B. Zeman, Chicago, Ill.

Three-in-One (3 in 1) Tonic, Fialla & Eppler (Inc.), New York City.

Tokay Quinine Iron Wine, Burger & Erdeky, Chicago, Ill.

Tolu Rock and Rye.

Tolu Rock Candy Cordial, Meyer Bros. Drug Co., St. Louis, Mo.

True's Magnetic Cordial, Standard Chemical Co., Fort Smith, Ark.

U-Go, Fritz T. Schmidt & Sons, Davenport, Iowa.

Uncle Josh's Dyspepsia Cure, Dr. Worthington's Drug Co., Birmingham, Ala.

Underberg's Bonekamp Maag Bitters, Imported by Luyties Bros., New York City.

Vermouth Stomach Bitters, Lobin Distilling Co., St. Louis, Mo.

Vigo Bitters, F. C. Altmeier & Co., Chicago, Ill.

Vigor-lix, Greenbaum Bros., Louisville, Ky.

Vin de Michael, Imported.

Vin Mariani, Mariani & Co., New York City.

Walker's Tonic, Dreyfuss, Veil & Co., Paducah, Ky.

Webb's A No. 1 Tonic, Webb's Coöperative Co., Sacramento, Cal.

Westphalia Stomach Bitters, E. R. Behlers, St. Louis, Mo.

White Cross Bitters, V. Gautier, New York City.

Williams' Kidney Relief, Parker, Blake & Co., New Orleans, La.

Wine of Chenstohow, Skarzynski & Co., Buffalo, N. Y.

Wine of Pomelo, with Beef and Iron, Irondequoit Wine Co., Rochester, N. Y.

Wine Zdrowia, American Bitter Wine Co., Chicago, Ill.

Woodbury Brand Bitters, Steinhart Bros. & Co., New York City.

Zeman's Medicinal Bitter Wine, B. Zeman, Chicago, Ill.

Zien Stomach Bitters, Zien Bros., Milwaukee, Wis.

Zig-Zag, Walker's Tonic Co., Paducah, Ky.

Manufacturers of the preparations listed herein cannot legally use in such manufacture distilled spirits produced from materials fermented after September 8, 1917, nor distilled spirits taxable at the rate of $2.20 per gallon.

DANIEL C. ROPER,
Commissioner of Internal Revenue.

Coming and Going.—The list of so-called remedies in which alcohol largely enters is constantly changing. Old preparations are losing their vogue and dropping out. Increasing numbers of new preparations are coming forward. Their virtues are set forth in elaborate advertisements carried by newspapers and magazines. Natural scenery bordering the great arteries of travel is defaced by them. Desperate humanity turns to these nostrums with a ray of hope on which the cunning advertiser relies to induce the sufferer to buy. The blasted hopes sadden the last hours of the victim. If he gets well he has a thirst for alcohol that often leads him to ruin.

PART XVI

BEVERAGES CONTAINING COCAIN

Source of Cocain.—The plant yielding cocain is the coca shrub *Erythroxylon coca*. It grows on the eastern slopes of the Andes to a great distance both north and south of the equator. The alkaloid is found in the leaves, which resemble the leaves of the tea plant. Cocain was separated from the leaves by Wöhler in 1860.

Length of Life.—The coca plant or shrub is long-lived. A well-cultivated orchard will remain in excellent condition from 50 to 75 years. Generally, however, the older or injured plants are replaced from time to time with young plants, so that the orchard is kept green and vigorous.

As is the case with ordinary fruit trees, the young plants are usually grown in a nursery. They are grown from seeds which are planted during the rainy season, and they are also propagated from cuttings. When seeds are used great care must be taken that the moisture in the soil is kept high enough to secure a proper germination; and, in the conservatory, this is done by covering the soil where seeds are planted with glass. Coca seeds are rather difficult to get, because of their being so prized by birds.

Transplanting.—When plants are about a foot and a half high they are at the proper size for transplanting. They are usually put out in rows. They are put rather close together, and the rows are not very wide apart.

Harvesting.—The leaves are of a proper quality to be harvested at about two years after the plants have been set. The period of harvesting is determined by the change of the character of the leaf, the chlorophyl becoming oxidized to xanthophyl, producing a slight yellowish tint. The leaf should be harvested before it would fall naturally, since at the time it drops off of its own weight it is too mature to produce the best results.

Size.—The coca shrub, if not pruned, will usually attain to

405

the height of about 12 feet. As this, however, would be inconvenient for picking purposes it is kept at a lower height by means of frequent pruning.

The orchards are weeded and cultivated at least after each harvest, but after this and until the next harvest not much is done to them. Nevertheless, if the orchards are kept free from weeds and grass it hastens the ripening of the leaves.

The leaves are gathered by women and children, just as in the days of the Incas. The women, however, do not take part in the cultivation. The harvester sits down in front of a shrub, and, taking the branch in her hands, strips the leaves off with both hands by a dexterous movement, being careful not to injure in any way the shrub itself. The leaves are collected in an apron, made usually of wool, and are thence transferred to larger sacks, and then conveyed to the drying sheds. Care is taken not to undertake the harvesting during a rain or when rain is threatening, since rain exerts a very injurious effect upon the leaves when freshly picked.

Drying.—The leaves, in order to dry, are spread in thin layers about three inches deep usually upon slate. They are carefully watched during the process of drying so that they can be protected from any rain, which would work great injury upon the partially dried leaves. If the sun be bright, the drying may be concluded inside of a day, and the quickly dried leaves, called day leaves, obtain the highest prices. When a coca leaf is properly dried it has a greenish tint, is easily bent, is free of dirt, is not rough to the touch, and it has a slightly glossy appearance. When very old or when they are dried too much the leaves take on a brownish tint, and do not bring so high prices in the market.

The dried leaves are brought together and undergo the usual sweating or fermentation as practised with coffee, tea and tobacco. After the sweating is over the leaves are spread out again for a short time in the sun, and are then ready for packing and shipment.

Marketing.—When the coca leaves reach the markets they are usually classed as either Bolivian or Peruvian in origin. The Bolivian leaf is heavier and thicker than the Peruvian leaf, which is more delicate and of a lighter color. In general the manufacturers of cocain use chiefly the Bolivian leaf, since it contains the highest percentage of the alkaloid and the less quantity of the other alkaloids which are present in the leaf.

The natives prefer for their use the Peruvian coca, although it contains less cocain. The Peruvian coca is richer in the other alkaloids aside from cocain than the Bolivian.

Method of Use.—The coca leaf, among the natives of South America, is used almost exclusively by chewing. The leaves are first dusted with an alkali (lime) which sets the cocain free from its natural combination with the plant acids. The natives do not undertake any great physical exertion without this adjuvant. The effect of the cocain is a high degree of exhilaration and an insensibility to the feeling of fatigue.

Forms in Which Coca Leaf is Used as a Beverage.—Attention has already been called to the fact that in South America, among the Indians, the coca leaf is used chiefly by chewing. In this form it cannot be regarded as a beverage. The use of the coca leaf in medicine is well known, or rather the alkaloid derived from it, namely: cocain. Cocain is a drug which is highly prized in the treatment of many diseases, and especially is it valued as a local anæsthetic. The application of cocain deadens the sensibilities of the nerves, and often in this way it relieves great suffering and distress. It also so diminishes the nervous sensibilities as to render possible small minor surgical operations without much pain. Thus when locally applied it takes the place of a general anæsthetic, such as chloroform or ether. In the form of a beverage it is nearly always offered in combination with wine, or diluted alcohol, called wine. A preparation of coca wine was recognized in the United States Pharmacopœia 8th decennial revision as a standard remedy, the strength of which is definitely prescribed. This preparation was omitted in the 9th decennial revision issued September 1, 1916.

It has been found that more than a hundred wines, bearing different names, have, in the past, been used as a vehicle for the administration of cocain without the knowledge on the part of the consumer. Among the most widely used of these preparations was Vin Mariani.

The injury from the use of a habit-forming drug, such as cocain in alcoholic solution, cannot be overestimated. In using any form of wine there is a double danger to be apprehended; first, on account of the wine itself being a pleasant and desirable beverage, and; second, on account of its other constituents and of the alcohol which it contains; and in addition thereto the preparation becomes dangerous by reason of the cocain extracted from the coca leaves.

Since the passage of the Foods and Drugs Act preparations of this kind, when used medicinally, as most of them are, are required to bear upon the label a statement of the quantity of cocain which they contain. This wise provision, in so far as protecting the innocent person is concerned from forming the habit, does not, however, protect those who have already formed the habit, but rather acts as a guide to enable them to secure the article desired. A better regulation to govern the consumption of cocain in this way is the federal law regulating commerce in opium and cocain, now in force. The use of beverages containing cocain, except for strictly medicinal purposes, is entirely prohibited. The decocainized leaves, by a special provision of the law may be used in the preparation of a beverage. Such a preparation is used in the manufacture of coca cola under the name of Merchandise No. 5.

INDEX

27